Studies in Computational Intelligence

Volume 665

Series editor

Janusz Kacprzyk, Polish Academy of Sciences, Warsaw, Poland
e-mail: kacprzyk@ibspan.waw.pl

About this Series

The series "Studies in Computational Intelligence" (SCI) publishes new developments and advances in the various areas of computational intelligence—quickly and with a high quality. The intent is to cover the theory, applications, and design methods of computational intelligence, as embedded in the fields of engineering, computer science, physics and life sciences, as well as the methodologies behind them. The series contains monographs, lecture notes and edited volumes in computational intelligence spanning the areas of neural networks, connectionist systems, genetic algorithms, evolutionary computation, artificial intelligence, cellular automata, self-organizing systems, soft computing, fuzzy systems, and hybrid intelligent systems. Of particular value to both the contributors and the readership are the short publication timeframe and the worldwide distribution, which enable both wide and rapid dissemination of research output.

More information about this series at http://www.springer.com/series/7092

Fabrice Guillet · Bruno Pinaud
Gilles Venturini
Editors

Advances in Knowledge Discovery and Management

Volume 6

Editors
Fabrice Guillet
Polytech Nantes
University of Nantes
Nantes Cedex 3
France

Gilles Venturini
Polytechnics Graduate School
University of Tours
Tours
France

Bruno Pinaud
University of Bordeaux
Talence Cedex
France

ISSN 1860-949X   ISSN 1860-9503   (electronic)
Studies in Computational Intelligence
ISBN 978-3-319-83368-2   ISBN 978-3-319-45763-5   (eBook)
DOI 10.1007/978-3-319-45763-5

Printed on acid-free paper

This Springer imprint is published by Springer Nature
The registered company is Springer International Publishing AG
The registered company address is: Gewerbestrasse 11, 6330 Cham, Switzerland

Preface

The recent and novel research contributions collected in this book are extended and reworked versions of a selection of the best papers that were originally presented in French at the EGC'2014 and EGC'2015 conferences respectively held in Rennes (France) in January 2014 and Luxembourg in January 2015. The papers have been selected among the papers accepted in long format at the conferences. For the conferences, the long papers are themselves the result of a double-blind peer-review process among the 106 papers initially submitted to the conference in 2013 and 83 papers in 2015 (conference acceptance rate for long papers of 26 % in 2014 and 27 % for 2015). These conferences were the 14th and 15th edition of this event, which takes place each year and which is now successful and well-known in the French-speaking community. This community was structured in 2003 by the Foundation of the International French-speaking EGC society (EGC in French stands for "Extraction et Gestion des Connaissances" and means "Knowledge Discovery and Management", or KDM). This society organizes every year its main conference (about 200 attendees) also workshops and other events with the aim of promoting exchanges between researchers and companies concerned with KDM and its applications in business, administration, industry or public organizations. For more details about the EGC society, please consult http://www.egc.asso.fr.

Structure of the Book

This book is a collection of representative and novel works done in Data Mining, Knowledge Discovery, Clustering and Classification. It is intended to be read by all researchers interested in these fields, including Ph.D. or M.Sc. students, and researchers from public or private laboratories. It concerns both theoretical and practical aspects of KDM.

This book has been structured into three parts. The first four chapters are related to optimization consideration while mining data. The second part presents four chapters dealing with specific quality measures, dissimilarities and ultrametrics. The five remaining chapters focus on semantics, ontologies and social networks.

Mining Data with Optimization

Chapter 1, *Online Learning of a Weighted Selective Naive Bayes Classifier with Non-convex Optimization*, is concerned with improving supervised classification for data streams with a high number of input variables. It focuses on direct estimation of weighted naïve Bayes classifiers using a sparse regularization of the model log-likelihood which takes into account knowledge relative to each input variable.

Chapter 2, *On Making Skyline Queries Resistant to Outliers*, aims to reduce the impact of exceptional points when computing skyline queries, so that outliers do not "hide" more interesting answers. The approach relies on the notion of fuzzy typicality and makes it possible to compute a graded skyline answers. A GPU-based parallel implementation is also described.

Chapter 3, *Adaptive Down-Sampling and Dimension Reduction in Time Elastic Kernel Machines for Efficient Recognition of Isolated Gestures*, addresses both the dimensionality reduction of the feature vector describing multidimensional motion time series and the dimensionality reduction along the time axis by the means of adaptive down-sampling used in conjunction with time Elastic Kernel Machines.

Chapter 4, *Exact and Approximate Minimal Pattern Mining*, presents a generic framework for exact and approximate minimal patterns mining by introducing the concept of minimizable set system, and it also demonstrates that minimal patterns mining is polynomial-delay and polynomial-space.

Quality Measures, Dissimilarities and Ultrametrics

Chapter 5, *Comparison of Proximity Measures for a Topological Discrimination*, proposes a methodology to make a clustering of proximity measures in the context of discrimination using a topological structure, and to choose the best discriminant measure for considered data.

Chapter 6, *Comparison of Linear Modularization Criteria Using the Relational Formalism, an Approach to Easily Identify Resolution Limit*, deals with the comparison of linear modularization criteria by using the Mathematical Relational analysis (MRA). MRA allows to compare numerous criteria on the same type of formal representation in order to facilitate their understanding and their usefulness in practical contexts.

Chapter 7, *A Novel Approach to Feature Selection Based on Quality Estimation Metrics*, proposes an adaptation of the Feature maximization (F-max) criterium in order to perform more efficient feature selection and feature contrasting within the framework of supervised classification. The comparison with other feature selection techniques shows a significant improvement of the performances, notably in the case of unbalanced, highly multidimensional and noisy textual data.

Chapter 8, *Ultrametricity of Dissimilarity Spaces and Its Significance for Data Mining*, evaluates the extent to which a dissimilarity is close to an ultrametric by introducing the notion of ultrametricity of a dissimilarity, and examines their influence on the accuracy of a classification or the quality of a clustering.

Semantics, Ontologies and Social Networks

Chapter 9, *SMERA: Semantic Mixed Approach for Web Query Expansion and Reformulation*, uses implicit and explicit concepts to automatically improve web queries. This approach handles several challenges related to query expansion, such as selective choice of expansion terms, named entities treatment, and concept-based query representation.

Chapter 10, *Multi-layer Ontologies for Integrated 3D Shape Segmentation and Annotation*, introduces an original framework where annotation and segmentation of 3D meshes are performed conjunctly. An expert's knowledge of the context is used while minimizing the use of geometric analysis, and a multi-layer ontology is designed to conceptualize 3D object features from the point of view of their geometry, topology, and possible attributes.

Chapter 11, *Ontology Alignment Using Web Linked Ontologies as Background Knowledge*, proposes an ontology matching method for aligning a source ontology with target ontologies already published and linked on the Linked Open Data (LOD) cloud. The evaluation was achieved on two well-known ontologies in the field of life sciences and environment: AgroVoc and Nalt.

Chapter 12, *LIAISON: reconciLIAtion of Individuals Profiles Across SOcial Networks*, describes an algorithm that uses the social network topology and the publicly available personal information to iteratively determine the profiles that belong to the same individuals across several social networks.

Chapter 13, *Clustering of Links and Clustering of Nodes: Fusion of Knowledge in Social Networks*, compares two network clustering approaches: the search for communities and the extraction of frequent conceptual links, in order to understand both the intersections that can exist between them and the knowledge that emerges from their fusion.

Acknowledgments

The editors would like to thank the chapter authors for their insights and contributions to this book.

The editors would also like to acknowledge the members of the review committee and the associated referees for their involvement in the review process of the book. Their in-depth reviewing, criticisms and constructive remarks have significantly contributed to the high quality of the slected papers.

Finally, we thank Springer and the publishing team, and especially T. Ditzinger and J. Kacprzyk, for their confidence in our project.

Nantes Cedex 3, France   Fabrice Guillet
Talence Cedex, France   Bruno Pinaud
Tours, France   Gilles Venturini
April 2016

Review Committee

All published chapters have been reviewed by two or three referees and at least one nonnative French speaker referee.

- Sadok Ben Yahia (Faculty of Sciences, Tunisia)
- Mohammed Bouguessa (Université du Québec à Montréal (UQAM), Canada)
- Hanen Brahmi (Faculty of Sciences, Tunisia)
- Francisco de A.T. De Carvalho (University Federal de Pernambuco, Brazil)
- Pascal Desbarats (University of Bordeaux, France)
- Gayo Diallo (University of Bordeaux, France)
- Carlos Ferreira (LIAAD INESC Porto LA, Portugal)
- Natalia Grabar (STL CNRS Université Lille 3, France)
- Antonio Irpino (Second University of Naples, Italy)
- Hoël Le Capitaine (University of Nantes, France)
- Daniel Lemire (LICEF Research Center, University of Québec, Canada)
- Paulo Maio (GECAD—Knowledge Engineering and Decision Support Research Group, Portugal)
- Fionn Murtagh (De Montfort University)
- Luiz Augusto Pizzato (Octosocial Labs, Australia)
- Jan Rauch (University of Economics, Czech Republic)
- Lorenza Saitta (University of Torino, Italy)
- Dan Simovici (University of Massachusetts Boston, USA)
- George Vouros (University of Piraeus, Greece)
- Jef Wijsen (University of Mons-Hainaut, Belgium)

Associated Reviewers

Nacéra Bennacer, Marc Boulé, Jean-Marie Favreau, Liliana Ibanescu, Nicolas Labroche, Jean-Charles Lamirel, Vincent Lemaire, Pierre-François Marteau, Josiane Mothe, Olivier Pivert, Gianluca Quercini, Alain Simac-Lejeune, Erick Stattner, François Rioult.

Contents

Part III Semantics, Ontologies, and Social Networks

Editors and Contributors

About the Editors

Fabrice Guillet is CS Professor at Polytech'Nantes, the graduate engineering school of University of Nantes, and a member of the "Knowledge and Decision" team (COD) of the LINA laboratory. He received a Ph.D. degree in CS in 1995 from the "École Nationale Supérieure des Télécommunications de Bretagne", and his Habilitation (HdR) in 2006 from Nantes university. He is a co-founder of the International French-speaking "Extraction et Gestion des Connaissances (EGC)" society. His research interests include knowledge quality and knowledge visualization in the frameworks of Data Mining and Knowledge Management. He has recently co-edited two refereed books of chapter entitled "Quality Measures in Data Mining" and "Statistical Implicative Analysis — Theory and Applications" published by Springer in 2007 and 2008.

Bruno Pinaud received the Ph.D. degree in Computer Science in 2006 from the University of Nantes. He is currently Assistant Professor at the University of Bordeaux in the Computer Science Department since September 2008. His current research interests are visual data mining, graph rewriting systems, graph visualization and experimental evaluation in HCI (Human Computer Interaction).

Gilles Venturini is CS Professor at François Rabelais University of Tours (France). His main researches interests concern visual data mining, virtual reality, 3D acquisition, biomimetic algorithms (genetic algorithms, artificial ants). He is co-editor in chief of the French New IT Journal (Revue des Nouvelles Technologies de l'Information) and was recently elected as President of the EGC society.

Contributors

Fatima-Zahra Aazi is Ph.D. student in joint supervision at the University Lumière of Lyon 2, France and the University Hassan 1er of Settat, Marocco. Her research interests include supervised learning methods applied in a topological context.

Rafik Abdesselam is Professor in Applied Mathematics at the University Lumière of Lyon 2. His research and teaching interests include statistics and data analysis, supervised classification, topological learning and methods for data mining. He is a member of various national and international program committees.

Bissan Audeh received her Ph.D. in Computer Science from Institut Fayol at the École Nationale Supèrieure des Mines de Saint-Étienne (France) in 2014. She has a Masters degree in Knowledge Discovery from Data from the university Lyon2 (France). Her research interests include semantic query expansion, information retrieval, recall evaluation, data extraction and semantic representation.

Marco Attene is studying geometry processing and analysis at CNR-IMATI since 1999, where has been the PI for regional, national and international projects. Marco collaborated with universities in Europe, USA, Asia and New Zealand and, in 2014, he was awarded the SGP Software Award for his "MeshFix" software. He is an associate editor of international journals in the area.

Philippe Beaune is a Graduate Engineer in Informatics and Applied Mathematics from ENSEEIHT (France) and he received his Ph.D. in Computer Science from the University of Montpellier in 1992. He is currently Associate Professor at Mines de Saint-Étienne where he teaches applied computer science and in particular Artificial Intelligence. He is also member of the Connected Intelligence Team of the Hubert Curien Laboratory. His main research is in cooperative engineering of ontologies.

Michel Beigbeder has been working as Assistant Professor at the École Nationale Supèrieure des Mines de Saint-Étienne (France) since he received his Ph.D. in Computer Science in 1988. His main interests currently are about text information retrieval, structured information retrieval and semantic information retrieval. With his team he has been participating in many Information Retrieval Campaigns such as TREC and INEX since 2004.

Nacéra Bennacer is Professor at the Computer Science Department of CentraleSupélec and member of LRI (Laboratoire de Recherche en Informatique). She obtained a Ph.D. in Computer Science from Conservatoire National des Arts et Métiers (CNAM-Paris) and received her accreditation to supervise research (HdR) from University Paris-Sud. Her research interests include semantic-based extraction information from Web resources, user-centric information retrieval by covering semi-structured data, linked open data and social networks data.

Marc Boullé was born in 1965 and graduated from Ecole Polytechnique (France) in 1987 and Sup Telecom Paris in 1989. Currently, he is a Senior Researcher in the data mining research group of Orange Labs. His main research interests include statistical data analysis, data mining, especially data preparation and modelling for large databases. He developed regularized methods for feature preprocessing, feature selection and construction, correlation analysis, model averaging of selective naive Bayes classifiers and regressors.

Patrice Buche received the Ph.D. degree in Computer Science from the University of Rennes, France, in 1990. He is a research engineer with INRA, Agricultural Research Institute of France. His research activities address data integration from heterogeneous sources multicriteria decision support based on imperfect data, stakeholders' preferences aggregation using argumentation, default reasoning.

Martine Collard is Full Professor in Computer Science at the University of the French West Indies. She is member of the laboratory of Mathematics and Computer Science at her university, where she is the head of the Data Analytics and bIg data gathering with Sensors team. She has a nice record of publications to her credit and is regularly appearing as a speaker at international conferences and workshops.

Patricia Conde-Céspedes is currently a postdoctoral researcher at L2TI, University Paris 13. In 2013, she defended her Ph.D. thesis about extensions of Mathematical Relational Analysis to graph clustering at University Paris 6. Before her Ph.D., she received a Master degree in statistics and a B.Sc. degree in industrial engineering. Her research interests include graph theory, data mining, machine learning, Mathematical Relational Analysis, and algorithmics.

Pascal Cuxac is Research Engineer at the INIST/CNRS (Institute for Scientific & Technical Information/National Center for Scientific Research) in Nancy, France. He obtained his Ph.D. in Geological and Mining Engineering from the Nancy School of Geology in 1991. In 1993, he joined the CNRS as Research Engineer where he is currently Project Manager ISTEX-RD for text mining activities. He currently collaborates to the RD part of the ISTEX French National Project.

Juliette Dibie is Professor in CS at AgroParisTech since 2013 and member of the INRA MIA research Computer Science team. She received her Ph.D. degree in Computer Science from the University Paris Dauphine, France, in 2000. Her research activity addresses knowledge representation using semantic Web languages, knowledge validation, flexible querying, semantic annotations, ontology building, ontology alignment and ontology evolution. She has published her work in 2 book chapters, 6 international journals, 2 national journals and 25 international conferences.

Thomas Dietenbeck received the Ph.D. degree in Image Processing in 2012 from INSA-Lyon, France. Since 2014, he is Assistant Professor at UPMC and conducts his research at the Laboratoire d'Imagerie Biomédicale (LIB). His research interests include medical image processing and segmentation.

Jean-Marie Favreau received the Ph.D. degree in computer science from the Université Blaise Pascal, Clermont- Ferrand, France, in 2009. After one year as postdoctoral fellow at the IMATI-Ge, Genova, Italy, he joined the ISIT unit at Université d'Auvergne, Clermont-Ferrand, France, as an Associate Professor in 2010. His current research interests include geometry and topology processing, with a strong focus on semantic identification.

Sylvie Gibet received her Ph.D. degree in Computer Science in 1987 from Institut National Polytechnique de Grenoble (Grenoble INP, France). She became Assistant Professor at Ecole Normale Supérieure de Cachan in 1991, and at University of Paris Sud, Orsay from 1992 to 1999 (LIMSI-CNRS Lab.), before joining in 2000 the Computer Science Lab. at University of Bretagne Sud where she is Professor since 2003. She joined the IRISA lab in 2012, and is currently involved in research on analysis and synthesis of gestures.

Mohammad Ghufran is a Ph.D. student in computer science at LRI—CentraleSupélec since 2014. He obtained a M.Sc. degree in Data Mining and Knowledge Management in 2012 within the Erasmus Mundus Program (University Lumiere Lyon 2, Ecole Polytechnique de Nantes, Universitat Politechnica de Catalunya).

Kafil Hajlaoui got his Ph.D. degree in CS in 2009. He was a postdoctoral student at INIST-CNRS (Institute for Scientific & Technical Information/National Center for Scientific Research) of Nancy (France) and was working on the exploitation of data mining techniques in the context of the OSEO QUAERO project. He continues his research in business intelligence systems and bank instrument.

Thomas Hecht received his MS degree in Computer Science from the University Paris Dauphine, France, in 2013. He is a Ph.D. Student at ENSTA ParisTech.

Kaixun Hua received the BS degree from Shanghai Jiao Tong University in 2012 and is currently working toward the Ph.D. degree in the CS Department at the University of Massachusetts, Boston where he is a member of Data Mining Research Lab at UMB. His research interests include the application of ultrametric spaces in clustering algorithm and classification of data with imbalanced classes.

Carine Hue received the Ph.D. degree in signal and image processing in 2003 from the University of Rennes 1, France. Since 2006, she worked in the data mining research group at Orange Labs in Lannion. Her recent research focuses on regularized methods for values grouping in the case of a large number of values and on sparse regularization methods for variables selection in large databases.

Liliana Ibanescu received her Ph.D. degree in CS from Institut National Polytechnique de Lorraine, Nancy, France, in 2004. Since 2008 she is Assistant Professor in Computer Science at AgroParisTech, Paris, France. Her current research activities in knowledge representation and data integration of heterogeneous data address ontology building, ontology alignment and ontology evolution, semantic annotations.

Hélène Jaudoin is a member of the research team Shaman at Irisa Lannion. Her research work concerns extraction and management of knowledge.

Coriane Nana Jipmo is a Ph.D. student in computer science at LRI—CentraleSupélec since 2014. She obtained a M.Sc degree in Databases and Decision-making Information Systems from Versailles Saint-Quentin-en-Yvelines University in 2013.

Jean-Charles Lamirel is Lecturer since 1997. He got his research accreditation in 2010. He currently takes part in the SYNALP team (ex. INRIA-TALARIS project) of the LORIA lab. In Nancy whose main concern is Automatic Language and Text Processing, his main domain of research is Textual Data Mining mainly based on Neural Networks with interests both in theoretical models and applications. He is the creator of the concepts of Data Analysis based on Multiple Viewpoints paradigm (MVDA) and of Feature Maximization metric (F-max).

Vincent Lemaire was born in 1968 and he obtained his undergraduate degree from the University of Paris 12 in signal processing and was in the same period an Electronic Teacher. He obtained a Ph.D. in Computer Science from the University of Paris 6 in 1999 and a Research Accreditation (HDR) in Computer Science from the University of Paris-Sud 11 (Orsay) in 2008. His recent research interests are the application of machine learning in various areas for telecommunication companies.

Jean-François Marcotorchino is currently Vice-President & Scientific Director at the Thales SIX Division. He is "Thales Technical Fellow" as well. Thanks to his "University Professor" position, JFM is also Research Director at LSTA (Lab of Theoretical and applied Statistics, Paris VI/UPMC University). Previously, he spent 30 years with IBM France and IBM EMEA, where he has performed the job of ECAM's Director (European Centre for Applied Mathematics, IBM EMEA).

Pierre-François Marteau received his Ph.D. degree in Computer Science in 1988 from Institut National Polytechnique de Grenoble (Grenoble INP). He is currently Professor at Universite Bretagne Sud (UBS) and at Institut de Recherche en Informatique et Systèmes Aléatoires (IRISA). His research interests include pattern recognition and machine learning with application in sequential (symbolic and digital) data processing.

Pierre Nerzic is a member of the research team Shaman at Irisa Lannion. His research interest is scalable implementations of preference query languages.

Alice Ahlem Othmani received the Ph.D. in 2014 from the University of Burgundy. In 2015, she was a postdoctoral researcher at the University of Auvergne at ISIT laboratory. Currently, she is postdoctoral researcher at Ecole Normale Supérieure of Paris. Her research interests include pattern recognition, texture analysis, 2D and 3D segmentation, classification, and object recognition. She is very keen to tackle fundamental and applied problems so that research results have impact on diverse real-life application domains including medical imaging, remote sensing and biological imaging.

Olivier Pivert heads the research team Shaman at Irisa Lannion. His research concerns fuzzy preference queries and uncertain data management. He has published over 300 papers on related topics in books, journals, and conference proceedings

Gianluca Quercini is Assistant Professor at the Computer Science Department of CentraleSupélec since 2012. He is also member of LRI (Laboratoire de Recherche en Informatique). He obtained a Ph.D. in computer science from the University of Genoa (Italy) in 2009 with a thesis titled "Optimizing and Visualizing Planar Graphs via Rectangular Dualization". He was a postdoctoral researcher at the University of Maryland from 2009 to 2011 and at the University of Paris-Sud from 2011 to 2012.

Clément Reverdy received his master degree in computer science at the Université Bretagne Sud (UBS) in 2014. He began a Ph.D. in September 2014 in the scope of analysis and synthesis of facial expressions in sign languages at the IRISA lab.

François Rioult received his Ph.D. in 2005 from the University of Caen Normandie. He is currently Associate Professor in computer science since 2006 at the University of Caen Normandie, and researcher at the CNRS UMR6072 GREYC laboratory. He is mainly interested in the founding principles of data mining and their applications to sports analytics.

Daniel Rocacher is a member of the research team Shaman at Irisa Lannion. His research interest is fuzzy preference queries.

Dan Simovici is Professor of Computer Science and Graduate Program Director at the University of Massachusetts Boston. He received his Ph.D. from the University of Bucharest, Romania and is the author of several books and more than 160 research papers. He serves as the Editor-in-Chief of the Journal for Multiple-Valued Logic and Soft Computing. His main research interests are in machine learning and data mining with a focus of information-theoretical techniques.

Arnaud Soulet received his Ph.D. in 2006 from the University of Caen Normandie. He is currently Associate Professor in computer science since 2007 at the computer science laboratory of University François Rabelais of Tours. His research interests include OLAP, data mining and machine learning.

Erick Stattner is Assistant Professor in CS at the University of the French West Indies. He is member of the laboratory of Mathematics and Computer Science and he has published more than 30 articles on the areas of Social Network Analysis and Mining, the study of diffusion phenomena and the social data collection.

Fakhri Torkhani received his Ph.D. degree (2015) from the University of Grenoble, France. Currently, he is a postdoctoral fellow in the University of Auvergne, France. His scientific interests include 3D mesh analysis, perceptual quality assessment for computer graphics, segmentation and semantic analysis.

Cassia Trojahn dos Santos received a Ph.D. degree in Computer Science from the University of Evora, Portugal, in 2009. Since 2012 she is Assistant Professor in Computer Science at the University of Toulouse 2, Toulouse, France. Her research interests are ontology matching and evaluation, knowledge representation, ontology construction and argumentation theory for reconciliating alignments.

Rosanne Vetro received a BS degree in Computer Science from the Federal University of Rio de Janeiro in Brazil, and a Ph.D. degree in Computer Science from the University of Massachusetts Boston in 2015. Her research areas include data mining and machine learning with application to bioinformatics, multimedia content and optimization. She currently works as a computational scientist at Seres Therapeutics in Cambridge, Massachusetts where she is responsible for the selection and development of machine learning and data mining algorithms for the analysis of microbiome data and characterization of states related to disease and health.

Emmanuel Viennet got his Ph.D. in 1993 on multi-layered neural networks for image recognition. Since 2008, he is Full Professor at L2TI, University Paris 13, where he leads a research group on data mining and social network analysis. This group recently developed social recommendation algorithms, and several new algorithms to detect and characterize local and global communities in very large complex networks.

Part I
Mining Data with Optimization

Online Learning of a Weighted Selective Naive Bayes Classifier with Non-convex Optimization

Carine Hue, Marc Boullé and Vincent Lemaire

Abstract We study supervised classification for data streams with a high number of input variables. The basic naïve Bayes classifier is attractive for its simplicity and performance when the strong assumption of conditional independence is valid. Variable selection and model averaging are two common ways to improve this model. This process leads to manipulate a weighted naïve Bayes classifier. We focus here on direct estimation of weighted naïve Bayes classifiers. We propose a sparse regularization of the model log-likelihood which takes into account knowledge relative to each input variable. The sparse regularized likelihood being non convex, we propose an online gradient algorithm using mini-batches and random perturbation according to a metaheuristic to avoid local minima. In our experiments, we first study the optimization quality, then the classifier performance under varying its parameterization. The results confirm the effectiveness of our approach.

Keywords Supervised classification · Naïve Bayes classifier · Non-convex optimization · Stochastic optimization · Variables selection · Sparse regularization

1 Introduction

Due to a continuous increase of storage capacities, data acquisition and processing have deeply evolved during the last decades. Henceforth, it is common to process data including a very large number of variables. Data amounts are so massive that it hardly seems possible to fully load them: online processing is then applied and data are seen only once. In this context, we consider the supervised classification problem where

C. Hue (✉) · M. Boullé · V. Lemaire
Orange Labs, 2 Avenue Pierre Marzin, 22300 Lannion, France
e-mail: Carine.Hue@orange.com

M. Boullé
e-mail: Marc.Boulle@orange.com

V. Lemaire
e-mail: Vincent.Lemaire@orange.com

© Springer International Publishing Switzerland 2017
F. Guillet et al. (eds.), *Advances in Knowledge Discovery and Management*,
Studies in Computational Intelligence 665, DOI 10.1007/978-3-319-45763-5_1

3

Y is a categorical target variable with J values $C_1, \ldots, C_J$ and $X = (X_1, \ldots, X_K)$ is the set of K explanatory variables, numerical or categorical.

Among the solutions to the problems of learning on data streams, the incremental learning algorithms are one of the most used techniques. These algorithms are able to update their model using just the new examples.

In this article we focus on one of the most used classifier in the literature i.e. the naïve Bayes classifier. A naive Bayes classifier is a simple probabilistic classifier based on applying Bayes' theorem with naive conditional independence assumption. The explanatory variables $(X_k)_{k=1,\ldots,K}$ are assumed to be independent given the target variable Y. Despite this strong assumption this classifier has proved to be very effective (Hand and Yu 2001) on many real applications and is often used on data stream for supervised classification (Gama 2010).

This "naïve" assumption allows us to compute the model directly from the univariate conditional estimates $P(X_k|C)$. For an instance denoted n, the probability of the target modality C conditionally to the value of the explanatory variables is computed according to the formulae[1]:

$$P_w(Y = C|X = x^n) = \frac{P(Y = C) \prod_{k=1}^{K} p(x_k^n|C)}{\sum_{j=1}^{J} P(C_j) \prod_{k=1}^{K} p(x_k^n|C_j)} \tag{1}$$

The literature shows that variable selection (Koller and Sahami 1996; Langley et al. 1992) or model averaging (Hoeting et al. 1999) can improve the classification results for batch learning. These two processes can be mixed iteratively. Moreover Boullé (2006) shows the close relation between weighting variables and averaging naive Bayes classifiers in the sense that, in the end, the two processes produce a similar single model where a weight is given to each explanatory variable. Equation (1) is just turned to the following equation:

$$P_w(Y = C|X = x^n) = \frac{P(Y = C) \prod_{k=1}^{K} p(x_k^n|C)^{w_k}}{\sum_{j=1}^{J} P(C_j) \prod_{k=1}^{K} p(x_k^n|C_j)^{w_k}} \tag{2}$$

In this paper, we particularly focus on weighing variables for data streams. We are not interested by learning ensemble of models which are then combined by fusion or selection (Kuncheva and Rodríguez 2007) or ensemble of models where individual classifiers do not share the same subset of used variables (Godec et al. 2010). One of the advantages of the classifier described by Eq. (2) in the context of data stream is its low complexity for deployment, which only depends on the number of explanatory variables: a weighted naïve Bayes classifier is completely described by its weight vector $w = (w_1, w_2, \ldots, w_K)$. The interpretation of the results is also simpler that in case of ensemble of models.

Within the 'weighted naive Bayes classifier' family, we can distinguish:

---

[1] We consider in this paper that estimates of prior probabilities $P(Y = C_j)$ and of conditional probabilities $p(x_k|C_j)$ are available. In our experiments, these probabilities are estimated using univariate discretization or grouping according to the MODL method (see Boullé 2007b).

- classifiers with weights equal to 1. It corresponds to the standard naïve Bayes classifier that uses all the explanatory variables.
- classifiers with boolean weights. It corresponds to the selective naïve Bayes classifiers which selects a subset of explanatory variables. The selection is generally done by optimizing a criteria over $\{0, 1\}^K$. However, when the variables number is high, such a browsing is infeasible and only a sub-optimal browsing of the space $\{0, 1\}^K$ can be completed.
- classifiers with continuous weights in $[0, 1]^K$. Such classifiers can be obtained by averaging classifiers with boolean weights with a weighting proportional to the posterior probability of the model (Hoeting et al. 1999) or proportionally to their compression rate (Boullé 2007a). However, for datasets with a very high number of variables, we observe that the models issued from averaging keep a lot of variables, which make the obtained classifiers both costly to deploy and difficult to interpret.

In the work presented in this paper, we are interested in direct estimation of the weight vector by optimization of the regularized log-likelihood in $[0, 1]^K$. Our main expectation is to obtain parcimonious robust models with less variables and equivalent performance. Preliminary works have shown the interest of such a direct estimation of the weights (Guigourès and Boullé 2011).

Moreover, the purpose of this work has been also to focus on (i) a proposition of a sparse regularization of the log-likelihood in Sect. 2 consistent with previous offline approach (Boullé 2007a) (ii) the setup of an online and anytime algorithm with limited budget dedicated to the optimization of the regularized criterion in Sect. 3 (iii) an evaluation of the obtained models in terms of parcimony, predictive performance and robustness. Experiments are presented in Sect. 4, before the conclusion and future work statement.

2 Construction of a Regularized Criterion

Given a dataset $D_N = (x_n, y_n)_{n=1}^N$, we are looking for the minimization of the negative log-likelihood, which is given by:

$$ll(w, D_N) = -\sum_{n=1}^{N} \left(\log P(Y = y^n) + \sum_{k=1}^{K} \log p(x_k^n | y^n)^{w_k} - \log \left(\sum_{j=1}^{J} P(C_j) \prod_{k=1}^{K} p(x_k^n | C_j)^{w_k} \right) \right)$$
(3)

Considered as a classical optimization problem, the regularization of the log-likelihood is performed by the addition of a regularization term, also called prior term, which expresses constraints on the weight vector w. The regularized criterion is:

$$CR^{D_N}(w) = -\sum_{n=1}^{N} ll(w, (x_n, y_n)) + \lambda f(w, X_1, \ldots, X_K) \tag{4}$$

where ll refers to the log-likelihood, f is the regularization function, and λ the regularization weight.

Several objectives have guided our choice of the regularization function:

1. Its sparsity, i.e. it favors the weight vectors composed of as much null components as possible. The L^p norm functions are usually employed with the addition of a regularization term of the form $\sum_{k=1}^{K} |w_k|^p$. All these functions are increasing and hence favor the weight vectors with low components. For $p >= 1$, the norm function L^p is convex, which makes the optimization easier and renders this function attractive. This explain the success of L^2 regularization in many contexts. For ill-posed linear problems, the ridge regression also called Tikhonov regularization (Hoerl and Kennard 1970) uses the L^2 norm. However, the minimization of the regularization terms for $p > 1$ does not necessarily lead to variables elimination whereas the choice $p \leq 1$ favors sparse weight vectors. The Lasso method and its variants (Trevor et al. 2015) exploit the advantages of the value $p = 1$, which enables sparsity and convex optimization. For $p < 1$, the L^p regularization more exploits the sparsity effect of the norm but conducts to non convex optimization.
2. Its ability to take into account a B_k coefficient associated to each explanatory variable so that, for equivalent likelihoods, the "simple" variables are preferred to "complex" ones. By weighting the term with L^p norm by such a coefficient, we obtain a penalization term of the form: $\sum_{k=1}^{K} B_k * |w_k|^p$. This coefficient is supposed to be known before the optimization. If no knowledge is available, this coefficient is fixed to 1. It can be used to include expert knowledge. In our case, this coefficient translates the preparation cost of the variable, i.e. the discretization cost for a numerical variable, resp. the grouping cost for a categorical variable described in equations (2.4), resp. (2.7), of Boullé (2007b).
3. Its consistency with the regularized criterion of the MODL naïve Bayes classifier with binary selection of variables (Boullé 2007a). In order that the two criteria coincide for $\lambda = 1$ and w_k with boolean values, we finally use the regularization term:

$$f(w, X_1, \ldots, X_K) = \sum_{k=1}^{K} (\log K - 1 + B_k) * w_k^p$$

3 Optimization Algorithm: Gradient Descent with Mini-Batches and Variable Neighborhood Search

Let $p_n = P(Y = y^n)$, $p_j = P(Y = C_j)$, $a_{k,n} = p(x_k^n|y^n)$, $a_{k,j} = p(x_k^n|C_j)$ be all constant quantities in this optimization problem.

The regularized criterion to minimize can be written:

$$CR^{D_N}(w) = -\sum_{n=1}^{N}\left\{\log p_n + \sum_{k=1}^{K}\left(w_k * \log a_{k,n}\right) - \log\left(\sum_{j=1}^{J}p_j\prod_{k=1}^{K}(a_{k,j})^{w_k}\right)\right\}$$

$$+ \lambda\sum_{k=1}^{K}(\log K - 1 + B_k)*w_k{}^p \tag{5}$$

Let us note that for $w = \{0\}^K$, i.e. without using explanatory variables, the criterion value is equal to

$$CR^{D_N}(\{0\}^K) = -\sum_{n=1}^{N}\log p_n = -N\sum_{j=1}^{J}p_j\log p_j \tag{6}$$

that is to say N times the Shannon entropy. For each n there is actually a j_n such that $y_n = C_{j_n}$. If we denote by N_j the number of instances among n such that $y_n = C_j$, then

$$-\sum_{n=1}^{N}\log p_n = -\sum_{j=1}^{J}N_j\log p_j = -\sum_{j=1}^{J}N*\frac{N_j}{N}\log p_j = -N\sum_{j=1}^{J}p_j\log p_j \tag{7}$$

We want to optimize the criterion $CR^{D_N}(w)$ subject to the constraint that w takes its values in $[0, 1]^K$ in order to obtain interpretable models. Our objective function consists of two terms. The first term is a convex function of w. In order to see this, let us represent its partial term $LL_n(w)$ in the following form :

$$LL_n(w) = \alpha_n + <c_n, w> + \log\left(\sum_{j=1}^{J}\exp^{-<b_{n,j},w>-\beta_j}\right), \tag{8}$$

where $\alpha_n = -\log p_n$, $c_n^{(k)} = -\log a_{k,n}$, $k = 1, \ldots, K$, vectors $b_{n,j} \in R^K$ have components $b_{n,j}^{(k)} = -\log a_{k,j}^n$, $k = 1, \ldots, K$, and $\beta_j = -\log P(C_j)$. The first and second term of LL_n are resp. constant and linear in w and then are both convex. The third term is convex because of the log-convexity of $\exp(x)$ (see for instance (Lange 2004) for definition and property of log-convexity). The second term (regularization term) is more complicated : it is not convex for $p < 1$ and its partial derivative are unbounded at the points with zero components. This makes impossible to establish theoretical guarantees even for convergence to a local solution. Efficient approaches have recently been proposed, which exploit sparsity and stochastic algorithms (Bach and Moulines 2013; Pilanci et al. 2015). However, these approaches rely on convex optimization criteria. In our case of a non convex optimization problem, the main available approach is the simplest gradient method (Nesterov 2004). This criterion is not convex but differentiable at each weight vector with partial derivative:

Algorithm 1: Projected gradient descent with mini-batches (PGDMB)

   Inputs : D : data stream
               N : historical depth to evaluate the criterion
               L : batch size used for weights update with $L << N$
               w^0 : initial weight vector
               η_0 : initial step vector
               Max : maximal number of iterations
               Tol : tolerated number of successive degradations
   Outputs: $w^* = argmin CR^D(w)$
             t_{total} = performed iterations number

1 **Initialization:**
2 Iteration index : $t = 0$;
3 Number of successive criterion degradations : $N_{deg} = 0$;
4 **while** *(Criterion improvement* **or** $N_{deg} < Tol$*)* **and** $t < Max$ **do**
5 | $D_{t,L}$=t-th batch of size L
6 | $D_{t,N}$=data historical of size N ending at the end of $D_{t,L}$
7 | $w^{t+1} = P_{[0,1]^K}\left(w^t - \eta_t \frac{1}{L} \nabla CR^{D_{t,L}}(w^t)\right)$
8 | Compute η_{t+1}
9 | Compute the criterion value on data historical of size N: $CR^{D_{t,N}}(w^{t+1})$
10 | **if** $CR^{D_{t,N}}(w^{t+1}) < CR^{D_{t-1,N}}(w^t)$ *(i.e. criterion improvement)* **then**
11 | | Best value storage: $w^* = w^{t+1}$
12 | **else**
13 | | Increment the counter of successive degradations : $N_{deg} = N_{deg} + 1$
14 | $t = t + 1$;

$$\frac{\partial CR^{D_N}(w)}{\partial w_\gamma} = -\sum_{n=1}^{N}\left\{\log a_{\gamma,n} - \frac{\sum_{j=1}^{J} p_j \log a_{\gamma,j} \prod_{k=1}^{K}(a_{k,j})^{w_k}}{\sum_{j=1}^{J} p_j \prod_{k=1}^{K}(a_{k,j})^{w_k}}\right\}$$
$$+ \lambda(\log K - 1 + B_k) * p * w_\gamma^{p-1} \tag{9}$$

The gradient $\nabla CR^{D_N}(w^t)$ is the vector of partial derivatives for $\gamma = 1, \ldots, K$.

To respect the constraint that w takes its values in $[0, 1]^K$, we have been interested in projected gradient descent algorithm (Bertsekas 1976) i.e. a gradient descent algorithm for which, at each iteration, the obtained w vector is projected on $[0, 1]^K$.

Several objectives have guided our choice for the algorithmic structure:

1. online algorithm: the algorithm structure is adapted to data stream processing and it does not need the processing of the entire dataset;
2. anytime algorithm: the algorithm is interruptible and is able to return the best optimization given a budgeted computational time.

Within a classical batch gradient descent algorithm, the weight vector is updated at each iteration t according to the gradient computed on all the instances. If the weight vector obtained at iteration t is denoted by w^t, the projected update at $t + 1$ iteration is performed according to the equation:

$$w^{t+1} = P_{[0,1]^K}[w^t - \eta_t \nabla CR^{D_N}(w^t)] \tag{10}$$

Algorithm 2: Projected gradient descent with variable neighbor search (PGDMB-VNS)

Inputs : T : total maximal number of iterations
           NeighSize: initial neighborhood size
Inputs : (PGDMB): D : data stream
           N : historical depth to evaluate the criterion
           L : batch size used for weights update
           η_0 : initial step vector
           Max : maximal number of iterations for one PGDMB optimization
           Tol: tolerated number of successive degradations
Outputs: $w^* = argmin CR^D(w)$

1 **Initialization :**
2 Initial weight vector for the first projected gradient descent with mini-batches (PGDMB) :
   $w_1^0 = \{0.5\}^K$;
3 Initial optimal weight vector : $w^* = w_1^0$;
4 Initial iterations sum $SumT = 0$;
5 **while** $SumT < T$ **do**
6   Compute $(w_m^*, t_{total}^m) = PGDMB(D, N, L, w_m^0, \eta_0, \text{Max}, \text{Tol})$
7    $SumT = SumT + t_{total}^m$
8   **if** *Improvement on w^** **then**
9 | Storage of $w^* = w_m^*$
10   **else**
11 | NeighSize $= min(2 * \text{NeighSize}, 1)$
12    $w_{m+1}^0 = P_{[0,1]^K}(w_m^* + Random([-\text{NeighSize}, \text{NeighSize}]))$

---

where the η step may, according to the variants, be a scalar constant or vary across the iterations and/or vary according to the weight vector components. We have chosen to compute η according to the Rprop method detailed later in this section. The projection $P_{[0,1]^K}$ on $[0, 1]^K$ just consists in bounding obtained values in interval $[0, 1]$. This batch approach assumes that the entire dataset is available to start the optimization.

In its stochastic version, the update is done using the gradient computed on one single instance. The gradient descent may turn out to be chaotic if the variance of the gradient from one instance to another one is high.

Aiming for an online approach, we have retained a variant mixing batch and stochastic, namely mini-batch approach (Dekel et al. 2012) which consists in directing the descent according to gradients computed on successive data batches of length L. To be able to compare descent paths when the size of mini-batches varies, we used a gradient standardized by the size of the mini-batches. The projected gradient descent with mini-batches is summarized in Algorithm 1.

The optimal value for step η_t has been the subject of several studies leading to more or less costly algorithms. We turned to the Rprop method (Riedmiller and Braun 1993): the step computation is specific for each vector component i.e. η is a step vector of dimension K, and each vector component is multiplied by a factor which is bigger, resp. smaller than 1, if the partial derivative sign change, resp. does not change from one iteration to another.

As far as the computational complexity is concerned, each iteration needs a criterion evaluation on a sample of size N that is to say a $O(K * N)$ complexity. The classical batch algorithm is obtained for $L = N$ and the stochastic one for $L = 1$.

As the criterion to be optimized is non convex, it often shows many local minima. It is then common to start several gradient descents with distinct random initializations (multi-start approach) hoping that one of these descent paths converges to a lower minimum. In order to make the optimization efficient and not to waste computational time at the beginning of each descent, it is also possible to modify the solution obtained after a given number of iterations in order to get out of a potential local minimum. We propose to use a metaheuristic so that the current solution is regularly randomized within a neighborhood of variable size. This randomization is inspired from the Variable Neighborhood Search (Hansen and Mladenovic 2001). Our approach denoted PGDMB-VNS is described in Algorithm 2. The projected gradient descent with mini-batches is runned several times with different initialization for the weight vector w. The initial weight vectors are generated in a neighborhood of the current optimal weight vector. If the last projected gradient descent improves the optimal weight vector, then the size of the neighborhood is reduced. Otherwise it is increased. This neighborhood variation enables either to exploit promising areas or to explore new areas of the weight vector space.

It can be noticed that, for a neighborhood that completely covers $[0, 1]^K$, the PGDMB-VNS algorithm is equivalent to a multi-start algorithm with random initializations. Besides, we stress that the random perturbations can lead to a non-null component for a weight set to zero after a precedent run. One variable can re-appear during the data stream reading.

The PGDMB-VNS algorithm is anytime in the sense that an estimation of the criterion argmin is available at the end of the first gradient descent and that it is improved afterwards according to the available budget and interruptible at any time. Its entire complexity is $O(T * K * N)$ where T is the total number of budgeted iterations.

4 Experiments

The purpose of the first experiments is to evaluate the optimization quality obtained with PGDMB-VNS according to the size L of the mini-batches and to the total number of iterations T. To study the intrinsic quality regardless of the associated classifier predictive performance, we have set the λ weight value to 0, which means that we directly optimize the non regularized likelihood. The second part of the experiments deals with predictive performance of the classifier obtained by optimization of regularized criterion ($\lambda \neq 0$).

For the whole experiments, the parameters for PGDMB algorithm are set to the following values:

- $w_0 = \{0.5\}^K$
- $\eta_0 = \{10^{-2}\}^K$ with a multiplication by 0.5, resp. 1.2, in case of sign change, resp. no sign change, between two successive gradients
- Max $= 100$ the iteration maximal number (i.e. the number of treated mini-batches). We have checked that this threshold had never been reached for the 36 tested datasets.
- Tol $= 5$ the number of authorized successive degradations

Improvement of the criterion is considered for a decreasing of at least $\epsilon = 10^{-4}$ with regard to the precedent criterion value. The weights smaller than 10^{-3} are set to 0.

When a VNS metaheuristic is applied, the initial neighborhood size is set to $1/16$.

The whole experiments have been done in 10-fold-cross-validation on the 36 UCI datasets described in Table 1. According to the values of L and N, it can be necessary to use the instances in several mini-batches. In this case, the datasets are randomly shuffled between two mini-batches.

In the results, 'SNB' stands for the performance of a selective naïve Bayes classifier with model averaging (Boullé 2007a).

4.1 Experiments on Optimization Quality

First of all, we have studied the PGDMB algorithm performance, that is to say, the performance of the projected gradient descent algorithm according to the mini-batch size denoted L, without using MS or VNS metaheuristic. We have chosen the compression rate as optimization quality indicator. It measures the complement to 1 of the negative logarithm of the model likelihood normalized by the Shannon entropy. As noticed in Sect. 3, for the "random" classifier with only null weights, the negative logarithm of the likelihood is equal to the Shannon entropy, which leads to a compression rate equal to 0. The closer the rate is to 1, the higher is the model likelihood. For model less competitive than the random model, compression rate is negative. The value of the compression rate on train data is then a good indicator of the optimization quality as the non regularized criterion is reduced to the negative log-likelihood.

Figure 1 presents the train and test compression rate averaged on 36 UCI datasets for various mini-batches sizes $L = 100, 1000, N$. In the last case, the choice $L = N$ corresponds to a batch algorithm. The train and test compression rates obtained with SNB classifier (Boullé 2007a) serve as a baseline. The obtained results indicate that, the larger the mini-batches size is, the better is the optimization quality. Moreover, the results obtained for $L = 1000$ and $L = N$ are very similar. The train compression rate is significantly better for batch mode than for $L = 1000$ for 8 of the 16 datasets with $N > 1000$.

Figure 2 presents as an example the serie of the criterion values obtained during optimization according to the mini-batches size $L = 100, 1000, N$ for the Phoneme dataset. For all the 36 datasets, when the mini-batches size decreases, the convergence is faster but more chaotic.

Table 1 Description of the 36 UCI datasets: Ni = instances number, Nv = initial number of variables, Nc = class number

Dataset	Ni	Nv	Nc	Dataset	Ni	Nv	Nc
Abalone	4177	8	28	Mushroom	8416	22	2
Adult	48842	15	2	PenDigits	10992	16	10
Australian	690	14	2	Phoneme	2254	256	5
Breast	699	10	2	Pima	768	8	2
Bupa	345	6	2	Satimage	768	8	6
Crx	690	15	2	Segmentation	2310	19	7
Flag	194	29	8	Shuttle	58000	9	7
German	1000	24	2	SickEuthyroid	3163	25	2
Glass	214	10	6	Sonar	208	60	2
Heart	270	13	2	Soybean	376	35	19
Hepatitis	155	19	2	Spam	4307	57	2
Horsecolic	368	27	2	Thyroid	7200	21	3
Hypothyroid	3163	25	2	Tictactoe	958	9	2
Ionosphere	351	34	2	Vehicle	846	18	4
Iris	150	4	3	Waveform	5000	21	3
LED	1000	7	10	WaveformNoise	5000	40	3
LED17	10000	24	10	Wine	178	13	3
Letter	20000	16	26	Yeast	1484	9	10

Fig. 1 Train and test mean compression rate for 36 UCI datasets

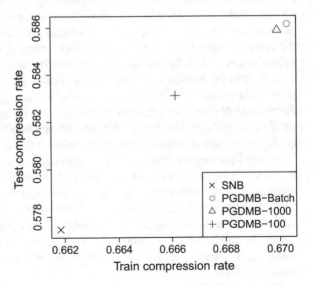

We have compared the optimization quality for the PGDMB algorithm without and with metaheuristic. Two metaheuristics have been tested: multi-start (PGDMB-MS) and variable neighborhood search (PGDMB-VNS).

Fig. 2 Criterion
convergence paths according
to the mini-batches size
(PGDMB) for the Phoneme
dataset

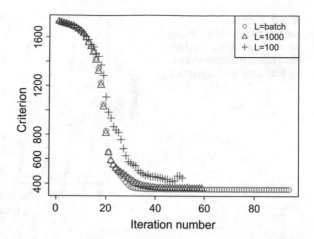

To get computational complexity equivalent to that of the univariate MODL pre-treatment, that is to say $O(K * N * \log(K * N))$, we have fixed the total number of authorized iterations T proportional to $\log(K * N)$. More precisely, we have chosen $T = \log(K * N) * 2^{\text{OptiLevel}}$ where OptiLevel is an integer which enables us to tune the desired optimization level.

For each of the two metaheuristics, we have studied the influence of the optimization level OptiLevel = 3, 4, 5. Since the obtained algorithm stores the best solution each time it is encountered, the metaheuristic can only improve the train compression rate. We have measured in a first step if the improvement was significant or not. For a MS metaheuristic, the train compression rate is significantly improved for resp. 7, 16, 18 of the 36 datasets with an optimization level equal resp. to 3, 4, 5. For a VNS metaheuristic, the train compression rate is significantly improved for resp. 18, 19, 23 of the 36 datasets with an optimization level equal resp. to 3, 4, 5. The VNS metaheuristic seems then better than the MS metaheuristic: the guided exploration within a variable sized neighborhood from the best minimum encountered enables a more fruitful exploration than a purely random exploration.

Figure 3 illustrates this iterations "waste" phenomenon with multi-start at the beginning of each start.

Experiments presented in this section have illustrated the effect of the mini-batches size on the optimization quality. They have also illustrated that the higher the size, the better the optimization quality, and that the VNS metaheuristic works better than the MS metaheuristic. For the rest of the experiments, we then retain a PGDMB-VNS algorithm with mini-batches size fixed to $L = 1000$ and an optimization level set to OptiLevel = 5.

Fig. 3 Criterion
convergence paths according
to the metaheuristic used for
the Phoneme dataset and an
optimization level equal to 5

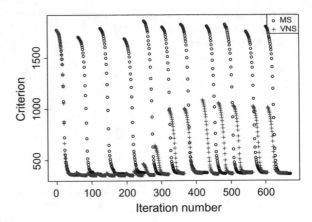

4.2 Regularized Classifier Performance

We present the classifier performance according to the setting of the regularization
weight λ and to the p exponent of the regularization function $|w_k|^p$. Three values
have been tested for λ, 0.01, 0.1, 0.5 and three values for p, 0.5, 1, 2. The AUC performance for the nine regularized classifiers are presented in Fig. 4. The performance
of the non-regularized classifier obtained with $\lambda = 0$ and of the SNB classifier have
been added as a baseline.

For the highest regularization weight, $\lambda = 0.5$ (in purple in the Figure), the AUC
performance are deteriorated with regards to the performance obtained without regularization (red circles in the Figure) whatever the p value. For the other weight values

Fig. 4 Train and test AUC
averaged for 36 UCI datasets
according to the weight and
the type of regularization

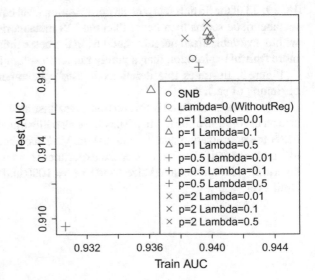

$\lambda = 0.01$ and $\lambda = 0.1$, the performance are similar for all p values and slightly greater or equal on average to those of the non-regularized classifier. These two regularization weights lead to statistical performance equivalent to those of non regularized classifier.

Figure 5 presents a study on the sparsity of the obtained classifiers. In this Figure, the number of kept variables and their weights sum are presented according to the weight and type of regularization. First, it shows that the smaller p, the smaller the non-null weights number. The quadratic regularization ($p = 2$) leads to non sparse classifiers. Among the regularization with absolute value ($p = 1$) and the squared

Fig. 5 Kept variable number and weight sum averaged for 36 UCI datasets according to the weight and the type of regularization

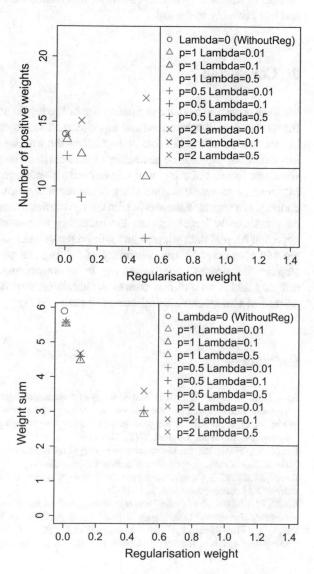

root one ($p = 0.5$), the second one enables the most important reduction of the number of kept variables.

As far as the weights sum is concerned, all the p regularization exponents enable to reduce the weights sum on average. Moreover, given a λ weight, the quadratic regularization has a less important impact on the weights sum reduction than the two other regularizations whose performance are very close for this indicator.

Considering both aspects of statistical performance and classifier sparsity, the compromise $p = 1$ and $\lambda = 0.1$ seems the most favorable. Without deteriorating the performance of the non regularized classifier, it enables a significant reduction of the number of selected variables. This reduction makes the classifier more interpretable and less complex to deploy.

5 Conclusion

We have proposed a sparse regularization of the log-likelihood for a weighted naïve Bayes classifier. We described and experimented a gradient descent algorithm, which treats online mini-batches data and optimizes the weights classifier through a more or less extensive exploration according to the iterations budget. The experiments have shown the interest of using mini-batches and a metaheuristic for deeper optimization. Moreover, a parameterization study of the regularization points out that the optimal choice was a regularization term with the L_1 norm and a weight $\lambda = 0.1$. Experiments on substantially larger datasets are necessary to evaluate the performance of our approach on real data streams and will be the subject of future work. Furthermore, we will also consider the possibility of solving our optimization problem in two steps : a first convex optimization step using convex approximation of the complete criterion and a second non convex optimization step that could be solved by the method of Composite Minimization (Nesterov 2013).

References

Bach, F., & Moulines, E. (2013). Non-strongly-convex smooth stochastic approximation with convergence rate O(1/n). In *Neural information processing systems (NIPS)* (pp. 773–781). USA

Bertsekas, D. P. (1976). On the goldstein-levitin-polyak gradient projection method. *IEEE Transactions on Automatic Control, 21*(2), 174–184.

Boullé, M. (2006). Regularization and averaging of the selective naive bayes classifier. In *International Joint Conference on Neural Network Proceedings* (pp. 1680–1688).

Boullé, M. (2007a). Compression-based averaging of selective naive bayes classifiers. *Journal of Machine Learning Research, 8*, 1659–1685.

Boullé, M. (2007b). *Recherche d'une représentation des données efficace pour la fouille des grandes bases de données*. PhD thesis, Ecole Nationale Supérieure des Télécommunications.

Dekel, O., Gilad-Bachrach, R., Shamir, O., & Xiao, L. (2012). Optimal distributed online prediction using mini-batches. *Journal of Machine Learning Research, 13*(1), 165–202.

Gama, J. (2010). *Knowledge discovery from data streams* (1st ed.). Boca Raton: Chapman & Hall, CRC.

Godec, M., Leistner, C., Saffari, A., & Bischof, H. (2010). On-line random naive bayes for tracking. In *International Conference on Pattern Recognition (ICPR)* (pp. 3545–3548). IEEE Computer Society.

Guigourès, R., & Boullé, M. (2011). Optimisation directe des poids de modèles dans un prédicteur bayésien naif moyenné. In *13èmes Journées Francophones "Extraction et Gestion de Connaissances" (EGC 2011)* (pp. 77–82).

Hand, D. J., & Yu, K. (2001). Idiot's bayes-not so stupid after all? *International Statistical Review, 69*(3), 385–398.

Hansen, P., & Mladenovic, N. (2001). Variable neighborhood search: Principles and applications. *European Journal of Operational Research, 130*(3), 449–467.

Hoerl, A. E., & Kennard, R. W. (1970). Ridge regression: Biased estimation for nonorthogonal problems. *Technometrics, 12*, 55–67.

Hoeting, J. A., Madigan, D., Raftery, A. E., & Volinsky, C. T. (1999). Bayesian model averaging: A tutorial. *Statistical Science, 14*(4), 382–417.

Koller, D., & Sahami, M. (1996). Toward optimal feature selection. In *International Conference on Machine Learning* (pp. 284–292).

Kuncheva, L. I., & Rodríguez, J. J. (2007). Classifier ensembles with a random linear oracle. *IEEE Transactions on Knowledge and Data Engineering, 19*(4), 500–508.

Lange, K. (2004). *Optimization*. Springer Texts in Statistics. New York: Springer.

Langley, P., Iba, W., & Thompson, K. (1992). An analysis of bayesian classifiers. In *National Conference on Artificial Intelligence* (pp. 223–228).

Nesterov, Y. (2004). *Introductory lectures on convex optimization: A basic course*. Applied optimization. Boston: Kluwer Academic Publishers.

Nesterov, Y. (2013). Gradient methods for minimizing composite functions. *Mathematical Programming, 140*(1), 125–161.

Pilanci, M., Wainwright, M. J., & Ghaoui, L. (2015). Sparse learning via boolean relaxations. *Mathematical Programming, 151*(1), 63–87.

Riedmiller, M., & Braun, H. (1993). A direct adaptive method for faster backpropagation learning: The RPROP algorithm. In *IEEE International Conference On Neural Networks* (pp. 586–591).

Trevor, H., Robert, T., & Martin, W. (2015). *Statistical learning with sparsity: The lasso and generalizations*. Boca Raton: Chapman and Hall/CRC.

On Making Skyline Queries Resistant to Outliers

Hélène Jaudoin, Pierre Nerzic, Olivier Pivert and Daniel Rocacher

Abstract This paper deals with the issue of retrieving the most preferred objects (in the sense of Skyline queries, i.e., of Pareto ordering) from a collection involving outliers. Indeed, many real-world datasets, for instance from ad sales websites, contain odd data and it is important to limit the impact of such odd data (outliers) on the result of skyline queries, and prevent them from hiding more interesting points. The approach we propose relies on the notion of fuzzy typicality and makes it possible to compute a graded skyline where each answer is associated with both a degree of membership to the skyline and a typicality degree. A GPU-based parallel implementation of the algorithm is described and experimental results are presented, which show the scalability of the approach.

1 Introduction

In this paper, a qualitative view of preference queries is considered, namely the *Skyline* approach introduced in Börzsönyi et al. (2001). Given a set of points in a space, a skyline query retrieves those points that are not dominated by any other in the sense of Pareto order. When the number of dimensions on which preferences are expressed gets high, many tuples may become incomparable. Several approaches have been proposed to define an order for two incomparable tuples, based on the number of other tuples that each of the two tuples dominates (notion of k-representative dominance proposed in Lin et al. (2007)), on a preference order over the attributes (see for instance the notions of k-dominance and k-frequency introduced in Chan et al. (2006a, b)),

H. Jaudoin (✉) · P. Nerzic · O. Pivert · D. Rocacher
Irisa, 6 rue Kerampont, Lannion, France
e-mail: hjaudoin@irisa.fr

P. Nerzic
e-mail: nerzic@univ-rennes1.fr

O. Pivert
e-mail: pivert@irisa.fr

D. Rocacher
e-mail: rocacher@irisa.fr

© Springer International Publishing Switzerland 2017
F. Guillet et al. (eds.), *Advances in Knowledge Discovery and Management*,
Studies in Computational Intelligence 665, DOI 10.1007/978-3-319-45763-5_2

or on a notion of representativity ((Tao et al. 2009) redefines the approach proposed by Lin et al. (2007) and proposes to return only the most representative points of the skyline, i.e., a point among those present in each cluster of the skyline points). Other approaches fuzzify the concept of skyline in different ways, see e.g. Goncalves and Tineo (2007) and Hadjali et al. (2011). See also Rojas et al. (2014) where the authors define soft skylines by relaxing the dominance relation. Here, we are concerned with a different problem, namely that of the possible presence of *exceptional* points, also known as outliers, in the dataset over which the skyline is computed. Such exceptions may correspond to noise or to the presence of *nontypical* points in the collection considered. The impact of such points on the skyline may obviously be important if they dominate some other, more representative ones.

At least two strategies can be considered to handle outliers. The former consists in removing anomalies by adopting cleaning procedures. However, the task of automatically distinguishing between odd points and simply exceptional points is not always easy. Another solution is to define an approach that is *tolerant to outliers*, that highlights representative points of the database and that points out the possible outliers. Literature about outlier detection is very abundant as shown by recent surveys (Hodge and Austin 2004; Niu et al. 2011; Zimek et al. 2012; Zhang 2013) and recent publications in the data mining area (Gupta et al. 2013; Gabel et al. 2013; Zimek et al. 2013; Ji et al. 2013). The approaches may be categorized into three classes: (i) those that aim to isolate data distant from another *normal* data, (ii) those that aim to detect if a new observation is normal or abnormal, (iii) those that exclusively focus on modelling normality and assess if a new observation is likely to be normal.

The estimation of the best approach for detecting outliers is out of the scope of this paper. In the tolerant skyline approach we propose, we use a simple detection technique based on the notions of frequency and distance. More precisely we adopt the fuzzy notion of typicality (Zadeh 1984) in order to identify non-typical, thus exceptional points. We revisit the definition of a skyline and show that it (i) makes it possible to retrieve the dominant points without discarding other potentially interesting ones, and (ii) constitutes a flexible tool for distinguishing between the answers.

The remainder of the paper is structured as follows. Section 2 provides a refresher about skyline queries and motivates the approach. Section 3 presents the principle of exception-tolerant skyline queries, based on the fuzzy concept of typicality. Section 4 deals with implementation aspects whereas Sect. 5 presents experimental results obtained on a real-world dataset. Finally, Sect. 6 recalls the main contributions and outlines perspectives for future work.

2 Refresher About Skyline Queries and Motivations

2.1 Skyline Queries

Let $\mathscr{D} = \{D_1, \ldots, D_d\}$ be a set of d dimensions. Let us denote by $dom(D_i)$ the domain associated with dimension D_i. Let $\mathscr{S}$ be a subset of $dom(D_1) \times \ldots \times dom(D_d)$, p and q two points of $\mathscr{S}$, and $\succ_i$ a preference relation on D_i. One says that p *dominates* q on $\mathscr{D}$ (p is better than q according to Pareto order), denoted by $p \succ_{\mathscr{D}} q$, iff

$$\forall i \in [1, d]: \ p_i \succeq_i q_i \text{ and } \exists j \in [1, d]: \ p_j \succ_j q_j$$

A skyline query on $\mathscr{D}$ applied to a set of points $\mathscr{S}$, whose result is denoted by $\text{SKY}_{\mathscr{D}}(\mathscr{S})$, according to preference relations $\succ_i$, produces the set of points that are not dominated by any other point of $\mathscr{S}$:

$$\text{SKY}_{\mathscr{D}}(\mathscr{S}) = \{p \in \mathscr{S} \mid \nexists q \in \mathscr{S} : q \succ_{\mathscr{D}} p\}$$

Depending on the context, one may try, for instance, to maximize or minimize the values of $dom(D_i)$, assuming that $dom(D_i)$ is a numerical domain.

In order to illustrate the principle of the approach we propose, let us consider the dataset *Iris* (Fisher 1936), graphically represented in Fig. 1.

The vertical axis corresponds to the attribute *sepal width* whereas the horizontal axis is associated with *sepal length*. The skyline query:

```
select * from iris
skyline of sepallength max, sepalwidth max
```

looks for those points that maximize the dimensions *length* and *width* of the sepals (the circled points in Fig. 1).

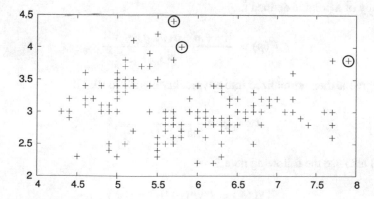

Fig. 1 The dataset *Iris*

In this dataset, the points form two groups that respectively correspond to the intervals [4, 5.5] and [5.5, 7] on attribute *length*. By definition, the skyline points are on the border of the region that includes the points of the dataset. However, these points are very distant from the areas corresponding to the two groups and are thus not very representative of the dataset. It could then be interesting for a user to be able to visualize the points that are "almost dominant", closer to the clusters, then more representative of the dataset. A way to make such points visible without discarding extrema, while allowing to discriminate them, is to use a gradual view of representativity. The notion of typicality discussed in the next section makes it possible to reach that goal.

2.2 Computing a Fuzzy Set of Typical Values

The concept of typicality has been studied by numerous authors, both in cognitive psychology (Rosch and Mervis 1975; Hampton 1988; Osherson and Smith 1997) and in fuzzy logic, see e.g. Zadeh (1982), Friedman et al. (1995) and Yager (1997). In the following, we consider Zadeh's interpretation (Zadeh 1987), but any other interpretation of typicality could be used without drastically altering the general principle of the approach. In Zadeh (1987), x is defined as a typical element of a fuzzy set A iff (i) x has a high degree of membership to A and (ii) most of the elements of A are similar to x. In the case where A is a classical set, this definition becomes: x belongs to A and most of the elements of A are similar to x. In the same spirit, in Dubois and Prade (1984), the authors define a typicality indice based on similarity and frequency. We adapt their definition as follows. Let us consider a set $\mathscr{E}$ of points. We say that a point is all the more typical as it is close to many other points. The proximity relation considered is based on Euclidean distance. We consider that two points p_1 and p_2 are close to each other if $d(p_1, p_2) \leq \tau$ where τ is a predefined threshold. In the experiment performed on the dataset *Iris*, we used $\tau = 0.5$. The frequency of a point is defined as:

$$F(p) = \frac{|\{q \in \mathscr{E}, \ d(p, q) \leq \tau\}|}{|\mathscr{E}|} \tag{1}$$

This degree is then normalized into a typicality degree in [0, 1]:

$$typ(p) = \frac{F(p)}{\max_{q \in \mathscr{E}}\{F(q)\}} \tag{2}$$

We will also use the following notations:

$$\text{TYP}(\mathscr{E}) = \{typ(p)/p \mid p \in \mathscr{E}\}$$

$$\text{TYP}_\gamma(\mathscr{E}) = \{p \mid p \in \mathscr{E} \text{ and } typ(p) \geq \gamma\}.$$

$\text{TYP}(\mathscr{E})$ represents the fuzzy set of points that are somewhat typical of the set $\mathscr{E}$ while $\text{TYP}_\gamma(\mathscr{E})$ gathers the points of $\mathscr{E}$ whose typicality is over the threshold γ.

Example 1 Let us consider the following multiset (of cardinality $n = 30$):

$$E = \langle 1/0,\ 1/3,\ 1/4,\ 4/5,\ 7/6,\ 5/7,\ 3/8,\ 5/9,\ 2/12,\ 1/23 \rangle$$

where k/e means that element e has k copies in E. Using Formulas (1) and (2) with $\tau = 2$, we get:

$$\text{TYP}(\mathscr{E}) = \{0.04/0,\ 0.25/3,\ 0.54/4,\ 0.75/5,\ 0.83/6,$$
$$1/7,\ 0.83/8,\ 0.54/9,\ 0.08/12,\ 0.04/23\}$$

and

$$\text{TYP}_{0.5}(\mathscr{E}) = \{4,\ 5,\ 6,\ 7,\ 8,\ 9\}.\diamond$$

An excerpt of the typicality degrees computed over the *Iris* dataset is presented in Table 1.

Remark 1 This typicality-based interpretation of outliers is close to the approach used in DBSCAN (Ester et al. 1996). Typical points (relative to thresholds τ and γ) correspond to *core* points in DBSCAN. However, as we will see, the fuzzy skyline definition introduced in Sect. 3.2 relies on a *gradual* view of outliers where no threshold γ is applied.

	Length	Width	Frequency	Typicality
Table 1 Excerpt of the *Iris* dataset with associated typicality degrees	7.4	2.8	0.0600	0.187
	7.9	3.8	0.0133	0.0417
	6.4	2.8	0.253	0.792
	6.3	2.8	0.287	0.896
	6.1	2.6	0.253	0.792
	7.7	3.0	0.0467	0.146
	6.3	3.4	0.153	0.479
	6.4	3.1	0.293	0.917
	6.0	3.0	0.320	1.000

3   Principle of the Exception-Tolerant Skyline

As explained in the introduction, our goal is to revisit the definition of the skyline so as to take into account the typicality of the points in the database, in order to control the impact of exceptions or anomalies. Thus, three variants of the classical skyline are defined hereafter:

- $\text{SKY}^1_{\mathscr{D}}$ that returns all sufficiently typical points of $\mathscr{S}$ that are not dominated by sufficiently typical points,
- $\text{SKY}^2_{\mathscr{D}}$ that returns all points of $\mathscr{S}$ that are not dominated by sufficiently typical points, and
- $\text{SKY}^3_{\mathscr{D}}$ that returns a fuzzy set of $\mathscr{S}$, where each point is associated with a membership degree which is a function of the typicality of the points that dominate it.

3.1   Boolean View

The first idea is to restrict the computation of the skyline to a subset of $\mathscr{E}$ that corresponds to sufficiently typical points. The corresponding definition is:

$$\text{SKY}^1_{\mathscr{D}}(\mathscr{S}) = \{p \in \text{TYP}_\gamma(\mathscr{S}) \mid \nexists q \in \text{TYP}_\gamma(\mathscr{S}) \text{ such that } q \succ_{\mathscr{D}} p\} \qquad (3)$$

Such an approach obviously reduces the cost of the processing since only the points that are typical at least to the degree γ are considered in the calculus. However, this definition does not make it possible to discriminate the points of the result since the skyline obtained is a crisp set. Figure 2 illustrates this behavior and shows the maxima (circled points) obtained when considering the points that are typical to a degree ≥ 0.7 (represented by crosses).

Another drawback of this definition is to exclude the nontypical points altogether, even though some of them could be interesting answers. A more cautious definition consists in keeping the nontypical points while computing the skyline and transform Eq. (3) into:

$$\text{SKY}^2_{\mathscr{D}}(\mathscr{S}) = \{p \in \mathscr{S} \mid \nexists q \in \text{TYP}_\gamma(\mathscr{S}) \text{ such that } q \succ_{\mathscr{D}} p\} \qquad (4)$$

Figure 3 illustrates this alternative solution. It represents (circled points) the objects from the *Iris* dataset that are not dominated by any item typical to the degree $\gamma = 0.7$ at least (represented by crosses).

With Eq. (3), the nontypical points are discarded, whereas with Eq. (4), the skyline is larger and includes nontypical extrema. This approaches relaxes skyline queries in such a way that the result obtained is not a polyline anymore but a stripe composed of the regular skyline elements completed with possible "substitutes". However, the main drawbacks of this definition are: (i) the potentially large number of points returned, (ii) the impossibility to distinguish, among the skyline points, those that are not at all dominated from those that *are* dominated (by more or less typical points).

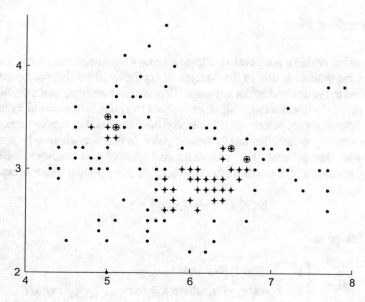

Fig. 2 Skyline of the Iris points whose typicality degree is ≥0.7

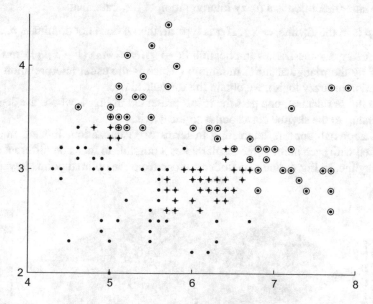

Fig. 3 Points that are not dominated by any other whose typicality degree is ≥0.7

3.2   Gradual View

A third version makes it possible to compute a *graded* skyline, seen as a fuzzy set, that preserves the gradual nature of the concept of typicality. By doing so, no threshold γ is applied to typicality degrees anymore. This variant considers that a point totally belongs to the skyline (membership degree equal to 1) if it is dominated by no other point. A point does not belong at all to the skyline (membership degree equal to 0) if it is dominated by at least one totally typical point. In the case where p is dominated by somewhat (but not totally) typical points, its degree of membership to the skyline is strictly positive and depends on the typicality of these points. The corresponding formula is:

$$\text{SKY}_{\mathscr{D}}^{3}(\mathscr{S}) = \{\mu/p \mid p \in \mathscr{S}\} \tag{5}$$

with μ defined as:

$$\mu = \begin{cases} 1 & \text{if } p \in \text{SKY}_{\mathscr{D}}(\mathscr{S}) \\ 1-t & \text{otherwise, where } t = \max_{q \in \mathscr{S} \mid q \succ_{\mathscr{D}} p} typ(q) \end{cases} \tag{6}$$

This latter definition is a fuzzy interpretation of the statement

p is in the skyline $\Leftrightarrow \forall q$, if q is typical, then q does not dominate p.

Indeed, using Kleene-Dienes implication ($x \rightarrow_{KD} y = max(1 - x, \ y)$) and translating $\forall$ by the triangular norm minimum (which is the usual interpretation of the conjunction in fuzzy logic), we obtain the formula above.

With the *Iris* dataset, one gets the result presented in Fig. 4, where the degree of membership to the skyline corresponds to the z axis.

This approach appears interesting in terms of visualization. Indeed, the score associated with each point of the result makes it possible to focus on different α-cuts of the skyline. In Fig. 4, one may notice a slope from the optimal points towards the

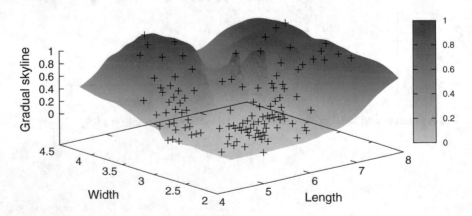

Fig. 4 Graded skyline obtained with the *Iris* dataset

less typical or completely dominated ones. The user may select points that are not necessarily optimal but that represent good alternatives to the regular skyline answers (in the case, for instance, where the latter look "too good to be true"). Finally, an element of the graded skyline is associated with two scores: a degree of membership to the skyline, and a typicality degree (that expresses the extent to which it is *not* exceptional). One may imagine different ways of navigating inside areas in order to explore the set of answers: a simple scan for displaying the characteristics of the points, the use of different filters aimed, for instance, at optimizing diversity on certain attributes, etc.

Remark 2 In Eq. (6), t denotes the typicality of the most typical point that dominates p. A possible refinement would be to take into account, not only the typicality of the most typical point that dominates p, but also the *number* of such points. Indeed, let p and p' be two points which have the same typicality r. Assume that p is dominated by a large number N of points, all of which have typicality r'. Assume further that p' is dominated by only one point, whose typicality is also r'. The approach described above cannot discriminate p and p'. A way to overcome this limitation would be to associate two values with each point p of the fuzzy skyline: a membership degree $\mu(p)$ as defined in Eqs. (5) and (6), and a number $n(p)$ defined as

$$n(p) = |\{q \in \mathscr{S} \mid q \succ_{\mathscr{D}} p \text{ and } typ(q) = \max_{u \in \mathscr{S} \mid u \succ_{\mathscr{D}} p} typ(u)\}| \qquad (7)$$

Finally, each element p of the result would be associated with three degrees: $typ(p)$, $\mu(p)$ and $n(p)$ and the selection of the most interesting points of this fuzzy skyline could be performed by means of a second (classical) skyline query aimed at maximizing both $typ(p)$ and $\mu(p)$, and involving a nested condition (*cascade* clause) minimizing $n(p)$ in order to break ties if any. The detailed study and implementation of such an extension is left for future work.

4 Implementation Aspects

Two steps are necessary for obtaining the graded result: (i) computation of the typicality degrees, and (ii) computation of the graded skyline. The typicality degrees are used to compute the degree of membership of all tuples to the skyline.

Let us first investigate the methods to process Boolean skylines. Lee et al. (2007) highlights two key points: (i) dominance tests are expensive, (ii) the organization of skyline candidates and their examination strategies are critical for the efficiency of the dominance test. This second point relates specifically to Boolean skyline queries. With regards to these points, three categories of algorithms have been proposed in the literature. The first category is based on Divide and Conquer (Börzsönyi et al. 2001). Processing is done separately on subsets of data, then the results are merged. The second category relies on exploiting indices such as B-trees or R-trees to organize the data in order to avoid many dominance tests in Papadias et al. (2005), or like bitmap indices in Tan et al. (2001). The third category gathers various methods based on sorting: Block-Nested-Loops (BNL) (Börzsönyi et al. 2001) and its improvement named Sort-Filter-Skyline (Chomicki et al. 2003, 2005), and a strategy proposed in Bartolini et al. (2008) that relies on a preordering of the tuples aimed to limit the number of elements to be accessed and compared.

We propose to implement the computation of both the typicality degrees and the skyline by parallelizing most of the calculations, so that they can be run on a CUDA processor. CUDA (Compute Unified Device Architecture) is a parallel application programming model defined by *nVidia* for their graphic cards. CUDA allows for developing programs that contain some functions to be executed on the GPU (Graphic Processing Unit) by taking advantage of the large number of computing cores. These functions, called *kernels*, are run simultaneously on different data. The number of parallel executions depends on the number of physical cores available on the GPU.

We can expect a major gain in computation speed with a large amount of cores, but only if the algorithm can be parallelized in the manner that CUDA expects. Indeed, CUDA has some drawbacks which can slow down the whole process, if they are not addressed properly. Generally, CUDA threads are designed to process an array in parallel, with one thread per cell. There are some recommandations for a thread to avoid random accesses to another cell, due to the very high latency delays of the global memory.

Only few research works deal with parallel implementation of skyline queries. Choi et al. (2012) proposes an adaptation of the Block-Nested-Loops algorithm to the CUDA architecture, taking the number of really parallel threads into consideration, in order to balance the workload. Park et al. (2009) compares two different approaches. The first approach parallelizes the BBS Branch-and-Bound algorithm proposed by Papadias et al. (2005). The second approach called *pskyline* for *parallel skyline*, aims at being much simpler, and is based on a *map-reduce* paradigm that we will describe later. Both approaches are implemented on OpenMP, and appear to have similar performances. However, in the second approach, the computation of the skyline itself is not fully parallelized. It uses a sequential function which computes a partial skyline by iterations on a subset of the tuples. Moreover, this function makes use of the memory in a way that could slow down CUDA a lot.

Bøgh et al. (2013) proposes a fast implementation of skyline queries on CUDA, taking finely into account all the particularities of the architecture. The first step is to sort the tuples according to the Manhattan distance, as the tuples at the top of the list may more likely belong to the skyline. This sorting also guarantees that a data point cannot be dominated by a successor. Then, the algorithm processes the tuples by subsets of fixed size at each major iteration. The authors have also carefully designed the computations to avoid branches in the CUDA kernels and minimize data transfers between GPU and CPU.

Unfortunately, in our problem, there is no motivation in efficiently selecting candidate tuples, by presorting the set and removing dominated tuples from the candidates. Indeed all tuples may belong to the graded skyline, but with different degrees, from 0^+ (very low membership to the skyline) to 1 (total membership to the skyline). We have to compare each tuple with every other tuples to determine its degree. So the main issue remains the efficient computation of typicality degrees and domination, as stated by Eqs. (2) and (5).

This leads to Algorithm 3, composed of two loops over the dataset $\mathscr{S}$. The outer loop computes the membership degree of all the tuples p. The inner loop compares p with all the other tuples and those who dominate p are used to get the highest dominant typicality.

Algorithm 3: Sequential Algorithm for computing the graded skyline

---

Require: dataset $\mathscr{S}$, typicality degree $Typ(p)$ available for all tuple p in $\mathscr{S}$
Ensure: graded skyline: $\forall\, p \in \mathscr{S}$, $Sky_{grad}(p)$
  for all $p \in \mathscr{S}$ **do**
    $best \leftarrow 0$
    for all $q \in \mathscr{S}$ **do**
      if $q \succ_{\mathscr{D}} p$ and $best < Typ(q)$ **then**
        $best \leftarrow Typ(q)$
      end if
    end for
    $Sky_{grad}(p) \leftarrow 1 - best$
  end for

Typicality degrees $Typ(p)$ are computed by Algorithm 4. Note that this second algorithm could be improved to avoid the loop that computes the maximum value of the frequency F, but this would hinder the parallelization.

Algorithm 4: Sequential Algorithm for computing the typicality degrees

---

Require: dataset $\mathscr{S}$ of cardinality n
Ensure: typicality degrees: $\forall\, p \in \mathscr{S}$, $Typ(p)$
  // compute the frequency
  for all $p \in \mathscr{S}$ **do**
    $neighbors \leftarrow 0$
    for all $q \in \mathscr{S}$ **do**
      if $d(p,q) \leq \tau$ **then**
        $neighbors \leftarrow neighbors + 1$
      end if
    end for
    $F(p) \leftarrow neighbors/n$
  end for
  // compute the maximum frequency
  $maxF \leftarrow 0$
  for all $p \in \mathscr{S}$ **do**
    if $F(p) > maxF$ **then**
      $maxF \leftarrow F(p)$
    end if
  end for
  // compute the typicality degrees
  for all $p \in \mathscr{S}$ **do**
    $Typ(p) \leftarrow F(p)/maxF$
  end for

The algorithm proposed in Tan et al. (2001) is well-suited to the calculation of the graded skyline, because it generates for each tuple p a Boolean vector D of the size of the database, and indicating for each cell $D[q]$ whether the tuple q dominates p. Then, one has just to replace each value equal to *true* in D by the typicality of the related tuple in order to find the most typical tuple which dominates p. However, the entire algorithm is not directly parallelizable on CUDA, both because of the data structure necessary (big bitmap index), and calculations such as Boolean operations on these bitmaps.

Therefore, we propose parallel algorithms based on general parallelization principles, that can be implemented efficiently with CUDA to evaluate Formulas (2) and (5) keeping the spirit of the approach of Tan and getting some inspiration from Bøgh et al. (2013) and Park et al. (2009).

4.1 Parallel Algorithm Principles

We presented the sequential version of the graded skyline computation in Algorithms 3 and 4. They contains several *for* loops that can easily be parallelized because they are kinds of *map* and *reduce* operations. In Skeletal Parallel Programming (Cole 1991), $map(f, C)$ is an operation that applies in parallel the same function f to every element of the collection C, and returns the resulting collection. If $C = \{c_1, c_2, \ldots, c_n\}$, $map(f, C)$ returns $\{f(c_1), f(c_2), \ldots, f(c_n)\}$. The other operation, $reduce(f, C)$ returns the aggregation of an associative binary operator f. $reduce(f, C)$ returns the value $c_1 \ f \ c_2 \ f \ldots f \ c_n$. For instance, $reduce(\lambda x, y : x + y, map(\lambda x : x^2, \{2, -3, 4\})) = 29$. Note that the *lambda* notation is necessary when functions have no name. The parameters come from the collection to process. The free variables, if any, come from the context.

Both operations, $map(f, C)$ and $reduce(f, C)$ are parallelizable with high performances in CUDA. Function f must be written as a CUDA kernel—this is a function which will run on the GPU. The dataset C shall be put into an array in the memory of the graphic card, called *device global memory*. Another array of appropriate size must be allocated to receive the results. Then, for *map*, the kernel is launched with as many instances as there are data to process. So, in theory, $map(f, C)$ has the same order of complexity than function f (multiplied by the collection size, divided by the number of parallel threads). For *reduce*, the kernel is launched hierarchically following a binary tree scheme. If f has a time complexity of $\theta(1)$, then $reduce(f, C)$ has a complexity of $\theta(\log_2 n)$ where n is the cardinality of C.

Using these principles, we propose Algorithm 5 that computes the degree of membership to the skyline of every tuple in the dataset. We use a convention in C/C++ about Booleans: *true* is equivalent to the integer value 1, and *false* to 0. This allows to replace a condition by a multiplication in the kernel. We can combine the typicality degree with the Boolean test $q \succ p$, and then get the degree by reducing with the *maximum* function.

Algorithm 5: Parallel algorithm for computing the graded skyline

Require: dataset $\mathscr{S}$, typicality degree $Typ(p)$ available for every tuple p in $\mathscr{S}$
Ensure: graded skyline: $\forall\ p \in \mathscr{S},\ Sky_{grad}(p)$
  for all $p\ \in \mathscr{S}$ **do**
    $best \leftarrow reduce(max,\ map(\lambda q : (q \succ_{\mathscr{D}} p) * Typ(q),\ \mathscr{S}))$
    $Sky_{grad}(p) \leftarrow 1 - best$
  end for

The computation of the test $q \succ_{\mathscr{D}} p$ is performed by Algorithm 6. It consists of a loop over the attributes of both tuples p and q. We use the technique proposed by Bøgh et al. (2013) to avoid then-else branches. Here, we handle two Boolean variables $p\_does\_not\_dominate\_q$ and $q\_does\_not\_dominate\_p$.

$p\_does\_not\_dominate\_q$ is changed to *true* when q is better than p on at least one attribute. Reciprocally, $q\_does\_not\_dominate\_p$ is changed to *true* when p is better than q. At the end of the loop, q dominates p $(q \succ_{\mathscr{D}} p)$ iff $p\_does\_not\_dominate\_q$ is *true* and $q\_does\_not\_dominate\_p$ is *false*. The loop on the attributes in Algorithm 6 has not been parallelized but has been unrolled, following CUDA guidelines (Harris 2007).

Algorithm 6: Determination of the domination $q \succ_{\mathscr{D}} p$ (Bøgh et al. 2013)

Require: tuples p and q, each one is a sub-array of attributes $attributes[1..number\ of\ attributes]$
Ensure: $q \succ_{\mathscr{D}} p$
  $p\_does\_not\_dominate\_q \leftarrow 0$
  $q\_does\_not\_dominate\_p \leftarrow 0$
  for all $i\ \in 1..number\ of\ attributes$ **do**
    $attribute\_p \leftarrow p.attributes[i]$
    $attribute\_q \leftarrow q.attributes[i]$
    $p\_does\_not\_dominate\_q \leftarrow p\_does\_not\_dominate\_q$ or $(attribute\_q \succ attribute\_p)$
    $q\_does\_not\_dominate\_p \leftarrow q\_does\_not\_dominate\_p$ or $(attribute\_p \succ attribute\_q)$
  end for
  return $(p\_does\_not\_dominate\_q > q\_does\_not\_dominate\_p)$

Algorithm 7 computes the typicality degree of every tuple with the same kind of operations. We also use the Boolean $d(p, q) \leq \tau$ as a number 0 or 1 that we accumulate along the tuples. Contrary to Algorithm 4, we compute the numbers of neighbors, instead of the frequencies, because $\frac{F(p)}{maxF} = \frac{nb\_neighbors(p)}{max\_nb\_neighbors}$.

Algorithm 7: Parallel algorithm for computing the typicality

Require: dataset $\mathscr{S}$ of cardinality n
Ensure: typicality degrees: $\forall\, p \in \mathscr{S},\ Typ(p)$
 // Compute the number of neighbors of all tuples
 for all $p \in \mathscr{S}$ **do**
  $nb\_neighbors(p) \leftarrow reduce(sum,\ map(\lambda q : d(p,q) \leq \tau,\ \mathscr{S}))$
 end for
 // Compute the highest number of neighbors
 $max\_nb\_neighbors \leftarrow reduce(max,\ nb\_neighbors)$
 // Compute the typicality of every tuple
 $Typ \leftarrow map(\lambda p :\ nb\_neighbors(p)/max\_nb\_neighbors,\ \mathscr{S})$

4.2 Implementation in CUDA

These algorithms are easy to code in CUDA, using the Thrust library.[1] Thrust is a CUDA Toolkit resembling the C++ Standard Template Library (STL). For instance, Thrust makes it simple to allocate an array on the GPU, to exchange data with the CPU and to launch *map* and *reduce* operations on predefined or custom kernels.

All the algorithms can be implemented this way. However, to obtain maximum speed, some computations have to be implemented more efficiently. With Thrust, each *reduce* operation brings back the result into the CPU memory. In some cases, it would be better to keep it in the GPU memory. For instance, in Algorithm 7, the instruction $F(p) \leftarrow reduce(sum,\ ...)$ causes many data transfers between CPU and GPU memories. For instance, the instruction $F(p) \leftarrow reduce(sum,\ map(...))$ in Algorithm 5 is not fast enough with Thrust because of the data transfers between the GPU and the CPU when implemented with Thrust. Such memory transfers are very slow. It is necessary to rewrite some parts of the program without Thrust.

Firstly, it appears very important to organize data in the global memory of the GPU in a way that memory accesses are coalesced between threads. In CUDA, threads are grouped by warps of 32, to work together. It is recommended that a thread number i shall access a memory cell i when the thread $i + 1$ accesses the cell $i + 1$. In other cases, memory accesses cannot be grouped and cause high latencies. The memory is accessed in two places: when computing the distance between two tuples $d(p,\ q) \leq \tau$ and when computing the domination between two tuples in Algorithm 4. Consecutive threads will deal with consecutive tuples, written as q in the algorithm, and each thread will try to compare the first attribute of p and q, then the second, and so on. So it is necessary to put all the values for the first attribute of all tuples in sequence, then all the values of the second attribute etc., instead of putting the first tuple, then the second tuple and so on.

Other optimizations can be thought of Harris (2007) shows how to efficiently design *reduce* functions on CUDA. It is better to group *map* and *reduce* operations when the latter is applied on the former. The *map* step shall be done inside the first step

---

[1] http://docs.nvidia.com/cuda/thrust/.

of *reduce*. To avoid waiting cycles in alternatives, it is worth to replace all branches by simple computations, when possible.

5 Experimental Results

We have experimented our implementation on both synthetic and real-world datasets. The first subsection shows the results obtained with data coming from an ad sales web site. The second subsection is devoted to the performances of the CUDA implementation of the graded skyline.

5.1 Application to a Real-World Dataset

The approach has been tested using a subset of the database of 845,810 ads about second hand cars from the website *Le bon coin*[2] from 2012. The skyline query used hereafter as an example aims at minimizing both the price and the mileage. In the query considered, we focus on small urban cars with a regular (non-diesel) engine, which corresponds to 441 ads. Figure 5 shows the result obtained. In dark grey are the points that belong the most to the skyline (membership degree between 0.8 and 1). These points are detailed in Table 2. According to the definition used, points dominated by others that are not totally typical belong to the result. It is the case for instance of ad number 916264 that is dominated by ads number 1054357 and 1229833. The identifiers in bold correspond to the points that belong to the regular skyline. One may observe that the points from Table 2 (area [0.8, 1]) are not very (or even not at all) typical. Moreover, certain features may not satisfy the user (the mileage can be very high, the price can be very low) and may look suspicious. On the other hand, Table 3, which shows an excerpt of the 0.6-cut of the graded skyline, contains more typical—thus more credible—points whose overall satisfaction remains high.

Let us mention that from such a result, we can devise different kinds of querying services such as sorting the answers by descending skyline degrees and then by descending typicality degrees or gathering answers by α-cuts over the skyline degree. As mentioned in Remark 2, it is also possible to obtain the most interesting points of the fuzzy skyline by means of a second (classical) skyline query aimed at maximizing both the membership degree and the typicality degree.

---

[2]www.leboncoin.fr.

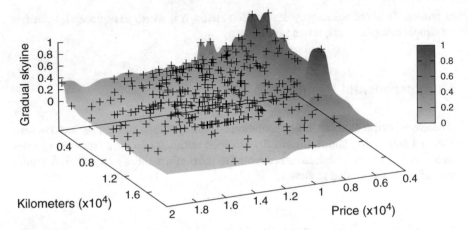

Fig. 5 3D representation of the graded skyline

Table 2 Excerpt of the database and associated degrees (skyline and typicality)

Id	Price	km	Skyline	Typicality
1211574	7000	500	1	0.352
1156771	6000	700	1	0.247
1229833	5990	10000	1	0.126
1596085	5800	162643	1	0.005
1054357	1800	118000	1	0
1333992	500	220000	1	0
1380340	800	190000	1	0
891125	1000	170000	1	0
1276388	1300	135000	1	0
916264	5990	2514000	0.874	0
1674045	6000	3500	0.753	0.315

Table 3 Excerpt of the area [0.6, 0.8]

Id	Price	km	Skyline	Typicality
1208620	6500	3300	0.716	0.363
870279	6900	1000	0.716	0.358
1334605	10500	500	0.647	0.642
1635437	9900	590	0.647	0.621
1529678	7980	650	0.647	0.458
1166077	7750	2214	0.642	0.532
1685854	7890	1000	0.642	0.458
1366336	7490	4250	0.637	0.516
981939	6500	4000	0.637	0.363
1022586	6500	7200	0.637	0.258
1267726	6500	100000	0.637	0

5.2 Application to Synthetic Data

Two kinds of measurements have been performed. We first compare the CUDA implementation to a sequential version on datasets of various cardinalities but a constant number of attributes. Then, we assess the efficiency of the CUDA version with a constant number of tuples, and an increasing number of attributes.

Our CUDA engine has four Tesla M2090, with 6 GB of memory on each. The data is very far from exhausting the memory. For instance, 1,000,000 tuples with 16 attributes represented as float numbers will occupy 64 Mbytes. To improve speed, we add padding bytes to align data, but the total size is many orders below the available memory.

Figure 6 shows a comparison between both programs, with an increasing number of tuples, from 1,000 to 1,000,000 and a constant number of 4 attributes, on logarithmic scales both in tuple number and in computation time, to show that time is a power of the number of tuples. The time is expressed in seconds. Broadly, we can say that the CUDA version is 40 times faster than the sequential version. The computation of the typicalities takes a bit less time, but has the same complexity.

We then studied the impact of the number of attributes with the same number of tuples: 100, 200 and 500 k. The result is shown in Fig. 7. The slope of the line is $\theta(n^2)$ where n is the number of tuples. These experimental results together confirm that the global computation time is $\theta(n^2 * a)$ where a the number of attributes for each tuple. Some long duration results are somewhat imprecise, due to other processes in the system.

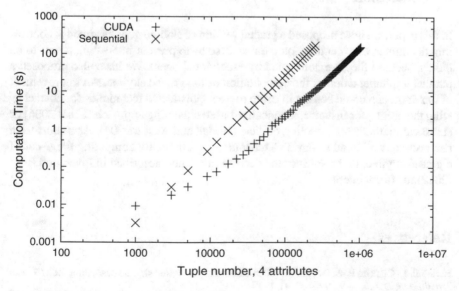

Fig. 6 CUDA compared to sequential algorithm

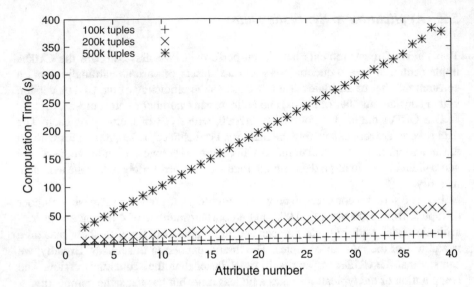

Fig. 7 Impact of the number of attributes

These results show that it is possible to calculate a graded Skyline within a reasonable time, even on a large number of tuples with many attributes.

6 Conclusion

In this paper, we have proposed a graded version of skyline queries aimed at controlling the impact of exceptions on the result (so as to prevent interesting points to be hidden because they are dominated by exceptional ones). We have also proposed a parallel implementation of the computation of the graded skyline. An improvement of our approach could consist in using more sophisticated techniques for characterizing the points, for instance, a typicality-based clustering approach (Lesot 2006) or statistical methods for detecting outliers (Hodge and Austin 2004). As a short-term perspective, we intend to devise a technique for efficiently computing the α-cut of a graded skyline, in the spirit of the derivation method described in Pivert and Bosc (2012) for fuzzy queries.

References

Bartolini, I., Ciaccia, P., & Patella, M. (2008). Efficient sort-based skyline evaluation. *ACM Transactions on Database Systems, 33*(4), 1–49.
Bøgh, K. S., Assent, I., & Magnani, M. (2013). Efficient GPU-based skyline computation. In *Proceedings of the Ninth International Workshop on Data Management on New Hardware*, DaMoN 2013 (pp. 5:1–5:6). New York: ACM.

Börzsönyi, S., Kossmann, D., & Stocker, K. (2001). The skyline operator. In *Proceedings of ICDE 2001* (pp. 421–430).

Chan, C., Jagadish, H., Tan, K., Tung, A., & Zhang, Z. (2006a). Finding k-dominant skylines in high dimensional space. In *Proceedings of SIGMOD 2006* (pp. 503–514).

Chan, C., Jagadish, H., Tan, K., Tung, A., & Zhang, Z. (2006b). On high dimensional skylines. In *Proceedings of EDBT 2006. LNCS* (Vol. 3896, pp. 478–495).

Choi, W., Liu, L., & Yu, B. (2012). Multi-criteria decision making with skyline computation. In *2012 IEEE 13th International Conference on Information Reuse and Integration (IRI)* (pp. 316–323).

Chomicki, J., Godfrey, P., Gryz, J., & Liang, D. (2003). Skyline with presorting. In *Proceedings of ICDE 2003* (pp. 717–719).

Chomicki, J., Godfrey, P., Gryz, J., & Liang, D. (2005). Skyline with presorting: Theory and optimizations. In *Proceedings of IIS 2005* (pp. 595–604).

Cole, M. (1991). *Algorithmic skeletons: Structured management of parallel computation.* Cambridge: MIT Press.

Dubois, D. & Prade, H. (1984). On data summarization with fuzzy sets. In *Proceedings of IFSA 1993* (pp. 465–468).

Ester, M., Kriegel, H., Sander, J., & Xu, X. (1996). A density-based algorithm for discovering clusters in large spatial databases with noise. In E. Simoudis, J. Han & U. M. Fayyad (Eds.), *Proceedings of the Second International Conference on Knowledge Discovery and Data Mining (KDD 1996), Portland, Oregon, USA* (pp. 226–231). AAAI Press.

Fisher, R. A. (1936). The use of multiple measurements in taxonomic problems. *Annals of Eugenics, 7*(2), 179–188.

Friedman, M., Ming, M., & Kandel, A. (1995). On the theory of typicality. *International Journal of Uncertainty, Fuzziness and Knowledge-Based Systems, 3*(2), 127–142.

Gabel, M., Keren, D., & Schuster, A. (2013). Communication-efficient outlier detection for scale-out systems. In *Proceedings of the First International Workshop on Big Dynamic Distributed Data, Riva del Garda, Italy, 30 August 2013* (pp. 19–24).

Goncalves, M., & Tineo, L. (2007). Fuzzy dominance skyline queries. In R. Wagner, N. Revell & G. Pernul (Eds.), *Proceedings 18th International Conference on Database and Expert Systems Applications, DEXA 2007, Regensburg, Germany, 3–7 September 2007* (Vol. 4653, pp. 469–478). Lecture notes in computer science. Heidelberg: Springer.

Gupta, M., Gao, J., and Han, J. (2013). Community distribution outlier detection in heterogeneous information networks. In H. Blockeel, K. Kersting, S. Nijssen & F. Železný (eds.), *Machine learning and knowledge discovery in databases* (Vol. 8188, pp. 557–573). *Lecture notes in computer science.* Berlin: Springer.

Hadjali, A., Pivert, O., & Prade, H. (2011). On different types of fuzzy skylines. In *Proceedings of ISMIS 2011* (pp. 581–591).

Hampton, J. (1988). Overextension of conjunctive concepts: evidence for a unitary model of concept typicality and class inclusion. *Journal of Experimental Psychology: Learning Memory and Cognition, 14*(1), 12–32.

Harris, M. (2007). Optimizing parallel reduction in cuda. Technical report, nVidia.

Hodge, V., & Austin, J. (2004). A survey of outlier detection methodologies. *Artificial Intelligence Review, 22*(2), 85–126.

Ji, T., Yang, D., & Gao, J. (2013). Incremental local evolutionary outlier detection for dynamic social networks. In H. Blockeel, K. Kersting, S. Nijssen & F. Železný (Eds.), *Machine learning and knowledge discovery in databases* (Vol. 8189, pp. 1–15). *Lecture notes in computer science.* Berlin: Springer.

Lee, K. C. K., Zheng, B., Li, H., & Lee, W.-C. (2007). Approaching the skyline in z order. In *VLDB* (pp. 279–290).

Lesot, M. (2006). Typicality-based clustering. *International Journal of Information technology and Intelligent Computing, 12*, 279–292.

Lin, X., Yuan, Y., Zhang, Q., & Zhang, Y. (2007). Selecting stars: The k most representative skyline operator. In *Proceedings of the ICDE 2007* (pp. 86–95).

Niu, Z., Shi, S., Sun, J., & He, X. (2011). A survey of outlier detection methodologies and their applications. In *Proceedings of Third International Conference Artificial Intelligence and Computational Intelligence, Part I, AICI 2011, Taiyuan, China, September 24–25, 2011* (pp. 380–387).

Osherson, D., & Smith, E. (1997). On typicality and vagueness. *Cognition, 64*, 189–206.

Papadias, D., Tao, Y., Fu, G., & Seeger, B. (2005). Progressive skyline computation in database systems. *ACM Transactions on Database Systems, 30*(1), 2005.

Park, S., Kim, T., Park, J., Kim, J., & Im, H. (2009). Parallel skyline computation on multicore architectures. In *ICDE* (pp. 760–771).

Pivert, O., & Bosc, P. (2012). *Fuzzy preference queries to relational databases*. London: Imperial College Press.

Rojas, W. U., Boizumault, P., Loudni, S., Crémilleux, B., and Lepailleur, A. (2014). Mining (soft-) skypatterns using dynamic CSP. In H. Simonis (ed.), *Proceedings of 11th International Conference on Integration of AI and OR Techniques in Constraint Programming, CPAIOR 2014, Cork, Ireland, 19–23 May 2014* (Vol. 8451, pp. 71–87). *Lecture notes in computer science*. Cham: Springer.

Rosch, E., & Mervis, C. (1975). Family resemblance: Studies in the internal structure of categories. *Cogn. Psychol., 7*, 573–605.

Tan, K.-L., Eng, P.-K., & Ooi, B. C. (2001). Efficient progressive skyline computation. In *Proceedings of VLDB 2001* (pp. 301–310).

Tao, Y., Ding, L., Lin, X., & Pei, J. (2009). Distance-based representative skyline. In Y. E. Ioannidis, D. L. Lee & R. T. Ng (Eds.), *ICDE* (pp. 892–903). IEEE.

Yager, R. (1997). A note on a fuzzy measure of typicality. *International Journal of Intelligent Systems, 12*(3), 233–249.

Zadeh, L. A. (1982). A note on prototype theory and fuzzy sets. *Cognition, 12*, 291–298.

Zadeh, L. A. (1984). A computational theory of dispositions. In Y. Wilks (Ed.), *COLING* (pp. 312–318). ACL.

Zadeh, L. (1987). A computational theory of dispositions. *International Journal of Intelligent Systems, 2*, 39–63.

Zhang, J. (2013). Advancements of outlier detection: A survey. *EAI Endorsed Transactions on Scalable Information Systems, 1*, e2.

Zimek, A., Gaudet, M., Campello, R. J. G. B., & Sander, J. (2013). Subsampling for efficient and effective unsupervised outlier detection ensembles. In *The 19th ACM SIGKDD International Conference on Knowledge Discovery and Data Mining, KDD 2013, Chicago, IL, USA, 11–14 August 2013* (pp. 428–436).

Zimek, A., Schubert, E., & Kriegel, H. (2012). A survey on unsupervised outlier detection in high-dimensional numerical data. *Statistical Analysis and Data Mining, 5*(5), 363–387.

Adaptive Down-Sampling and Dimension Reduction in Time Elastic Kernel Machines for Efficient Recognition of Isolated Gestures

Pierre-Francois Marteau, Sylvie Gibet and Clément Reverdy

Abstract In the scope of gestural action recognition, the size of the feature vector representing movements is in general quite large especially when full body movements are considered. Furthermore, this feature vector evolves during the movement performance so that a complete movement is fully represented by a matrix M of size DxT, whose element $M_{i,j}$ represents the value of feature i at timestamps j. Many studies have addressed dimensionality reduction considering only the size of the feature vector lying in $\mathbb{R}^D$ to reduce both the variability of gestural sequences expressed in the reduced space, and the computational complexity of their processing. In return, very few of these methods have explicitly addressed the dimensionality reduction along the time axis. Yet this is a major issue when considering the use of elastic distances which are characterized by a quadratic complexity along the time axis. We present in this paper an evaluation of straightforward approaches aiming at reducing the dimensionality of the matrix M for each movement, leading to consider both the dimensionality reduction of the feature vector as well as its reduction along the time axis. The dimensionality reduction of the feature vector is achieved by selecting remarkable joints in the skeleton performing the movement, basically the extremities of the articulatory chains composing the skeleton. The temporal dimensionality reduction is achieved using either a regular or adaptive down-sampling that seeks to minimize the reconstruction error of the movements. Elastic and Euclidean kernels are then compared through support vector machine learning. Two data sets that are widely referenced in the domain of human gesture recognition, and quite distinctive in terms of quality of motion capture, are used for the experimental assessment of the proposed approaches. On these data sets we experimentally show that it is feasible, and possibly desirable, to significantly reduce simultaneously the size of the feature vector and the number of skeleton frames to represent body movements while

P.-F. Marteau (✉) · S. Gibet · C. Reverdy
IRISA (UMR 6074), Université Bretagne Sud Campus de Tohannic,
56000 Vannes, France
e-mail: Pierre-Francois.Marteau@univ-ubs.fr

S. Gibet
e-mail: Sylvie.Gibet@univ-ubs.fr

C. Reverdy
e-mail: Clement.Reverdy@univ-ubs.fr

© Springer International Publishing Switzerland 2017
F. Guillet et al. (eds.), *Advances in Knowledge Discovery and Management*,
Studies in Computational Intelligence 665, DOI 10.1007/978-3-319-45763-5_3

maintaining a very good recognition rate. The method proves to give satisfactory results at a level currently reached by state-of-the-art methods on these data sets. We experimentally show that the computational complexity reduction that is obtained makes this approach eligible for *real-time* applications.

1 Introduction

Gesture recognition is a challenging task in the computer vision community with numerous applications using motion data such as interactive entertainment, human-machine interaction, automotive, or digital home. Recently, there is an increasing availability of large and heterogeneous motion captured data characterized by a various range of qualities, depending on the type and quality of the motion sensors. We thus separate (i) databases of high resolution and quality built from expensive capturing devices and requiring a particular expertise, (ii) and low resolution and noisier databases produced with cheap sensors that do not require any specific expertise. Such databases open new challenges for assessing the robustness and generalization capabilities of gesture recognition algorithms on diversified motion data sets. Besides the quality of recognition, the complexity of the algorithms and their computational cost is indeed a major issue, especially in the context of real-time interaction.

We address in this paper the recognition of isolated gestures from motion captured data. As motion data are generally represented by high-dimensional time series, many approaches have been developed to reduce their dimension, so that the recognition process is more efficient while being still accurate. Among them, low-dimensional embeddings of motion data have been proposed that enable to characterize and parameterize action sequences. Some of them are based on statistical descriptors (Hussein et al. 2013), rely on relevant meaningful trajectories (Ofli et al. 2013), or characterize the style (Hussein et al. 2013). In this paper we focus on dimensionality reduction along two complementary axes: the spatial axis representing the configuration of the skeleton at each frame, and the temporal axis representing the evolution over time of the skeletal joints trajectories. With such an approach, two main challenges are combined simultaneously:

- We use relevant trajectories (end-extremities) whose content may characterize complex actions;
- Considering that the temporal variability is of primary importance when recognizing skilled actions and expressive motion, we apply an adaptive temporal down-sampling to reduce the complexity of elastic matching.

These low-dimensional dual-based representations will be coupled with appropriate recognition algorithms that we expect to be more tractable and efficient.

Our recognition principle is based on a recent method that improves the performance of classical support vector machines when used with regularized elastic kernels dedicated to time series matching. Our objective is to show how the spatial and temporal dimensionality reductions, associated with such regularized elastic kernels significantly improve the efficiency, in terms of response time, of the algorithm while preserving the recognition rates.

The second section briefly presents the related works and state-of-the-art for isolated gesture recognition. In the third section we describe the nature of the motion data as well as the main pre-processing of the data. The fourth section gives the major keypoints of the method, positioning it in the context of multivariate sequential data classification. We present in the fifth part the evaluation of our algorithm carried out on two data sets with very distinct qualities and compare its performance with those obtained by some of the state-of-the-art methods. A final discussion is provided as well as some perspectives.

2 Related Work

In human gesture recognition, there are two main challenges to be addressed: dimension reduction closely linked to feature descriptors, and recognition models which cover different aspects of dynamic modeling, statistical and machine learning approaches. In this section, we give a brief and non-exhaustive overview of the literature associated with each challenge.

Dimension Reduction
The problem of dimensionality reduction (also called manifold learning) can be addressed with the objective to find a low-dimensional space that best represents the variance of the data without loosing too much information. Action descriptors have thus been defined for characterizing whole motion sequences, or punctual frames that need additional step of temporal modeling to achieve the recognition goal.

Numerous method are available to carry out such dimensionality reduction, the most popular being linear approaches such as Principal Component Analysis (PCA, Jolliffe 1986; Masoud and Papanikolopoulos 2003), Linear Discriminant Analysis (LDA, McLachlan 2004), or linear projections preserving locally neighborhoods (Locality Preserving Projection) (He and Niyogi 2003). Among non-linear approaches, Locally Linear Embeddings (LLE, Roweis and Saul 2000), Metric Multidimensional Scaling (MDS, Kruskal and Wish 1978) and variants like Laplacian Eigenmap (Belkin and Niyogi 2002), Isomap (Tenenbaum et al. 2000) have been implemented to embed postures in low dimensional spaces in which a more efficient time warp (DTW, see Sect. 5) algorithm can be used to classify movements (Blackburn and Ribeiro 2007). An extension of this method, called ST-Isomap, considers temporal relationships in local neighborhoods that can be propagated globally via a shortest-path mechanism (Jenkins and Matarić 2004). Models based on Gaussian processes with latent variables are also largely used, for instance a hierarchical version has been recently exploited for gesture recognition (Han et al. 2010).

Other methods define discriminative features that best classify motion classes. This is the case in the work of Yu and Aggarwal (2009) that reduces the motion data to only five end-extremities of the skeleton (two feet, two hands and the head), thus giving some meaningful insight of the motion related to the action task. Fothergill et al. (2012) and Zhao et al. (2012) have in particular applied random forests to recognize actions, using a Kinect sensor, while Ofli et al. (2013) recently proposed to automatically select the most informative skeletal joints to explain the current action. In the same line, Hussein et al. (2013) use the covariance matrix for skeleton joint locations over time as a discriminative descriptor to characterize a movement sequence. Multiple covariance matrices are deployed over sub-sequences in a hierarchical fashion in order to encode the relationship between joint movement and time. In Li et al. (2010) a simple bag of 3D points is used to represent and recognize gestural action. Similarly, in Wang et al. (2012), *actionlets* are defined from Fourier coefficients to characterize the most discriminative joints. Finally, it can be mentioned, among many existing applications that address the use of elastic distances into a recognition process, the recent work described in Sempena et al. (2011), as well as the hardware acceleration proposed in Hussain and Rashid (2012). However, to our knowledge, no work exploiting this type of distance has directly studied the question of data reduction along the time axis.

Gesture Recognition
Recognition methods essentially aim at modeling the dynamics of gestures. Some approaches Veeraraghavan et al. (2004), based on linear dynamic models, have used autoregressive (AR) and autoregressive moving-average (ARMA) models to characterize the kinematics of movements, while other approaches, based on nonlinear dynamic models (Bissacco et al. 2007), have developed movement analysis and recognition scheme based on dynamical models controlled by Gaussian processes. Mitra and Acharya (2007) propose a synthesis of the major gesture recognition approaches relying on Hidden Markov Models (HMM). Histograms of oriented 4D normals have also been proposed in Oreifej and Liu (2013) for the recognition of gestural actions from sequences of depth images. Wang et al. (2006) have exploited conditional random fields to model joint dependencies and thus increase the discrimination of HMM-like models. Recurrent neural network models have also been used (Martens and Sutskever 2011); among them, conditional restricted Boltzmann's machines (Larochelle et al. 2012) have been studied recently in the context of motion captured data modeling and classification.

In this paper, we propose a new representation of human actions that results from a dual-based reduction method that occurs both spatially and temporally. We couple this representation to a SVM classification method associated with regularized elastic kernels.

3 Motion Representation

We are working on isolated human motions acquired through motion captured databases. In recent years, there is an increasing availability of these databases, some of them being captured by high resolution devices (infrared marker-tracking system), such as those provided by CMU (2003), HDM05 (Müller et al. 2007), and other ones captured by low-cost devices, such as MSRAction3D captured with the Microsoft Kinect system. With the first type of sensors, the acquisition process is expensive, as it necessitates for capturing skilled motion with a good accuracy several cameras with many markers located on an actor, and a post-processing pipeline which is costly in time and expertise. Instead, with the second type of sensors, the data acquisition is affordable and necessitates less time and expertise, but with a loss of accuracy which remains acceptable for some tasks.

After the recording, the captured data is filtered and reconstructed so that to eliminate most of the noise, data loss and labelling errors (markers inversion), and the output of this acquisition pipeline is generally a set of 3D-trajectories of the skeleton joints determined from the positions of markers. This kind of data is inherently noisy, mainly due to the quality of the sensors and the acquisition process, but also to approximations made during the reconstruction process. The modeling of the skeleton is indeed subject to some variation for different reasons: in particular the markers being positioned on cloths or on the skin of the actor's body, they can move during the capturing process, also the determination of the segment lengths, of the joints' centers and their associated rotation axis are not trivial and lead to modeling errors. To overcome these difficulties, the skeleton model is obtained through an optimization process such as the ones described in de Aguiar et al. (2006), O'Brien et al. (2000), or Shotton et al. (2011). The techniques based on a skeleton model

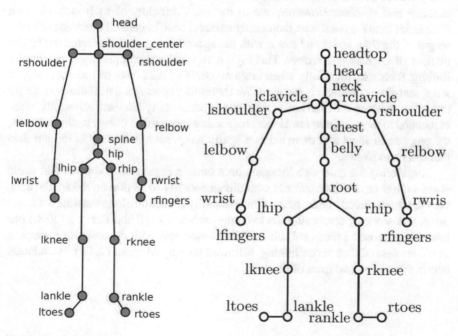

Fig. 1 Examples of skeletons reconstructed from motion data captured from the Kinect (*left*) and from the Vicon-MX device used by the Max Planck Institute (*right*)

hence convert 3D sensor data into Cartesian or angular coordinates that define the state of the joints over time with various accuracies.

Figure 1 presents two skeletons reconstructed from two very distinct capture systems. On the right, the skeleton is reconstructed from data acquired via the Microsoft Kinect, on the left from the Vicon-MX device used by the Max Planck Institute to produce the HDM05 datasets.

Thus, any skeletal-based model can be represented by a hierarchical tree-like structure composed of rigid segments (bones) connected by joints. A motion can be defined by a sequence of postures over time, formalized as a multivariate state vector describing a trajectory over time, i.e. a time series: $\{Y_t \in \mathbf{R}^k\}_1^T = [Y_1, \ldots, Y_T]$, where the k spatial dimension ($k = 3 \cdot N$, with N the number of joints) typically varies between 20 and 100 according to the capture devices and the considered task. As this state vector is obviously not composed of independent scalar dimensions, the spatio-temporal encoded redundancies open solutions for dimension reduction as well as noise reduction approaches. This is particularly relevant for motion recognition, as the objective is to aim at improving computation time and error rate.

4 Dimension Reduction of Motion Capture Data

Using elastic distances or kernels for recognition problems has proved to be very accurate and efficient. However, one of the main difficulty of such methods with time series data is to deal with their computational cost, in general quadratic with the length of the time series and linear with the *spatial* dimension characterized by the number of degrees-of-freedom. This high computational complexity is potentially limiting their use, especially when large amounts of data have to be processed, or when real-time constraint is required. We therefore expect that a dual dimensionality reduction, both on the time and spatial axis is particularly relevant, especially when associated to these techniques. Hence, for motion recognition using elastic distances, we propose to show that there exists a spatio-temporal redundancy in motion data that can be exploited.

Considering the quite rich literature on motion recognition, it appears that while some studies have shown success with dimensionality reduction on the spatial axis, very few have directly addressed a dimensionality reduction along the time axis, and much less work by combining the two approaches. Keogh and Pazzani (2000) has however proposed a temporal sub-sampling associated with dynamic time warping in the context of time series mining, followed later by Marteau (2009). We address herein after these two lines of research.

4.1 Dimension Reduction Along the Spatial Axis

Dimension reduction (or manifold learning), is the process consisting of mapping high dimensional data to representations lying in a lower-dimensional manifold. This can be interpreted as mapping the data into a space characterized by a smaller dimension, from which most of the variability of the data can be reproduced.

We consider in this paper a more direct approach based on the knowledge of the mechanism underlying the production of motion data and the way human beings perceive and discriminate body movements. We make the assumption that the perception of human motion is better achieved in the so-called *task-space* represented by a selection of significant 3D joint trajectories. This hypothesis is supported by Giese et al. (2008) who show that visual perception of body motion closely reflects physical similarities between joint trajectories. This is also consistent with the motor theory of motion perception presented in Gibet et al. (2011). Besides, we may reasonably accept that these joint trajectories embed sufficient discriminative information as inverse kinematics (widely used in computer animation and robotics) has shown to be very efficient and robust to reconstruct the whole skeleton movement from the knowledge of the end effector trajectories (hands, feet, head), possibly with the additional knowledge of mid-articulated joints trajectories (such as elbows and knees) and constraints. Hence a straightforward approach consists in constructing a motion descriptor that discards all joints information but the 3D positions of the mid and end effectors extremities. We thus select the 3D positions for the two wrists, the two ankles (the fingers and toes markers are less reliable in general), the two elbows, the two knees and the head. This leads to a time-varying descriptor lying in a 18D space, while a full body descriptor is embedded in a 60D, space for the Kinect sensor, significantly more for vicon settings in general.

4.2 Dimension Reduction Along the Time Axis

The straightforward approach we are developing to explicitly reduce dimensionality along the time axis consists in sub-sampling the movement data such that each motion trajectories takes the form of a fixed-size sequence of L skeletal postures. Then it becomes easy to perform a classification or recognition task by using elastic kernel machines on such fixed-size sequences. With such an approach, performance rates depend on the chosen degree of sub-sampling. This approach seems coarse since long sequences are characterized with the same number of skeletal poses than short sequences. For very short sequences, whose lengths are shorter than L, if any, we over-sample the sequence to meet the fixed-size requirement. But we consider this case as very marginal since we seek a sub-sampling rate much lower than the average sequence of movement length. In the following, we will experiment and compare uniform and adaptive down-sampling.

4.2.1 Uniform Down-Sampling

In order to explicitly reduce dimensionality along the time axis, our first straight-forward approach here consists in down-sampling the motion data so that each motion trajectory takes the form of a fixed-size sequence of L skeletal postures, evenly distributed along the time axis. We refer to this approach as uniform down-sampling (UDS).

4.2.2 Adaptive Down-Sampling

The second approach is the so-called adaptive down-sampling (ADS) approach. Similarly to UDS, each motion trajectory takes the form of a fixed-size sequence of L skeletal postures, but these postures are not evenly distributed anymore along the time axis. They are selected such as to minimize a trajectory reconstruction criteria. Basically we follow the previous work by Marteau and Gibet (2006). A data modeling approach is used to handle the adaptive sampling of the $\{Y_t\}$ multidimensional and discrete time series. More precisely, we are seeking an approximation Y_{θ^*} of Y such as:

$$\theta^* = \underset{\theta}{ArgMin}(E(\{Y_t\}, \{Y_\theta, t\}))\tag{1}$$

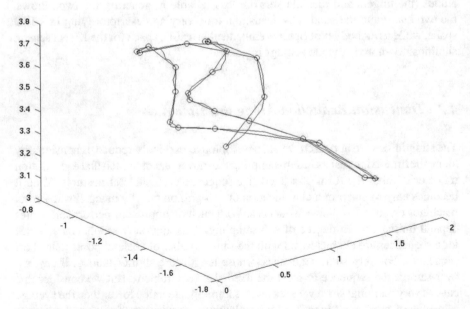

Fig. 2 Trajectory of the human wrist in the 3D Cartesian space adaptively down-sampled with the localization of the 25 selected samples (*red circles*): motion capture data (*blue*) and reconstructed data by linear interpolation (*red*)

where E is the RMS error between Y and Y_θ selected among the set of piecewise linear approximations defined from Y. Since the optimal solution of the optimization problem defined by Eq. 1 is $O(L.n^2)$, where n is the length of the input time series, we adopt a near to optimal solution as developed in Marteau and Ménier (2009) whose complexity is $O(n)$. As an example, in Fig. 2 the human wrist 3D trajectory is down-sampled using 25 samples positioned along the trajectory by minimizing the piecewise linear approximation.

5 Elastic Kernels and Their Regularization

Dynamic Time Warping (DTW) (Velichko and Zagoruyko 1970; Sakoe and Chiba 1971), by far the most used elastic measure, is defined as

$$d_{dtw}(X_p, Y_q) = d_E^2(x(p), y(q)) \tag{2}$$
$$+ \text{Min} \begin{cases} d_{dtw}(X_{p-1}, Y_q) & sup \\ d_{dtw}(X_{p-1}, Y_{q-1}) & sub \\ d_{dtw}(X_p, Y_{q-1}) & ins \end{cases}$$

where $d_E(x(p), y(q)$ is the Euclidean distance (possibly the square of the Euclidean distance) defined on $\mathbb{R}^k$ between the two postures in sequences X and Y taken at times p and q respectively.

When performed by a support vector machine (SVM) model, the optimization problem inherent to this type of learning algorithm is no longer quadratic. Moreover, the convergence towards the *optimum* is no longer guaranteed, which, depending on the complexity of the task may be considered as detrimental.

Besides the fact that the DTW measure does not respect the triangle inequality, it is furthermore not possible to directly define a positive definite kernel from it. Hence, the optimization problem, inherent to the learning of a kernel machine, is no longer quadratic which could, at least on some tasks, be a source of limitation.

Regularized DTW: recent works (Cuturi et al. 2007; Marteau and Gibet 2014) allowed to propose new guidelines to regularize kernels constructed from elastic measures such as DTW. A simple instance of such regularized kernel, derived from Marteau and Gibet (2014) for time series of equal length, takes the following form, which relies on two recursive terms:

$$K_{rdtw}(X_p, Y_q) = K_{rdtw}^{xy}(X_p, Y_q) + K_{rdtw}^{xx}(X_p, Y_q)$$

$$K_{dtw}^{xy}(X_p, Y_q) = \frac{1}{3}e^{-vd_E^2(x(p),y(q))}$$
$$\sum \begin{cases} h(p-1, q)K_{rdtw}^{xy}(X_{p-1}, Y_q) \\ h(p-1, q-1)K_{rdtw}^{xy}(X_{p-1}, Y_{q-1}) \\ h(p, q-1)K_{rdtw}^{xy}(X_p, Y_{q-1}) \end{cases} \tag{3}$$

$$K_{rdtw}^{xx}(X_p, Y_q) = \frac{1}{3}$$
$$\sum \begin{cases} h(p-1, q)K_{rdtw}^{xx}(X_{p-1}, Y_q)e^{-vd_E^2(x(p),y(p))} \\ \Delta_{p,q}h(p, q)K_{rdtw}^{xx}(X_{p-1}, Y_{q-1})e^{-vd_E^2(x(p),y(q))} \\ h(p, q-1)K_{rdtw}^{xx}(X_p, Y_{q-1})e^{-vd_E^2(x(q),y(q))} \end{cases}$$

where $\Delta_{p,q}$ is the Kronecker's symbol, $v \in \mathbb{R}^+$ is a *stiffness* parameter which weights the local contributions, i.e. the distances between locally aligned positions, and $d_E(.,.)$ is a distance defined on $\mathbb{R}^k$.

The initialization is simply $K_{rdtw}^{xy}(X_0, Y_0) = K_{rdtw}^{xx}(X_0, Y_0) = 1$.

The main idea behind this line of regularization is to replace the operators min and max (which prevent the symmetrization of the kernel) by a summation operator ($\sum$). This leads to consider, not only the best possible alignment, but also all the best (or nearly the best) paths by summing up their overall cost. The parameter v is used to control what we call nearly-the-best alignment, thus penalizing more or less alignments too far from the optimal ones. This parameter can be easily optimized through a cross-validation.

5.1 Normalization

As K_{rdtw} evaluates the sum on all possible alignment paths of the products of local alignment costs $e^{-d_E^2(x(p),y(p))/(2.\sigma^2)} \leq 1$, its values can be very small depending on the size of the time series and the selected value for σ. Hence, K_{DTW} values tend to 0 when σ tends towards 0, except when the two compared time series are identical (the corresponding Gram matrix suffers from a diagonal dominance problem). As proposed in Marteau and Gibet (2014), a manner to avoid numerical troubles consists in using the following *normalized* kernel:

$$\tilde{K}_{rdtw}(.,.) = exp\left(\alpha \frac{log(K_{rdtw}(.,.)) - log(min(K_{rdtw}))}{log(max(K_{rdtw})) - min(K_{rdtw})}\right)$$

where $max(K_{rdtw})$ and $min(K_{rdtw})$ respectively are the max and min values taken by the kernel on the learning data set and $\alpha > 0$ a positive constant ($\alpha = 1$ by default). If we forget the proportionality constant, this leads to take the kernel K_{rdtw} at a power $\tau = \alpha/(log(max(K_{rdtw})) - log(min(K_{rdtw})))$, which shows that the normalized kernel $\tilde{K}_{rdtw} \propto K_{rdtw}^{\tau}$ is still positive definite (Berg et al. 1984, Proposition 2.7).

We consider in this paper the non definite exponential kernel (Gaussian or Radial Basis Function (RBF) types) $K_{dtw} = e^{-d_{dtw}(.,.)/(2.\sigma^2)}$ constructed directly from the elastic measures d_{dtw}, the normalized regularized elastic kernel K_{rdtw}^{τ}, and the non-elastic kernel obtained from the Euclidean distance,[1] i.e., $K_E(.,.) = e^{-d_E^2(.,.)/\sigma}$.

6 Experimentation

To estimate the robustness of the proposed approaches, we evaluate them on two motion capture databases of opposite quality, the first one, called *HDM05*, developed at the Max Planck Institute, the other one, called *MSR-Action3D*, at Microsoft research laboratories.

HDM05 Data Set (Müller et al. 2007) consists of data captured at 120 Hz by a Vicon MX system composed of a set of reflective optical markers followed by six high-definition cameras and configured to record data at 120 Hz. The movement sequences are segmented and transformed into sequences of skeletal poses consisting of N = 31 joints, each associated to a 3D position (x, y, z). In practice the position of the root of the skeleton (located near its center of mass) and its orientation serving as referential coordinates, only the relative positions of the remaining 30 joints are used, which leads to represent each position by a vector $Y_T \in \mathbb{R}^k$, with $k = 90$. We consider two recognition/classification tasks: HDM05-1 and HDM05-2 that are respectively those proposed in Ofli et al. (2012) (also exploited in the work of Hussein et al. (2013)) and Ofli et al. (2013). For both tasks, three subjects are involved during learning and two separate subjects are involved during testing. For task HDM05-1, 11 gestural actions are processed: *{deposit floor, elbow to knee, grab high, hop both legs, jog, kick forward, lie down floor, rotate both arms backward, sneak, squat, and throw basketball}*. This constitutes 249 motion sequences. For task HDM05-2, the subjects are the same, but five additional gestural actions are considered in addition to the previous 11: *{jump, jumping jacks, throw, sit down, and stand up}*. For this task, the data set includes 393 movement sequences in total. For both tests, the lengths of the gestural sequences are between 56 and 901 postures (corresponding to a movement duration between 0.5 and 7.5 s).

MSR-Action3D data set: This database (Li et al. 2010) has recently been developed to provide a Kinect data *benchmark*. It consists of 3D depth image sequences (*depth map*) captured by the Microsoft Kinect sensor. It contains 20 typical interaction gestures with a game console that are labeled as follows *high arm wave, horizontal arm wave, hammer, hand catch, forward punch, high throw, draw x, draw tick, draw circle, hand clap, two hand wave, side-boxing, bend, forward kick, side kick, jogging, tennis swing, tennis serve, golf swing, pickup and throw*. Each action was carried out

---

[1]The Euclidean distance is usable only because a fixed number of skeletal positions is considered to characterize each movement, and this, irrespectively of their initial length.

by 10 subjects facing the camera, 2 or 3 times. This data set includes 567 motion sequences whose lengths vary from 14 to 76 skeletal poses. The 3D images of size 640×480 were captured at a frequency of 15 Hz. From each 3D image a skeletal posture has been extracted with $N = 20$ joints, each one being characterized by three coordinates. As for the previous data set, we characterize postures relatively to the referential coordinates located at the root of the skeleton, which leads to represent each posture by a vector $Y_t \in \mathbb{R}^k$, with $k = 3 \times 19 = 57$. The task is to provide a cross-validation on the subjects, i.e. 5 subjects participating in learning and 5 subjects participating in testing, considering all possible configurations which represent 252 learning/testing pairs in total.

Hence, we perform three classification tasks: HDM05-1, HDM05-2 and MSR Action3D, with or without a spatial dimensionality reduction while simultaneously considering a down-sampling on the time axis:

- The spatial dimensionality reduction is obtained by constructing a frame (skeletal pose) descriptor composed only with the end-effector trajectories in 3D (EED) comparatively to a full-body descriptor (FBD) that integrates all the joints trajectories that compose the skeleton. The FBD rests in a 90D space for HDM05 and in a 60D space for MSRAction3D. The EED rests in a 24D space (3D positions for 2 elbows, 2 hands, two knees and two feet) for the three tasks, leading to a data compression of 73 % for HDM05 and 55 % for MSRAction3D.
- The dimensionality reduction on the time axis is obtained through either a uniform down-sampling (UDS) or an adaptive down-sampling (ADS) based on a piecewise approximation of the FBD or EED trajectories. The number of skeletal poses varies from 5 to 30 for each trajectories, leading to an average data compression of 97 % for HDM05 and 74 % for MSRAction3D.

6.1 Results and Analysis

For the three considered tasks, we present the results obtained using a SVM classifier built from the LIBSVM library (Chang and Lin 2011), the elastic non definite kernel $K_{dtw} = e^{-d_{dtw}(.,.)/(2.\sigma^2)}$, the elastic definite kernel K_{rdtw}^τ, and as a baseline, the Euclidean RBF kernel, $K_E = e^{-d_E^2(.,.)/(2.\sigma^2)}$.

Figures 3, 4 and 5 present the classification accuracies for respectively the HDM05-1, HDM05-2 and MSRAction3D tasks for the test data when the number of skeletal postures selected after down-sampling varies between 5 and 30. For these three figures, the top sub-figure presents classification accuracies when the FBD (Full Body) descriptor associated to a uniform down sampling (UDS) are used, the middle sub-figure classification accuracies when the FBD (Full Body) descriptor associated to an adaptive down sampling (ADS) are used, and the bottom sub-figure gives classification accuracies when the EED (End Extremities) descriptor associated to an adaptive down sampling (ADS) is used.

Fig. 3 Classification accuracies for HDM05-1 task as defined in Ofli et al. (2012) uniform down-sampling, full body (*top*), adaptive down-sampling, full body (*middle*), adaptive down-sampling, end effector extremities (*bottom*), when the number of skeletal poses varies: K_E (*red, circle, dash*), K_{dtw} (*black, square, plain*), K_{rdtw}^{τ} (*blue, star, dotted*)

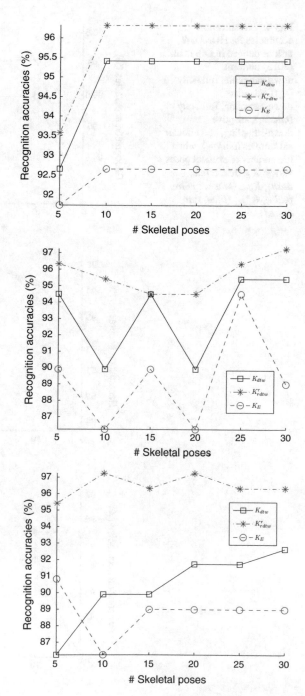

Fig. 4 Classification accuracies for HDM05-2 task as defined in Ofli et al. (2013) uniform down-sampling, full body (*top*), adaptive down-sampling, full body (*middle*), adaptive down-sampling, end effector extremities (*bottom*), when the number of skeletal poses varies: K_E (*red, circle, dash*), K_{dtw} (*black, square, plain*), K_{rdtw}^{τ} (*blue, star, dotted*)

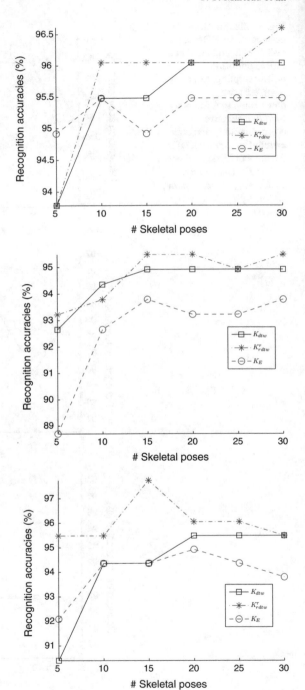

Fig. 5 Classification accuracies for the MSRAction3D data set, uniform down-sampling, full body (*top*), adaptive down-sampling, full body (*middle*), adaptive down-sampling, end effector extremities (*bottom*), when the number of skeletal poses varies: K_E (*red, circle, dash*), K_{dtw} (*black, square, plain*), K_{rdtw}^{τ} (*blue, star, dotted*). Additionally, the cross validation on subjects (252 tests) allows to show the variance of the results

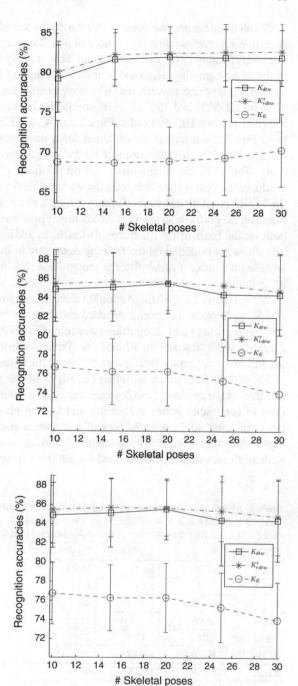

On all three figures, we observe that the down-sampling does not degrade the classification accuracies when the number of poses is over 10, and for some tasks it may even significantly improves the accuracy. This is likely due to the noise filtering effect of the down-sampling. High levels of down-sampling (e.g. when 10–15 postures are retained to describe movements, which represents an average compression ratio of 97 % for HDM05 and 70 % on MSRAction3D) lead to very satisfactory results (90–98 % for the two HDM05 tasks: Figs. 3 and 4, and 82–86 % for the MSRAction3D task: Fig. 5). Best results are obtained for a number of skeletal poses between 15 and 20, when the EED descriptor is used in conjunction with an adaptive down sampling. The SVM classifier constructed on the basis of the regularized kernel K_{rdtw}^{τ} produces the best recognition rates (>=96 % for the two HDM05 tasks). We note that the MSRAction3D task is much more difficult since it consists in a cross validation based on the performing subjects. Much lower performance are obtained for the SVM built on the basis of the Euclidean distance; in addition, if very good classification rate (96 %) is obtained on the training data, due to the noisy nature of Kinect data and the inter subject variability, the recognition rate on the test data falls down from 82 to 86 %.

Table 1 gives for the MSRAction3D data set and for the SVM based on K_E, K_{dtw} and K_{rdtw}^{τ} kernels, means and standard deviations, obtained on the training data (L) and testing data (T), of recognition rates (classification accuracies) when performing the cross-validation over the 10 subjects (5 train—5 test splits leading to 252 tests), for the full body descriptor (FBD) and the end extremities descriptor (EED) associated either to a uniform down-sampling (UDS) or an adaptive Down Sampling (ADS). For this test, movements are represented as sequences of 15 skeletal postures. The drop of accuracies between learning and testing phases is due, on this dataset, to the large inter subjects variability of movement performances. Nevertheless, our experiment shows that the best average classification accuracies (obtained in general with minimal variance) are obtained for the most compact movement representation,

Table 1 Means and standard deviations of classification accuracies on the MSRAction3D data set obtained according to a cross-validation on the subjects (5-5 splits, 252 tests) (L): on the training data, (T): on test data for a number of skeletal postures equal to 15

FBD-UDS	K_E (L)	K_E (T)	K_{dtw} (L)	K_{dtw} (T)	K_{rdtw}^{τ} (L)	K_{rdtw}^{τ} (T)
Mean	87,71	69,73	96,04	81,41	96,65	82,50
Stand. dev.	2,34	5,73	1,36	5,04	1,13	3,22
FBD-ADS	K_E (L)	K_E (T)	K_{dtw} (L)	K_{dtw} (T)	K_{rdtw}^{τ} (L)	K_{rdtw}^{τ} (T)
Mean	85,06	76,48	92,13	84,72	91,89	85,09
Stand. dev.	3,01	3,18	2,74	3,31	2,66	3,58
EED-ADS	K_E (L)	K_E (T)	K_{dtw} (L)	K_{dtw} (T)	K_{rdtw}^{τ} (L)	K_{rdtw}^{τ} (T)
Mean	92.46	76.27	97.14	85.16	**97.19**	85.72
Stand. dev.	3.45	0.96	3.57	5.04	0.93	3.07

Table 2 Comparative study on the MSRAction3D dataset, according to a cross-validation on the subjects (5-5 splits, 252 tests)

HDM05-1	Accuracy (%)
SMIJ (Ofli et al. 2012)	84.40
Cov3DJ, L = 3 (Hussein et al. 2013)	95.41
SVM K_{rdtw}^{τ}, EED, ADS with 15 poses	**96.33**
HDM05-2	Accuracy (%)
SMIJ (Ofli et al. 2013), 1-NN	91.53
SMIJ (Ofli et al. 2013), SVM	89.27
SVM K_{rdtw}^{τ}, EED, ADS with 15 poses	**97.74**
MSR-Action3D	Accuracy (%)
Cov3DJ, L = 3 (Hussein et al. 2013)	72.33 ± 3.69[a]
HON4D, (Oreifej and Liu 2013),	82.15 ± 4.18
SVM K_{rdtw}^{τ} 15 poses,	**85.72 ± 3.07**

[a] According to our own implementation of Cov3DJ

i.e. when the EED descriptor is used associated to an adaptive down-sampling. This is true both for the training and testing datasets.

For comparison, Table 2 gives results obtained by different methods of the state-of-the-art and compare them with the performance of an SVM that exploits the regularized DTW kernel (K_{rdtw}^{τ}) associated to the end extremity descriptor (EED) and an adaptive down-sampling (ADS) of 15 skeletal poses. To that end, we have reimplemented the Cov3DJ approach (Hussein et al. 2013) to get, for the MSR Action3D data set, the average result given by a 5-5 cross-validation on the subjects (252 tests). This comparative analysis shows that the SVM constructed from regularized DTW kernel provides results slightly above the current state-of-the-art for the considered data sets and tasks.

Fig. 6 Elapsed time as a function of the number of skeletal poses (10–30 poses): (i) RBF Euclidean Kernel, *Red/round/dashed line*, (ii) RBF DTW kernel, *Black/square/plain line*, (iii) normalized regularized DTW kernel(K_{rdtw}^{τ}), *blue/star/dotted line*

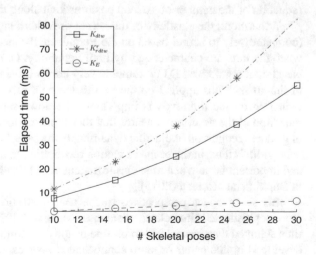

Finally, in Fig. 6, we give the average CPU elapsed time for the processing of a single gestural MSRAction3D action when varying the number of retained skeletal poses. The test has been performed on an Intel Core i7-4800MQ CPU, 2.70 GHz. Although the computational cost for the elastic kernel is quadratic, the latency for the classification of a single gestural action using a SVM K^{τ}_{rdtw} is less than 25 ms when 15 poses are considered, which effectively meets easily *real-time* requirements. The speed-up is, as expected, quadratic with the reduction of the number of skeletal poses since the elapsed time is roughly 100 ms when 30 poses are considered.

7   Conclusion and Perspectives

In the context of isolated action recognition, where few studies explicitly consider dimension reduction along both the spatial and time axes simultaneously, we have presented a recognition model based on the dimensionality reduction of the skeletal pose descriptor and the down-sampling of motion sequences coupled to elastic kernel machines. Two ways of down-sampling have been considered: a uniform down-sampling that evenly selects samples along the time axis and an adaptive down-sampling based on a piecewise linear approximation model. The dimensionality reduction of the skeletal pose descriptor is straightforwardly obtained by considering only end effector trajectories, which is consistent with some sensorimotor perceptual evidence about the way human beings perceive and interpret motion. On the data sets and tasks that we have addressed, we have shown that, even when quite important down-sampling is considered, the recognition accuracy only slightly degrades. In any case, best accuracies are obtained when an adaptive down-sampling is used on the end effector 3D trajectories. The temporal redundancy is therefore high and apparently not critical for the discrimination of the selected movements and tasks. In return, the down-sampling benefits in terms of computational complexity is quadratic with the reduction of the number of skeletal postures kept along the time axis.

Furthermore, the elasticity of the kernel provides a significant performance gain (comparatively to kernel based on the Euclidean distance) which is very important when the data are characterized by high variability. Our results show that a SVM based on a regularized DTW kernel is very competitive comparatively to the state-of-the-art methods applied on the two tested data sets, even when the dimension reduction on the time axis is important. The down-sampling and dimensionality reduction of the descriptor ensures that this approach meets the real-time constraint of gesture recognition. Adjusting dynamically the sampling rate to the current signal bandwidth is thus, in the scope of motion recognition in data stream, quite a feasible and important issue when time elastic distances or kernels are involved as proposed in Dupont and Marteau (2015).

This study opens perspectives to the use of elastic kernels constructed from more sophisticated time elastic distances (Marteau 2009) that cope explicitly with time stamped data, associated to on-line adaptive sampling techniques such the one developed in this paper or more sophisticated ones capable of extracting the most significant and discriminant skeletal poses in movement sequences, based on semantic segmentation. We also aim at testing these powerful tools to more complex tasks, where skilled gestures are studied, or/and expressive variations are considered.

References

Belkin, M., & Niyogi, P. (2002). Laplacian eigenmaps for dimensionality reduction and data representation. *Neural Computation, 15*, 1373–1396.

Berg, C., Christensen, J. P. R., & Ressel, P. (1984). *Harmonic analysis on semigroups: Theory of positive definite and related functions* (Vol. 100). Graduate texts in mathematics. New York: Springer.

Bissacco, A., Chiuso, A., & Soatto, S. (2007). Classification and recognition of dynamical models: the role of phase, independent components, kernels and optimal transport. *IEEE Transactions on Pattern Analysis and Machine Intelligence, 29*(11), 1958–1972.

Blackburn, J., & Ribeiro, E. (2007). Human motion recognition using isomap and dynamic time warping. In A. Elgammal, B. Rosenhahn, & R. Klette (Eds.), *Human motion—understanding, modeling, capture and animation* (Vol. 4814, pp. 285–298). Lecture notes in computer science. Berlin: Springer.

Chang, C. C., & Lin, C.-J. (2011). LIBSVM: A library for support vector machines. *ACM Transactions on Intelligent Systems and Technology, 2*, 27:1–27:27.

CMU. (2003). Motion capture database, Carnegie Mellon University. http://mocap.cs.cmu.edu/.

Cuturi, M., Vert, J.-P., Birkenes, O., & Matsui, T. (2007). A kernel for time series based on global alignments. In *Proceedings of ICASSP 2007* (pp. II-413–II-416). Honolulu: IEEE.

de Aguiar, E., & Theobalt, C. (2006). Automatic learning of articulated skeletons from 3D marker trajectories. In G. Bebis, et al. (Eds.), *ISVC* (Vol. 4291, pp. 485–494). Lecture notes in computer science. Berlin: Springer.

Dupont, M., & Marteau, P.-F. (2015). Coarse-DTW: exploiting sparsity in gesture time series. In A. Douzal-Chouakria, et al. (Eds.), *Advanced Analytics and Learning on Temporal Data (AALTD), Proceedings of the 1st International Workshop on Advanced Analytics and Learning on Temporal Data (AALTD)* (Vol. 1425). Porto, Portugal: CEUR Workshop Proceedings.

Fothergill, S., Mentis, H., Kohli, P., & Nowozin, S. (2012). Instructing people for training gestural interactive systems. In *Proceedings of the SIGCHI Conference on Human Factors in Computing Systems CHI 2012*, (pp. 1737–1746). New York: ACM.

Gibet, S., Marteau, P. -F., & Duarte, K. (2011). Toward a motor theory of sign language perception. In E. Efthimiou, G. Kouroupetroglou, & S. -E. Fotinea (Eds.), *Gesture Workshop* (Vol. 7206, pp. 161–172). Lecture notes in computer science Berlin: Springer.

Giese, M. A., Thornton, I., & Edelman, S. (2008). Metrics of the perception of body movement. *Journal of Vision, 8*(9), 1–18. Reviewed.

Han, L., Wu, X., Liang, W., Hou, G., & Jia, Y. (2010). Discriminative human action recognition in the learned hierarchical manifold space. *Image and Vision Computing, 28*(5), 836–849.

He, X., & Niyogi, P. (2003). *Locality preserving projections* (Vol. 16). Advances in neural information processing systems. Cambridge: MIT Press.

Hussain, S., & Rashid, A. (2012). User independent hand gesture recognition by accelerated DTW. In *International Conference on Informatics, Electronics Vision (ICIEV)* (pp. 1033–1037).

Hussein, M. E., Torki, M., Gowayyed, M. A., & El-Saban, M. (2013). Human action recognition using a temporal hierarchy of covariance descriptors on 3D joint locations. In *IJCAI*.

Jenkins, O. C. & Matarić, M. J. (2004). A spatio-temporal extension to isomap nonlinear dimension reduction. In *The International Conference on Machine Learning (ICML 2004)*, (pp. 441–448).

Jolliffe, I. (1986). *Principal component analysis*. Springer series in statistics. New York: Springer.

Keogh, E. J. & Pazzani, M. J. (2000). Scaling up dynamic time warping for datamining applications. In *Proceedings of the Sixth ACM SIGKDD KDD 2000* (pp. 285–289). New York.

Kruskal, J., & Wish, M. (1978). *Multidimensional scaling*. Beverly Hills: Sage Publications.

Larochelle, H., Mandel, M., Pascanu, R., & Bengio, Y. (2012). Learning algorithms for the classification restricted Boltzmann machine. *Journal of Machine Learning Research, 13*, 643–669.

Li, W., Zhang, Z., & Liu, Z. (2010). Action recognition based on a bag of 3D points. In *Proceedings of IEEE International Workshop on CVPR for Human Communicative Behavior Analysis* (pp. 9–14). In Press.

Marteau, P. F. (2009). Time warp edit distance with stiffness adjustment for time series matching. *IEEE Transactions on Pattern Analysis and Machine Intelligence, 31*(2), 306–318.

Marteau, P.-F., & Gibet, S. (2006). Adaptive sampling of motion trajectories for discrete task-based analysis and synthesis of gesture. In S. Gibet, N. Courty, & J.-F. Kamp (Eds.), *Gesture in human-computer interaction and simulation* (Vol. 3881, pp. 224–235). Lecture notes in computer science. Springer: Berlin.

Marteau, P.-F. & Gibet, S. (2014). On recursive edit distance kernels with application to time series classification. *IEEE Transactions on Neural Networks and Learning Systems*, 1–14.

Marteau, P.-F., & Ménier, G. (2009). Speeding up simplification of polygonal curves using nested approximations. *Pattern Analysis and Applications, 12*(4), 367–375.

Martens, J. & Sutskever, I. (2011). Learning recurrent neural networks with hessian-free optimization. In *ICML* (pp. 1033–1040).

Masoud, O., & Papanikolopoulos, N. (2003). A method for human action recognition. *Image and Vision Computing, 21*(8), 729–743.

McLachlan, G. (2004). *Discriminant analysis and statistical pattern recognition*. Probability and statistics. New York: Wiley.

Mitra, S., & Acharya, T. (2007). Gesture recognition: a survey. *Transactions on Systems, Man, and Cybernetics, Part C, 37*(3), 311–324.

Müller, M., Röder, T., Clausen, M., Eberhardt, B., Krüger, B., & Weber, A. (2007). Documentation mocap database HDM05. Technical report CG-2007-2, Universität Bonn.

O'Brien, J. F., Bodenheimer, R. E., Brostow, G. J., & Hodgins, J. K. (2000). Automatic joint parameter estimation from magnetic motion capture data. In *Proceedings of Graphics Interface* (Vol. 2000, pp. 53–60).

Ofli, F., Chaudhry, R., Kurillo, G., Vidal, R., & Bajcsy, R. (2012). Sequence of the most informative joints (SMIJ): A new representation for human skeletal action recognition. In *CVPR Workshops* (pp. 8–13). IEEE.

Ofli, F., Chaudhry, R., Kurillo, G., Vidal, R., & Bajcsy, R. (2013). Sequence of the most informative joints (SMIJ): A new representation for human skeletal action recognition. *Journal of Visual Communication and Image Representation*, 1–20.

Oreifej, O. & Liu, Z. (2013). HON4D: Histogram of oriented 4D normals for activity recognition from depth sequences. In *2013 IEEE CVPR* (pp. 716–723).

Roweis, S. T., & Saul, L. K. (2000). Nonlinear dimensionality reduction by locally linear embedding. *Science, 290*, 2323–2326.

Sakoe, H. & Chiba, S. (1971). A dynamic programming approach to continuous speech recognition. In *Proceedings of the 7th International Congress of Acoustic* (pp. 65–68).

Sempena, S., Maulidevi, N., & Aryan, P. (2011). Human action recognition using dynamic time warping. In *International Conference on Electrical Engineering and Informatics (ICEEI)* (pp. 1–5).

Shotton, J., Fitzgibbon, A., Cook, M., Sharp, T., Finocchio, M., Moore R., et al. (2011). Real-time human pose recognition in parts from single depth images. In *Conference on Computer Vision and Pattern Recognition CVPR 2011* (pp. 1297–1304). IEEE.

Tenenbaum, J. B., de Silva, V., & Langford, J. C. (2000). A global geometric framework for nonlinear dimensionality reduction. *Science, 290*(5500), 2319.

Veeraraghavan, A., Chowdhury, A. K. R., & Chellappa, R. (2004). Role of shape and kinematics in human movement analysis. In *CVPR* (Vol. 1, pp. 730–737).

Velichko, V. M., & Zagoruyko, N. G. (1970). Automatic recognition of 200 words. *International Journal of Man-Machine Studies, 2*, 223–234.

Wang, J., Liu, Z., Wu, Y., & Yuan, J. (2012). Mining actionlet ensemble for action recognition with depth cameras. In *IEEE International Conference CVPR* (pp. 1290–1297).

Wang, S. B., Quattoni, A., Morency, L., Demirdjian, D., & Darrell, T. (2006). Hidden conditional random fields for gesture recognition. In *IEEE International Conference CVPR* (Vol. 2, pp. 1521–1527).

Yu, E. & Aggarwal, J. (2009). Human action recognition with extremities as semantic posture representation. In *2012 IEEE Computer Society Conference on Computer Vision and Pattern Recognition Workshops* (pp. 1–8).

Zhao, X., Song, Z., Guo, J., Zhao, Y., & Zheng, F. (2012). Real-time hand gesture detection and recognition by random forest. In M. Zhao & J. Sha (Eds.), *Communications and information processing* (Vol. 289, pp. 747–755). Berlin: Springer.

Exact and Approximate Minimal Pattern Mining

Arnaud Soulet and François Rioult

Abstract Condensed representations have been studied extensively for 15 years. In particular, the maximal patterns of the equivalence classes have received much attention with very general proposals. In contrast, the minimal patterns remained in the shadows in particular because they are too numerous and they are difficult to extract. In this paper, we present a generic framework for *exact* and *approximate* minimal patterns mining by introducing the concept of minimizable set system. This framework based on set systems addresses various languages such as itemsets or strings, and at the same time, different metrics such as frequency. For instance, the free, δ-free and the essential patterns are naturally handled by our approach, just as the minimal strings. Then, for any minimizable set system, we introduce a fast minimality checking method that is easy to incorporate in a depth-first search algorithm for mining the δ-minimal patterns. We demonstrate that it is polynomial-delay and polynomial-space. Experiments on traditional benchmarks complete our study by showing that our approach is competitive with the best proposals.

1 Introduction

Minimality is an essential concept of pattern mining. Given a function f and a language $\mathscr{L}$, a minimal pattern X is one of the smallest pattern with respect to the set inclusion in $\mathscr{L}$ satisfying the property $f(X)$. Interestingly, the whole set of minimal patterns forms a condensed representation of $\mathscr{L}$ adequate to f: it is possible to retrieve $f(Y)$ for any pattern of Y in $\mathscr{L}$. Typically, the set of free itemsets (Boulicaut et al. 2000) (also called generators or key itemsets (Pasquier et al. 1999)) is a condensed representation of all itemsets (here, f and $\mathscr{L}$ are respectively the frequency and the itemset languages). Of course, it is often more efficient to extract minimal

A. Soulet (✉)
LI, Université François Rabelais Tours, Tours, France
e-mail: arnaud.soulet@univ-tours.fr

F. Rioult
GREYC, Université de Caen Normandie, Caen, France
e-mail: francois.rioult@unicaen.fr

© Springer International Publishing Switzerland 2017
F. Guillet et al. (eds.), *Advances in Knowledge Discovery and Management*,
Studies in Computational Intelligence 665, DOI 10.1007/978-3-319-45763-5_4

61

patterns rather than all patterns because they are less numerous. In addition, minimal patterns have a lot of useful applications including higher KDD tasks: producing the most relevant association rules (Zaki 2000; Crémilleux and Boulicaut 2003), building classifiers (Liu et al. 1998) or generating minimal traversals (Eiter and Gottlob 2002). Minimality has been studied in the case of different functions (like frequency (Calders et al. 2004) and condensable functions (Soulet and Crémilleux 2008)) and different languages (e.g., itemsets (Boulicaut et al. 2000) and sequences (Lo et al. 2008)). Although the minimality has obvious advantages (Li et al. 2006), very few studies are related to the minimality while maximality (i.e., closed patterns) has been widely studied. Indeed, 59 % of publications about condensed representations benefit from closed patterns and only 13 % rely on free patterns (Giacometti et al. 2013). In particular, to the best of our knowledge, there is no framework as general as those proposed for maximality (Arimura and Uno 2009).

1.1 Depth-First Search Mining

We think that a current major drawback of minimal patterns lies in their inefficient extraction. This low efficiency comes mainly from the fact that most existing algorithms use a levelwise approach (Boulicaut et al. 2000; Soulet and Crémilleux 2008; Calders and Goethals 2005; Hébert and Crémilleux 2005) (i.e., breadth-first search/generate and test method). As they store all candidates in memory during the generation phase, the extraction may fail due to memory lack. To tackle this memory pitfall, it seems preferable to adopt a depth-first traversal which often consumes less memory and is still very fast. However, check whether the minimality is satisfied or not is very difficult in a depth-first traversal. In the case of frequency with itemsets, the best way for evaluating the minimality for a pattern (saying abc) is to compare its frequency with that of all its direct subsets (here, ab, ac and bc). But, when the pattern abc is achieved by a depth-first traversal, only frequencies of a and ab have previously been calculated. As the frequency of ac and bc are unknown, it is impossible to check whether the frequency of abc is strictly less than that of ac and bc. To cope with this problem, (Calders and Goethals 2005; Liu et al. 2008; Szathmary et al. 2009) have adopted a different traversal with reordered items. For instance, when the itemset abc is reached by this new traversal, c, b, bc, a, ac and bc were previously scanned and their frequency are known for checking whether abc is minimal. Unfortunately, such a method requires to store all the patterns in memory (here, c, b, bc and so on) using a trie (Calders and Goethals 2005) or an hash table (Liu et al. 2008; Szathmary et al. 2009). For this reason, existing DFS proposals (Calders and Goethals 2005; Liu et al. 2008; Szathmary et al. 2009) do not solve the memory consumption issue as expected.

1.2   Approximate Minimality

The large size of condensed representations based on minimal patterns is often mentioned as a major drawback of the minimality. For instance, the free patterns are always more numerous than closed patterns by definition (Calders et al. 2004). Using an approximate minimality is a way to reduce the cardinality of representations. With such representations, it is not possible to exactly regenerate the information about all the patterns. Now we can only get bounds on it. In other words, the whole set of approximate minimal patterns forms a condensed representation of $\mathscr{L}$ adequate to f: it is possible to retrieve $\tilde{f}(Y)$ for any pattern of Y in $\mathscr{L}$ with a bounded error: $\varepsilon(\tilde{f}(Y), f(Y)) \leq \delta$. Nevertheless, in many situations, these bounds are sufficient to approximate queries (Boulicaut et al. 2003). More evidences of the practical interest of such representations have been discussed in (Mannila and Toivonen 1996). To the best of our knowledge, this paper is the first work to generalize the notion of approximate representation of minimal patterns to measure other than frequency.

1.3   Contributions

The main goal of this paper is to present a generic and efficient framework for exact and approximate minimal pattern mining by providing a depth-first search algorithm. We introduce the notion of ε-*minimizable set system* which is at the core of the definition of this framework. This latter covers a broad spectrum of minimal patterns including all the languages and measures investigated in (Soulet and Crémilleux 2008; Arimura and Uno 2009). Fast minimality checking in a depth-first traversal is achieved thanks to the notion of *critical objects* which depends on the ε-minimizable set system. Based on this new technique, we propose the DEFME algorithm. It mines the δ-minimal patterns for any minimizable set system using a depth-first search algorithm. To the best of our knowledge, this is the first algorithm that enumerates exact and approximate minimal patterns in polynomial delay and in linear space with respect to the dataset. This paper extends our previous work (Soulet and Rioult 2014) by adding the ability to extract approximate patterns. Furthermore, this paper presents the demonstrations of all our theoretical results and it is also enriched with an extensive study concerning free itemset mining.

The outline of this paper is as follows. In Sect. 2, we discuss some related work about minimality in the landscape of condensed representations. In Sect. 3, we propose our generic framework for minimal pattern mining based on set systems. We introduce our fast minimality checking method in Sect. 4 and we indicate how to use it by sketching the DEFME algorithm. Section 5 shows that ε-MSS framework enable us to deal with various pattern mining problem related with minimality. Section 6 provides experimental results.

2 Related Work

As discussed in the introduction, we propose the first algorithm for minimal pattern mining with moderate memory usage. Beyond algorithmic aspects, we think that that our framework also has the advantage of extending the notion of minimality in three orthogonal aspects: flexible function, flexible language and exact/approximated (see Sect. 2.1). The expected benefits of this generalization are important because minimality is a central task of Knowledge Discovery in Databases (see Sect. 2.2).

2.1 Minimality for Exact and Approximate Condensed Representations

The collection of minimal patterns is a kind of condensed representations. Let us recall that a condensed representation of the frequent patterns is a set of patterns that can regenerate all the patterns that are frequent with their frequency. The success of the condensed representations stems from their undeniable benefit to reduce the number of mined patterns by eliminating redundancies. A large number of condensed representations have been proposed in literature (Calders et al. 2004; Hamrouni 2012): closed itemsets (Pasquier et al. 1999), free itemsets (Boulicaut et al. 2000), essential itemsets (Casali et al. 2005), Non-Derivable Itemsets (Calders and Goethals 2005), k-free itemsets (Calders and Goethals 2003), itemsets with negation (Kryszkiewicz 2005) and so on. Besides, approximate condensed representations have also been proposed to approximate the frequency of itemsets as it is the case with δ-free itemsets (Boulicaut et al. 2003). Two ideas are at the core of the condensed representations: the closure operator (Hamrouni 2012) that builds equivalence classes and the principle of inclusion-exclusion. As the inclusion-exclusion principle only works for the frequency, this paper exclusively focuses on minimal patterns considering equivalence classes.

The *exact* condensed representations of minimal patterns are not limited to frequency or itemsets. First, there are a few extensions of condensed representations to measure other than frequency. In Soulet et al. (2004), it is shown that the former condensed representations are adequate to any frequency-based measure (e.g., lift, growth rate). Soulet and Crémilleux (2008) addresses all the condensable measures dealing with aggregate measures, bond and so on. As the notion of closure operator is at the core of these approaches, it is also possible to mine the minimal patterns adequate to classical aggregate functions such as min, max or sum (Soulet and Crémilleux 2008). In the rest of this paper, we show that our unifying framework deals with free (Boulicaut et al. 2000), δ-free (Boulicaut et al. 2003), essential (Casali et al. 2005) and adequate free (Soulet and Crémilleux 2008) itemsets. More interestingly, many other forms of minimal patterns never discussed in the literature fit into this framework. In particular, we generalize the concept of approximate condensed representation to a wide range of measures.

In parallel, several studies have extended the notion of generators to address other languages such as sequences (Lo et al. 2008; Gao et al. 2008), negative itemsets (Gasmi et al. 2007), graphs (Zeng et al. 2009). Unfortunately no work proposed a generic framework extending the condensed representations based on minimality to a broad spectrum of languages, as it was done with the closed patterns (Arimura and Uno 2009). For instance, Boulicaut et al. (2000), Pasquier et al. (1999), Calders and Goethals (2005), Liu et al. (2008) only address itemsets or Lo et al. (2008), Gao et al. 2008 focus exclusively on sequences. In this paper, we make the connection between the set systems and only two languages: itemsets and strings due to space limitation. Numerous other languages can be represented using this set system framework. In particular, all the languages depicted by Arimura and Uno (2009) are suitable.

2.2 Interest of Minimality

Minimal pattern mining has a lot of applications and their use is not limited to obtain frequent patterns more efficiently. Their properties are useful for higher KDD tasks. For instance, minimal patterns are used in conjunction of closed patterns to produce non-redundant (Zaki 2000), informative rules (Pasquier et al. 1999) or simplest rules (Crémilleux and Boulicaut 2003). The sequential rules also benefit from minimality (Lo et al. 2009). It is also possible to exploit the minimal patterns for mining the classification rules that are the key elements of associative classifiers (Liu et al. 1998; Rioult et al. 2010).

In addition, the essential patterns are useful for deriving minimal traversals that exactly corresponds to the largest essential patterns with respect to the inclusion. Let us recall that the minimal transversal generation is a very important problem which has many applications in Logic (e.g., satisfiability checking), Artificial Intelligence (e.g., model-based diagnosis) and Machine Learning (e.g., exact learning) (Eiter and Gottlob 2002; Murakami and Uno 2013). For instance, minimal traversal are useful for the discovery of key actors in social networks that belong to several communities (Jelassi et al. 2014). Thus, the efficient extraction of minimal patterns as proposed in this paper is a very crucial stage at the core of many tasks.

3 ε-Minimizable Set System Framework

3.1 Basic Definitions

A *set system* $(\mathcal{F}, E)$ is a collection $\mathcal{F}$ of subsets of a *ground set* E (i.e. $\mathcal{F}$ is a subset of the power set of E). A member of $\mathcal{F}$ is called a *feasible set*. A *strongly accessible* set system $(\mathcal{F}, E)$ is a set system where for every feasible sets X, Y

satisfying $X \subset Y$, there is an element $e \in Y \setminus X$ such that $Xe \in \mathscr{F}$.[1] Obviously, itemset language fits this framework with the set system $(2^{\mathscr{I}}, \mathscr{I})$ where $\mathscr{I}$ is the set of items. $(2^{\mathscr{I}}, \mathscr{I})$ is even strongly accessible. But the notion of set system allows considering more sophisticated languages. For instance, it is easy to build a family set $\mathscr{F}_S$ denoting the collection of substrings of $S = abracadabra$ by encoding each substring $s_{k+1} s_{k+2} \ldots s_{k+n}$ by a set $\{(s_{k+1}, 1), (s_{k+2}, 2), \ldots, (s_{k+n}, n)\}$. The set sytem $(\mathscr{F}_S, E_S = \bigcup \mathscr{F}_S)$ is also strongly accessible. The set system formalism has already been used to describe pattern mining problems (see for instance Arimura and Uno 2009).

Intuitively, a pattern always describes a set of objects. This set of objects is obtained from the pattern by means of a *cover operator* formalized as follows:

Definition 1 (*Cover operator*) Given a set of objects $\mathscr{O}$, a cover operator $cov : 2^E \rightarrow 2^{\mathscr{O}}$ is a function satisfying $cov(X \cup Y) = cov(X) \cap cov(Y)$ for every $X \in 2^E$ and $Y \in 2^E$.

This definition indicates that the coverage of the union of two patterns is exactly the intersection of their two covers. For itemsets, a natural cover operator is the extensive function of an itemset X that returns the set of tuple identifiers supported by X: $cov_{\mathscr{I}}(X) = \{o \in \mathscr{O} \mid X \subseteq o\}$. But, in general, the cover is not the final aim: the cardinality of $cov_{\mathscr{I}}(X)$ corresponds to the frequency of X. In the context of strings, the index list of a string X also defines a cover operator: $cov_S(X) = \{i \mid \forall (s_j, j) \in X, \ (s_j, j+i) \in S\}$. Continuing our example with the string $S = abracadabra$, it is not difficult to compute the index lists $cov_S(\{(a, 1)\}) = \{0, 3, 5, 7, 10\}$ and $cov_S(\{(b, 2)\}) = \{0, 7\}$ and then, to verify $cov_S(\{(a, 1), (b, 2)\}) = cov_S(\{(a, 1)\}) \cap cov_S(\{(b, 2)\}) = \{0, 7\}$.

For some languages, the same pattern is described by several distinct sets and then it is necessary to have a canonical form. For example, consider the set $\{(a, 1), (b, 2), (r, 3)\}$ corresponding to the string abr. Its suffix $\{(b, 2), (r, 3)\}$ encodes the same string br as $\{(b, 1), (r, 2)\}$. The latter is the canonical form of the string br. To retrieve the canonical form of a pattern, we introduce the notion of canonical operator:

Definition 2 (*Canonical operator*) Given two set systems $(\mathscr{F}, E)$ and $(\mathscr{G}, E)$, a canonical operator $\phi : \mathscr{F} \cup \mathscr{G} \rightarrow \mathscr{F}$ is a function satisfying (i) $X \subset Y \Rightarrow \phi(X) \subset \phi(Y)$ and (ii) $X \in \mathscr{F} \Rightarrow \phi(X) = X$ for all sets $X, Y \in \mathscr{G}$.

In this definition, the property (i) ensures us that the canonical forms of two comparable sets with respect to the inclusion remain comparable. The property (ii) means that the set system $(\mathscr{F}, E)$ includes all canonical forms. Continuing our example about strings, it is not difficult to see that $\phi_S : \{(s_k, k), (s_{k+1}, k+1), \ldots, (s_{k+n}, n)\} \mapsto \{(s_k, 1), (s_{k+1}, 2), \ldots, (s_{k+n}, n-k+1)\}$ satisfies the two desired properties (i) and (ii). For instance, $\phi_S(\{(b, 2), (r, 3)\})$ returns the canonical form of the string $\{(b, 2), (r, 3)\}$ which is $\{(b, 1), (r, 2)\}$.

---

[1] We use the notation Xe instead of $X \cup \{e\}$.

3.2   ε-Minimizable Set System

Rather than considering an entire set system, it is wise to select a smaller part that provides the same information (w.r.t. a cover operator). For this, it is necessary that this set system plus the cover operator form a *minimizable* set system:

Definition 3 (*ε-Minimizable set system*) A ε-minimizable set system is a tuple $\langle (\mathscr{F}, E), \mathscr{G}, cov, \phi, \varepsilon \rangle$ where:

- $(\mathscr{F}, E)$ is a finite, strongly accessible set system. A feasible set in $\mathscr{F}$ is called a pattern.
- $(\mathscr{G}, E)$ is a finite, strongly accessible set system satisfying for every feasible set $X, Y \in \mathscr{F}$ such that $X \subseteq Y$ and element $e \in E$, $X \setminus \{e\} \in \mathscr{G} \Rightarrow Y \setminus \{e\} \in \mathscr{G}$. A feasible set in $\mathscr{G}$ is called a generalization.
- $cov : 2^E \to 2^\mathcal{O}$ is a cover operator.
- $\phi : \mathscr{F} \cup \mathscr{G} \to \mathscr{F}$ is a canonical operator such that for every feasible set $X \in \mathscr{G}$, it implies $cov(\phi(X)) = cov(X)$.
- $\varepsilon : \mathcal{O} \to \mathfrak{R}^+$ is an error function.

Let us now illustrate the role of $\mathscr{G}$ compared to $\mathscr{F}$ in the case of strings. In fact, $\mathscr{G}_S$ gathers all the suffixes of any pattern of $\mathscr{F}_S$. Typically, $\{(b, 2), (r, 3)\} \in \mathscr{G}_S$ is a generalization of $\{(a, 1), (b, 2), (r, 3)\} \in \mathscr{F}_S$. As said above, $\{(b, 2), (r, 3)\}$ has an equivalent form in $\mathscr{F}_S$: $\phi_S(\{(b, 2), (r, 3)\}) = \{(b, 1), (r, 2)\}$. By convention, we extend the definition of cov_S to $\mathscr{G}_S$ by considering that $cov_S(\phi_S(X)) = cov_S(X)$. In addition, it is not difficult to see that $\mathscr{G}_S$ satisfies the desired property with respect to $\mathscr{F}_S$: for every feasible set $X, Y \in \mathscr{F}_S$ such that $X \subseteq Y$ and element $e \in E_S, X \setminus \{e\} \in \mathscr{G}_S \Rightarrow Y \setminus \{e\} \in \mathscr{G}_S$. Indeed, if $X \setminus \{e\}$ is a suffix of X, it means that e is the first letter. If we consider a specialization of X and we again remove the first letter, we also obtain a suffix belonging to $\mathscr{G}_S$. Therefore, $\langle (\mathscr{F}_S, E_S), \mathscr{G}_S, cov_S, \phi_S, \varepsilon_S \rangle$ is a minimizable set system (where ε_S returns 1 for any object as done for δ-free (Boulicaut et al. 2000)). Note that Sect. 5 provides other examples of minimizable set systems.

In comparison with Soulet and Rioult (2014), Definition 3 introduces a new parameter ε to be able to extract minimal approximate patterns. This error function ε gives the weight of each object. Indeed, suppose that we have two feasible sets $X \subseteq Y$ whose covers are respectively O_X and O_Y. To determine whether X is sufficient to approximate Y, we propose to see if the objects $O_X \setminus O_Y$ are important. More precisely, the cover difference $O_X \setminus O_Y$ is measured by $\sum_{o \in O_X \setminus O_Y} \varepsilon(o)$. In the case of strings, ε_S returns 1 for any object saying that all the objects have the same importance.

To simplify notations, we extends ε to any subsets of $\mathcal{O}$ as follows: $\varepsilon : O \mapsto \sum_{o \in O} \varepsilon(o)$. Then, $\varepsilon(O_X \setminus O_Y)$ equals to $\sum_{o \in O_X \setminus O_Y} \varepsilon(o)$.

3.3 δ-Minimal Patterns

Obviously, a minimizable set system can be reduced to a system of smaller cardinality of which the patterns are called the δ-*minimal patterns*:

Definition 4 (δ-*Minimal pattern*) A pattern X is δ-minimal for $\langle (\mathcal{F}, E), \mathcal{G}, cov, \phi, \varepsilon \rangle$ iff $X \in \mathcal{F}$ and for every generalization $Y \in \mathcal{G}$ such that $Y \subset X$, $\varepsilon(cov(Y) \setminus cov(X)) > \delta$. $\mathcal{M}_\delta(\mathcal{S})$ denotes the set of all δ-minimal patterns.

Definition 4 means that a pattern is δ-minimal whenever its cover differs from that of any generalization with an error ε greater than δ. The higher the threshold δ is, the higher the approximation error of non-minimal patterns by δ-minimal patterns will be. If the threshold δ is set to 0, the extracted patterns are said to be *exact* and instead of writing 0-minimal patterns, we write minimal pattern. For example, for the cover operator cov_S with $\delta = 0$, the minimal patterns have a *strictly* smaller cover than their generalizations. The string ab is not minimal due to its suffix b because $cov_S(\{(b, 2)\}) = cov_S(\{(a, 1), (b, 2)\}) = \{0, 7\}$. For our running example, the whole collection of 0-minimal strings is $\mathcal{M}_0(\mathcal{S}_S) = \{a, b, r, c, d, ca, ra, da\}$.

Given a minimizable set system $\mathcal{S} = \langle (\mathcal{F}, E), \mathcal{G}, cov, \phi, \varepsilon \rangle$**, the δ-minimal pattern mining problem consists in enumerating all the δ-minimal patterns for** $\mathcal{S}$**.**

4 Enumerating the δ-Minimal Patterns

This section aims at effectively mining all the minimal patterns in a depth-first search manner (Sect. 4.3). To do this, we rely on two key ideas: the pruning of the search space (Sect. 4.1) and the fast minimality checking (Sect. 4.2).

Before, 0-minimal patterns are sufficient to induce the cover of any pattern (including non-minimal ones). From now, we consider a ε-minimizable set system $\mathcal{S} = \langle (\mathcal{F}, E), \mathcal{G}, cov, \phi, \varepsilon \rangle$. The 0-minimal patterns $\mathcal{M}_0(\mathcal{S})$ is a lossless representation of all patterns of $\mathcal{F}$ in the sense we can find the cover of any pattern. More importantly, we can even bound the approximation error of an approximate condensed representation stemming from the δ-minimal patterns $\mathcal{M}_\delta(\mathcal{S})$:

Theorem 1 (Exact and approximate condensed representation) *The set of δ-minimal patterns is a concise representation of $\mathcal{F}$ adequate to cov: for any pattern $X \in \mathcal{F}$, there exists $Y \subseteq X$ such that $\phi(Y) \in \mathcal{M}_\delta(\mathcal{S})$ and $\varepsilon(cov(\phi(Y)) \setminus cov(X)) \leq \delta$.*

Proof Let X be a pattern. There are two cases:

- X is δ-minimal. By Definition 4, X belongs to $\mathcal{M}_\delta(\mathcal{S}) \subseteq \mathcal{F}$. In such case, as $\phi(X) = X$, Thereom 1 is correct by considering $Y = X$.
- X is not δ-minimal. There exists $Y \in \mathcal{G}$ such that $Y \subset X$ with $\varepsilon(cov(Y) \setminus cov(X)) \leq \delta$ (see Definition 4). Definition 2 gives that $\phi(Y) \subset \phi(X)$ and even,

$\phi(Y) \subset X$ (because $X \in \mathscr{F}$ implies that $\phi(X) = X$). Note that the cardinality of $\phi(Y)$ is strictly smaller than that of X.

By induction, it is sure that there exists Y such that $\phi(Y)$ is a smaller feasible set than X verifying $\phi(Y) \in \mathscr{M}_\delta(\mathscr{S})$ and $\varepsilon(cov(\phi(Y)) \setminus cov(X)) \leq \delta$ (since the most general feasible sets of $\mathscr{F}$ are δ-minimal by definition). $\qquad\square$

Theorem 1 means that $\mathscr{M}_\delta(\mathscr{S})$ is really a condensed representation of $\mathscr{S}$ because the δ-minimal pattern mining enables us to approximate the cover of any pattern in $\mathscr{S}$ with δ as upper bound. For instance, the cover of the non-0-minimal pattern $\{(a, 1), (b, 2)\}$ equals to that of the 0-minimal pattern $\phi(\{(b, 2)\}) = \{(b, 1)\}$: $cov_S(\{(a, 1), (b, 2)\}) = cov_S(\{(b, 1)\}) = \{0, 7\}$.

It is preferable to extract $\mathscr{M}_0(\mathscr{S})$ instead of $\mathscr{S}$ because its size is lower (and, in general, much lower) than the total number of patterns. As indicated in introduction, the number of 0-minimal patterns may remain too important and it is often interesting to replace the exact regeneration by an approximate one. Indeed, the size of the approximate condensed representation decreases with δ: $\mathscr{M}_{\delta_1}(\mathscr{S}) \subseteq \mathscr{M}_{\delta_2}(\mathscr{S})$ whenever $\delta_1 \geq \delta_2$.

4.1 Search Space Pruning

The first problem we face is fairly classical. Given a ε-minimizable set system $\mathscr{S} = \langle(\mathscr{F}, E), \mathscr{G}, cov, \phi\rangle$, the number of patterns $|\mathscr{F}|$ is huge in general (in the worst case, it reaches $2^{|E|}$ patterns). So, it is absolutely necessary not to completely scan the search space for focusing on the minimal patterns. Effective techniques can be used to prune the search space due to the downward closure of $\mathscr{M}_\delta(\mathscr{S})$:

Theorem 2 (Independence system) *If a pattern X is δ-minimal for $\mathscr{S}$, then any pattern $Y \in \mathscr{F}$ satisfying $Y \subseteq X$ is also δ-minimal for $\mathscr{S}$.*

The proof of this theorem strongly relies on a key lemma saying that a non-minimal pattern (whatever δ) has a direct generalization having approximately the same cover with respect to ε:

Lemma 1 *If X is not δ-minimal, there exists $e \in X$ such that $X \setminus \{e\} \in \mathscr{G}$ and $\varepsilon(cov(X \setminus \{e\}) \setminus cov(X)) \leq \delta$.*

Proof Let X be a non-minimal pattern of $\mathscr{F}$, there exists $Y \in \mathscr{G}$ such that $Y \subset X$ and $\varepsilon(cov(Y) \setminus cov(X)) \leq \delta$ by Definition 4. As $\mathscr{G}$ is strongly accessible, there exists a chain of generalizations $Y_0 \subset Y_1 \subset \cdots \subset Y_n$ such that $Y_0 = Y$, $Y_n = X$ and $|Y_i| = |Y_{i-1}| + 1$ for any $i \in \{1, \ldots, n-1\}$. Thereby, Y_{n-1} is a generalization in $\mathscr{G}$ and the isotony of cov gives that $cov(Y) \supseteq cov(Y_{n-1})$. So, $cov(Y_{n-1}) \setminus cov(X) \subseteq cov(Y) \setminus cov(X)$. As $\varepsilon(O)$ increases with O, it means that $\varepsilon(cov(Y_{n-1}) \setminus cov(X)) \leq \varepsilon(cov(Y) \setminus cov(X)) \leq \delta$. Considering $e = X \setminus Y_{n-1}$, it means that $X \setminus \{e\} \in \mathscr{G}$ and $\varepsilon(cov(X \setminus \{e\}) \setminus cov(X)) \leq \delta$. $\qquad\square$

Now we prove Theorem 2:

Proof Let X be a pattern such that $X \in \mathscr{F} \setminus \mathscr{M}_\delta(\mathscr{S})$, there exists $e' \in E$ such that $\varepsilon(cov(X \setminus e') \setminus cov(X)) \leq \delta$ (Lemma 1). Now let us consider $e \in E$ such that $Xe \in \mathscr{F}$.

- $\varepsilon(cov(Xe \setminus e') \setminus cov(Xe))$ equals to $\varepsilon((cov(X \setminus e') \cap cov(e)) \setminus (cov(X) \cap cov(e)))$ due to the definition of cov. As set intersection is distributive over set difference, $\varepsilon(cov(Xe \setminus e') \setminus cov(Xe))$ can be rewritten as follows: $\varepsilon((cov(X \setminus e') \setminus cov(X)) \cap cov(e))$. As $\varepsilon(O)$ increases with O and $(cov(X \setminus e') \setminus cov(X)) \cap cov(e) \subseteq (cov(X \setminus e') \setminus cov(X))$, we obtain that $\varepsilon((cov(X \setminus e') \setminus cov(X)) \cap cov(e)) \leq \varepsilon(cov(X \setminus e') \setminus cov(X)) \leq \delta$.
- As $X \in \mathscr{F}$ and $X \setminus e' \in \mathscr{G}$, it gives $(Xe) \setminus e' \in \mathscr{G}$ due to the property of $\mathscr{G}$ with respect to $\mathscr{F}$ (see Definition 3) considering that Xe is a superset of X.

As the set $(Xe) \setminus \{e'\}$ is a generalization of Xe where $\varepsilon(cov((Xe) \setminus \{e'\}) \setminus cov(Xe)) \leq \delta$, the pattern Xe is not δ-minimal. In other words, a specialization of a non-δ-minimal pattern is also non-δ-minimal. The contrapositive of this implication proves that Theorem 2 is right.   □

For instance, as the string da is 0-minimal, the substrings d and a are also 0-minimal. More interestingly, as ab is not 0-minimal, the string abr is not 0-minimal. It means that the string ab is a cut-off point in the search space. In practice, anti-monotone pruning is recognized as a very powerful tool whatever the traversal of the search space (level by level or in depth).

4.2   Fast δ-Minimality Checking

The main difficulty in extracting the δ-minimal patterns is to test whether a pattern is δ-minimal or not. As we mentioned earlier, this is particularly difficult in a depth-first traversal because all subsets have not yet been enumerated. Indeed, depth-first approaches only have access to the first parent branch contrary to level-wise approaches. To overcome this difficulty, we introduce the concept of *critical objects* inspired from critical edges in case of minimal traversals (Murakami and Uno 2013). Intuitively, the critical objects of an element e for a pattern X are objects that are not covered by X due to the element e. We now give a formal definition of the critical objects derived from any cover operator:

Definition 5 (*Critical objects*) For a pattern X, the critical objects of an element $e \in X$, denoted by $\widehat{cov}(X, e)$ is the set of objects that belong to the cover of X without e and not to the cover of e: $\widehat{cov}(X, e) = cov(X \setminus e) \setminus cov(e)$.

Let us illustrate the critical objects with our running example. For $\{(a, 1), (b, 2)\}$, the critical objects $\widehat{cov}(ab, a)$ of the element $(a, 1)$ correspond to $\emptyset$ ($= \{0, 7\} \setminus \{0, 3, 5, 7, 10\}$). It means that the addition of a to b has no impact on the cover of

ab. At the opposite, for the same pattern, the critical objects of $(b, 2)$ are $\{3, 5, 10\}$ $(= \{0, 3, 5, 7, 10\} \setminus \{0, 7\})$. It is due to the element b that ab does not cover the objects $\{3, 5, 10\}$.

The critical objects are central in our proposition for the following reasons: (1) the critical objects easily characterize the minimal patterns; and (2) the critical objects can efficiently be computed in a depth-first search algorithm.

4.2.1 Minimal Pattern Characterization

The converse of Lemma 1 says that a pattern is δ-minimal if its cover differs from that of its generalization with an error ε exceeding δ. We can reformulate this definition thanks to the notion of critical objects as follows:

Property 1 (δ-Minimality) $X \in \mathscr{F}$ is δ-minimal iff $\forall e \in X$ such that $X \setminus e \in \mathscr{G}$, $\varepsilon(\widehat{cov}(X, e)) > \delta$.

Proof Let X be a δ-minimal and e be an element in X such that $X \setminus e \in \mathscr{G}$. The following equality is satisfied:

$$
\begin{aligned}
cov(X \setminus e) \setminus cov(X) &= cov(X \setminus e) \setminus cov((X \setminus e) \cup e) \\
&= cov(X \setminus e) \setminus (cov(X \setminus e) \cap cov(e)) \\
&= (cov(X \setminus e) \setminus cov(X \setminus e)) \cup (cov(X \setminus e) \setminus cov(e)) \\
&= \emptyset \cup \widehat{cov}(X, e)
\end{aligned}
$$

First, as $X \setminus e$ is a generalization, Definition 4 implies that $\varepsilon(cov(X \setminus e) \setminus cov(X)) > \delta$ and then, $\varepsilon(\widehat{cov}(X, e)) > \delta$ due to the above equality.

Now, assume that $\varepsilon(\widehat{cov}(X, e)) > \delta$ is satisfied by any element $e \in X$. Let us consider Y a generalization of X with $Y \subset X$. Due to strong accessibility, it is sure that there exists at least e' such that $Y \subseteq X \setminus e' \subset X$. We have that $\varepsilon(\widehat{cov}(X, e')) > \delta$ and even, $\varepsilon(cov(X \setminus e') \setminus cov(X)) > \delta$ (see the above equality). As $cov(X \setminus e') \subseteq cov(Y)$, we conclude that $\varepsilon(cov(Y) \setminus cov(X)) > \delta$ and then, X is δ-minimal according to Definition 4. $\qquad\square$

Typically, as b is a generalization of the string ab and at the same time, $\widehat{cov}(ab, a)$ is empty, ab is not 0-minimal. Property 1 means that checking whether a candidate X is δ-minimal only requires to know the critical objects of all the elements in X. Unlike the usual definition, no information is required on the subsets. Therefore, the critical objects allow us to design a depth-first algorithm if (and only if) computing the critical objects does not also require information on the subsets.

4.2.2 Efficiently Critical Object Computation

In a depth-first traversal, we want to update the critical objects of an element e for the pattern X when a new element e' is added to X. In such case, we now show that the critical objects can efficiently be computed by intersecting the old set of the critical objects $\widehat{cov}(X, e)$ with the cover of the new element e':

Property 2 *The following equality holds for any pattern* $X \in \mathscr{F}$ *and any two elements* $e, e' \in E$: $\widehat{cov}(Xe', e) = \widehat{cov}(X, e) \cap cov(e')$.

Proof Given a feasible set $X \in \mathscr{F}$ and two elements $e, e' \in E$:

$$
\begin{aligned}
\widehat{cov}(Xe', e) &= cov(Xe' \setminus e) \setminus cov(e) \\
&= (cov(X \setminus e) \cap cov(e')) \setminus cov(e) \\
&= (cov(X \setminus e) \setminus cov(e)) \cap cov(e') \\
&= \widehat{cov}(X, e) \cap cov(e') \qquad\qquad \square
\end{aligned}
$$

For instance, Definition 5 gives $\widehat{cov}_S(a, a) = \{1, 2, 4, 6, 8, 9\}$. As $cov_S(b) = \{0, 7\}$, we obtain that $\widehat{cov}_S(ab, a) = \widehat{cov}_S(a, a) \cap cov_S(b) = \{1, 2, 4, 6, 8, 9\} \cap \{0, 7\} = \emptyset$. Interestingly, Property 2 allows us to compute the critical objects of any element included in a pattern X having information on a single branch. This is the ideal situation for a depth-first search algorithm.

4.3 Algorithm DEFME

The algorithm DEFME takes as inputs the current pattern, the current tail (the list of the remaining items to be checked) and a maximal error threshold. It returns all the δ-minimal patterns containing X (based on $tail$). More precisely, Line 1 checks whether X is δ-minimal or not. If X is δ-minimal, it is output (Line 2). Lines 3–14 explores the subtree containing X based on the tail. For each element e where Xe is a pattern of $\mathscr{F}$ (Line 4) (Property 1), the branch is built with all the necessary information. Line 7 updates the cover and Lines 8–11 updates the critical objects using Property 2. Finally, the function DEFME is recursively called at Line 12 with the updated tail (Line 5) (Algorithm 1).

Theorems 3 and 4 demonstrate that the algorithm DEFME has an efficient behavior both in space and time. This efficiency mainly stems from the inexpensive handling of covers/critical objects as explained by the following property:

Algorithm 1: DEFME($X, tail, \delta$)

Input: X is a pattern, $tail$ is the set of the remaining items to be used in order to generate the candidates. δ is the maximal error threshold. Initial values: $X = \emptyset, tail = E$.
Output: polynomially incrementally outputs the minimal patterns.

  if $\forall e \in X, \ \varepsilon(\widehat{cov}(X, e)) > \delta$ **then**
    print X
    for all $e \in tail$ **do**
      if $Xe \in \mathscr{F}$ **then**
        $tail := tail \setminus \{e\}$
        $Y := Xe$
        $cov(Y) := cov(X) \cap cov(e)$
        $\widehat{cov}(Y, e) := cov(X) \setminus cov(e)$
        for all $e' \in X$ **do**
          $\widehat{cov}(Y, e') := \widehat{cov}(X, e') \cap cov(e)$
        end for
        DEFME($Y, tail, \delta$)
      end if
    end for
  end if

Property 3 *The following inequality holds for any pattern $X \in \mathscr{F}$:*

$$|cov(X)| + \sum_{e \in X} |\widehat{cov}(X, e)| \leq |cov(\emptyset)|$$

Proof Let X be a feasible set in $\mathscr{F}$ and e be an element in X. Assume that $o \in \widehat{cov}(X, e) = cov(X \setminus e) \setminus cov(e)$. It means that $o \in cov(X \setminus e)$ and $o \notin cov(e)$.

- We note that $cov(X) = cov(X \setminus e \cup e) = cov(X \setminus e) \cap cov(e)$. As $o \notin cov(e)$, o does not belong to $cov(X)$. In other words, there is no intersection between $cov(X)$ and any critical set $\widehat{cov}(X, e)$.
- Suppose now that there is also $e' \neq e$ such that o belongs to $\widehat{cov}(X, e')$ i.e., $o \in cov(X \setminus e')$ and $o \notin cov(e')$. Then, $o \in cov(X \setminus e) \cap cov(X \setminus e')$. As $cov(X \setminus e) \cap cov(X \setminus e') = cov(X \setminus e \cup X \setminus e')$ and $e \neq e'$, we obtain that $cov(X \setminus e \cup X \setminus e') = cov(X)$. As seen above, o cannot belong to $cov(X)$. We conclude that there is no intersection between $\widehat{cov}(X, e)$ and $\widehat{cov}(X, e')$.

Thus, an object cannot be at the same time in two sets among $cov(X), cov(X \setminus e_1)$, ..., $cov(X \setminus e_n)$. We proved that Property 3 is correct. $\square$

Property 3 means that for a pattern, the storage of its cover plus that of all the critical objects is upper bounded by the number of objects (i.e., $|cov(\emptyset)|$). Thus, it is straightforward to deduce the memory space required by the algorithm:

Theorem 3 (Polynomial-space complexity) *Given a minimal set sytem $\mathscr{S} = \langle (\mathscr{F}, E), \mathscr{G}, cov, \phi \rangle$ and $\delta \geq 0$, $\mathscr{M}_\delta(\mathscr{S})$ is enumerable in $O(|cov(\emptyset)| \times m)$ space where m is the maximal size of a feasible set in $\mathscr{F}$.*

Proof It has been shown in Property 3 that the storage of the cover of a feasible set X plus that of all its critical objects is upper bounded by the number of objects (i.e., $|cov(\emptyset)|$). Besides, the number of consecutive recursive calls is at most $m + 1$ where m is the maximal size of a feasible set in $\mathcal{F}$. We conclude that $\mathcal{M}_\delta(\mathcal{S})$ is enumerable in $O(|cov(\emptyset)| \times m)$ space. $\square$

In practice, the used memory space is very limited because m is small. In addition, the amount of time between each output pattern is polynomial:

Theorem 4 (Polynomial-delay complexity) *Given a minimal set sytem* $\mathcal{S} = \langle(\mathcal{F}, E), \mathcal{G}, cov, \phi\rangle$ *and* $\delta \geq 0$, $\mathcal{M}_\delta(\mathcal{S})$ *is enumerable in* $O(|E|^2 \times |cov(\emptyset)|)$ *time per minimal pattern.*

Proof It is not difficult to see that between two output patterns, DEFME requires a polynomial number of operations assuming that the membership oracle is computable in polytime (Line 4). Indeed, the computation of the cover and that of the critical objects (Lines 7–11) is linear with the number of objects due to Property 3; the loop in Line 3 does not exceed $|E|$ iterations and finally, the number of consecutive backtracks is at most $|E|$. $\square$

5 Scope of ε-MSS Framework

We now illustrate the flexibility of ε-MSS framework to model well-known tasks about itemset mining. Of course, the three mining problems below can be solved using DEFME.

Free, δ-free and essential patterns: As indicated above the system $\mathcal{S}_\mathcal{G} = \langle(2^\mathcal{I}, \mathcal{I}), 2^\mathcal{I}, cov_\mathcal{G}, Id, \varepsilon_\mathcal{G}\rangle$ is minimizable and $\mathcal{M}_\delta(\mathcal{S}_\mathcal{G})$ corresponds exactly to the δ-free itemsets (or generators). The frequency of each itemset is computed using the cardinality of the cover. Replace the cover operator $cov_\mathcal{G}$ by $\overline{cov_\mathcal{G}} : X \mapsto \{o \in \mathcal{O} \mid X \cap o = \emptyset\}$ leads to a new minimizable set system $\langle(2^\mathcal{I}, \mathcal{I}), 2^\mathcal{I}, \overline{cov_\mathcal{G}}, Id\rangle$ of which minimal patterns are essential itemsets (Casali et al. 2005). The disjunctive frequency of an itemset X is $|\mathcal{O}| - |\overline{cov_\mathcal{G}}(X)|$.

Classification rules: Our framework is well-adapted for mining all minimal classification rules that satisfy interestingness criteria involving frequencies as in (Crémilleux and Boulicaut 2003). Assuming that the set of objects $\mathcal{O}$ is divided into two disjoint classes $\mathcal{O} = \mathcal{O}_1 \cup \mathcal{O}_2$, the confidence of the classification rule $X \rightarrow class_1$ is $|\mathcal{O}_1 \cap cov_\mathcal{G}(X)|/|cov_\mathcal{G}(X)|$. More generally, it is easy to show that any frequency-based measure (e.g., lift, bond) can be derived from the positive and negative covers. In addition, the essential patterns are useful for deriving minimal traversals that exactly corresponds to the maximal patterns of $\mathcal{M}_\delta(\langle(2^\mathcal{I}, \mathcal{I}), 2^\mathcal{I}, \overline{cov_\mathcal{G}}, Id, \varepsilon_\mathcal{G}\rangle)$.

Condensed representations for aggregate functions: Minizable set systems are also well-adapted for aggregate functions such as *min*, *max* and *sum*

(Soulet and Crémilleux 2008). For instance, let us consider the function $cov_{min}(X) = \{val(i)|\exists i \in \mathcal{I}, val(i) \leq min(X.val)\}$ that returns all the possible values of val less than $min(X.val)$. This function is a cover operator and $\langle(2^{\mathcal{I}}, \mathcal{I}), 2^{\mathcal{I}}, cov_{min}, Id, \varepsilon_{\mathcal{I}}\rangle$ is even a minimizable set system. The minimal patterns adequate to min correspond to the minimal patterns of the previous set system. Furthermore, the value $min(X.val)$ could be obtained as follows $max(cov_{min}(X))$. A similar approach enables us to deal with max and sum.

ε-MSS framework not only unifies a lot of previous works about minimal pattern mining, but it also opens the way for more sophisticated condensed representations. Typically, it is possible to combine original cover operators (e.g., negative cover $\overline{cov_{\mathcal{I}}}$ or function cov_{min}) with advanced languages such as graphs and pictures using set systems proposed by Arimura and Uno (2009).

6 Experimental Study

The aim of our experiments is to quantify the benefit brought by DEFME both on effectiveness and conciseness. We show its effectiveness with the problem of free itemset mining for which several prototypes already exist in the literature. Then we instantiate DEFME to extract the collection of minimal strings and compare its size with that of closed strings. All tests were performed on a 2.2 GHz Opteron processor with Linux operating system and 200 GB of RAM memory.

6.1 Free Itemset Mining

We designed a prototype of DEFME for itemset mining as a proof of concept and we compared it with four other prototypes:

- two of them are based on a traditional levelwise traversal: ACMINER (Boulicaut et al. 2000), which explores the itemset space and FTMINER (Hébert and Crémilleux 2005), exploring the transaction space;
- the two others use a depth-first traversal with reordered items: GRGROWTH (Liu et al. 2008), NDI[2] (Calders and Goethals 2005) and the TALKYG2 algorithm of the CORON platform.

For this purpose, we conducted experiments on benchmarks coming from the FIMI repository and the 2004 PKDD Discovery Challenge.[3] The first three columns of Table 1 give the characteristics of these datasets. The fourth column gives the used

---

[2]As this prototype mines non-derivable itemsets, it enable us to compute free patterns when the depth parameter is set to 1.

[3]http://fimi.ua.ac.be/data/ and http://lisp.vse.cz/challenge/ecmlpkdd2004/.

Table 1 Characteristics of benchmarks, minimum support and time of free itemset mining, in seconds

Dataset	Objects	Items	Minsup (%)	ACMINER	FTMINER	GRGROWTH	NDI	TALKYG2	DEFME
74 × 822	74	822	88	fail	158	122	fail	5,920	**45**
90 × 27679	90	27,679	91	fail	270	206	fail	fail	**79**
chess	3,196	75	22	6,623	345	**56**	187	1,291	192
connect	67,557	129	7	34,943	fail	**37**	115	1,104	4,873
pumsb	49,046	2,113	51	70,014	fail	**64**	212	6,897	548
pumsb*	49,046	2,088	5	21,267	2971	**89**	202	fail	4,600

Table 2 Memory usage, in MBytes

Dataset	Minsup (%)	ACMINER	FTMINER	GRGROWTH	NDI	TALKYG2	DEFME
74 × 822	88	fail	12,467	1,990	fail	20,096	**3**
90 × 27679	91	fail	6,763	2,929	fail	fail	**13**
chess	22	3,914	7,990	914	1,684	12,243	**8**
connect	7	2,087	fail	684	1,181	12,305	**174**
pumsb	51	7,236	fail	916	1,818	30,941	**118**
pumsb*	5	5,175	51,702	1,330	2,523	fail	**170**

minimal support threshold. The next five columns report the running times. Table 2 indicates the memory consumption.

The best time performances are highlighted in bold in Table 1. Depth first approaches of GRGROWTH, NDI and DEFME clearly state their domination over the levelwise approach of ACMINER and FTMINER: level-wisely mining is both time and memory consuming. GRGROWTH is by far the fastest prototype, except on the genomic datasets 74 × 822 and 90 × 27679, where it is outperformed by of DEFME.

The right part of Fig. 1 details, for various minsup thresholds, the speed of DEFME. It plots the number of minimal patterns it extracted for each second of computing time.

Concerning memory consumption in Table 2, DEFME is (as expected) the most efficient algorithm. For the other prototypes, even increasing the storage memory would not be sufficient to treat the most difficult datasets. Here, GRGROWTH and NDI are not suitable to process genomic datasets even with 200 GB of RAM memory and relatively high thresholds. More precisely, the left part of Fig. 1 plots the ratio between GRGROWTH's and DEFME's memory use for various minsup thresholds. It is easy to notice that this ratio could quickly lead GRGROWTH to go out of memory. DEFME works with bounded memory and then is not minsup limited.

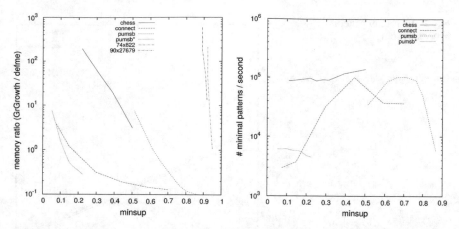

Fig. 1 Comparing NDI and DEFME: ratio of memory use (*left*) and mining speed (*right*)

6.2 Minimal String Mining

In this section, we adopt the formalism of strings stemming from our running example. We compared our algorithm for minimal string mining with the MAXMOTIF prototype provided by Takeaki Uno that mines closed strings (Arimura and Uno 2009). Our goal is to compare the size of condensed representations based on minimal strings with those based on all strings and all closed strings. We do not report the execution times because MAXMOTIF developed in Java is much slower than DEFME (developed in C++). Experiments are conducted on two datasets: chromosom[4] and msnbc coming from the UCI ML repository (http://www.ics.uci.edu/mlearn).

Figures 2 and 3 report the number of strings/minimal strings/closed strings mined in chromosom and msnbc. Of course, whatever the collection of patterns, the number of patterns increases with the decrease of the minimal frequency threshold. Interestingly, the two condensed representations become particularly useful when the frequency threshold is very small. Clearly the number of minimal strings is greater than the number of closed strings, but the gap is not as important as it is the case with free and closed itemsets.

---

[4]This dataset is provided with MAXMOTIF: http://research.nii.ac.jp/~uno/codes.htm.

Fig. 2 Number of patterns
in chromosom

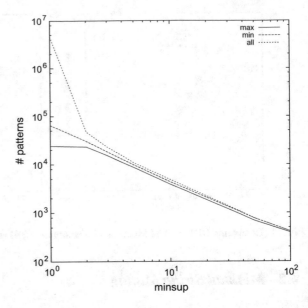

Fig. 3 Number of patterns
in msnbc

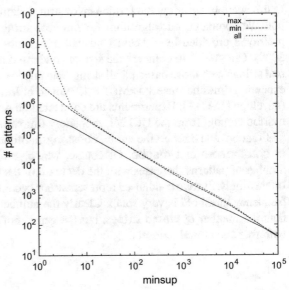

7 Conclusion

By proposing the new notion of ε-*minimizable set system*, this paper not only extended the paradigm of minimal patterns to a broad spectrum of functions and languages, but also introduced the ability of approximate condensed representation. This feature is important for mining smaller condensed representations. This framework unifies the current methods since the existing condensed representations (e.g., δ-free or essential itemsets) fit to specific cases of our framework. More importantly, new types of minimal patterns based on more sophisticated languages are also offered. Besides, DEFME efficiently mines such minimal patterns even in difficult datasets, which are intractable by state-of-the-art algorithms due to its low memory consumption. Experiments also showed on strings that the sizes of the minimal patterns are smaller than the total number of patterns.

Of course, we think that there is still room to improve our implementation even if it is difficult to find a compromise between generic method and speed. We especially want to test the ability of the minimal patterns for generating minimal classification rules with new types of data. For instance, the simplest rules (Crémilleux and Boulicaut 2003) which benefit from approximate free itemsets, can be naturally extended to strings. Similarly, it would be interesting to build associative classifiers from minimal patterns.

Acknowledgments This article has been partially funded by the Hybride project (ANR-11-BS02-0002).

References

Arimura, H., & Uno, T. (2009). Polynomial-delay and polynomial-space algorithms for mining closed sequences, graphs, and pictures in accessible set systems. In *SDM* (pp. 1087–1098). SIAM.

Boulicaut, J.-F., Bykowski, A., & Rigotti, C. (2000). Approximation of frequency queries by means of free-sets. In D. A. Zighed, J. Komorowski & J. Żytkow (Eds.), *PKDD*. LNCS (Vol. 1910, pp. 75–85). Heidelberg: Springer.

Boulicaut, J.-F., Bykowski, A., & Rigotti, C. (2003). Free-sets: A condensed representation of boolean data for the approximation of frequency queries. *Data Mining and Knowledge Discovery*, 7(1), 5–22.

Calders, T., & Goethals, B. (2003). Minimal k-free representations of frequent sets. In *Proceedings of the 7th European Conference on Principles and Practice of Knowledge Discovery in Databases (PKDD 2003)* (pp. 71–82). Heidelberg: Springer.

Calders, T., & Goethals, B. (2005). Depth-first non-derivable itemset mining. In *SDM* (pp. 250–261).

Calders, T., Rigotti, C., & Boulicaut, J. F. (2004). A survey on condensed representations for frequent sets. In J.-F. Boulicaut, L. De Raedt, & H. Mannila (Eds.), *Constraint-based mining and inductive databases*. Lecture notes in computer science (Vol. 3848, pp. 64–80). Heidelberg: Springer.

Casali, A., Cicchetti, R., & Lakhal, L. (2005). Essential patterns: A perfect cover of frequent patterns. In A. M. Tjoa & J. Trujillo (Eds.), *DaWaK*. Lecture notes in computer science (Vol. 3589, pp. 428–437). Heidelberg: Springer.

Crémilleux, B., & Boulicaut, J.-F. (2003). Simplest rules characterizing classes generated by δ-free sets. In M. Bramer, A. Preece, & F. Coenen (Eds.), *Research and development in intelligent systems XIX* (pp. 33–46). London: Springer.

Eiter, T., & Gottlob, G. (2002). Hypergraph transversal computation and related problems in logic and AI. In S. Flesca, S. Greco, G. Ianni, & N. Leone (Eds.), *JELIA. Lecture notes in computer science* (Vol. 2424, pp. 549–564). Heidelberg: Springer.

Gao, C., Wang, J., He, Y., & Zhou, L. (2008). Efficient mining of frequent sequence generators. In *WWW* (pp. 1051–1052). ACM.

Gasmi, G., Yahia, S. B., Nguifo, E. M., & Bouker, S. (2007). Extraction of association rules based on literalsets. In Y. Song, J. Eder, & T. M. Nguyen (Eds.), *DaWaK. Lecture notes in computer science* (Vol. 4654, pp. 293–302). Heidelberg: Springer.

Giacometti, A., Li, D. H., Marcel, P., & Soulet, A. (2013). 20 years of pattern mining: a bibliometric survey. *SIGKDD Explorations, 15*(1), 41–50.

Hamrouni, T. (2012). Key roles of closed sets and minimal generators in concise representations of frequent patterns. *Intelligent Data Analysis, 16*(4), 581–631.

Hébert, C., & Crémilleux, B. (2005). Mining frequent delta-free patterns in large databases. In A. Hoffmann, H. Motoda, & T. Scheffer (Eds.), *Discovery science. Lecture notes in computer science* (Vol. 3735, pp. 124–136). Heidelberg: Springer.

Jelassi, M. N., Largeron, C., & Yahia, S. B. (2014). Efficient unveiling of multi-members in a social network. *Journal of Systems and Software, 94*, 30–38.

Kryszkiewicz, M. (2005). Generalized disjunction-free representation of frequent patterns with negation. *Journal of Experimental and Theoretical Artificial Intelligence, 17*(1–2), 63–82.

Li, J., Li, H., Wong, L., Pei, J. & Dong, G. (2006). Minimum description length principle: Generators are preferable to closed patterns. In *AAAI* (pp. 409–414).

Liu, B., Hsu, W. & Ma, Y. (1998). Integrating classification and association rule mining. In *KDD* (pp. 80–86).

Liu, G., Li, J., & Wong, L. (2008). A new concise representation of frequent itemsets using generators and a positive border. *Knowledge and Information Systems, 17*(1), 35–56.

Lo, D., Khoo, S. -C., & Li, J. (2008). Mining and ranking generators of sequential patterns. In *SDM* (pp. 553–564). SIAM.

Lo, D., Khoo, S.-C., & Wong, L. (2009). Non-redundant sequential rules-theory and algorithm. *Information Systems, 34*(4–5), 438–453.

Mannila, H. & Toivonen, H. (1996). Multiple uses of frequent sets and condensed representations (extended abstract). In E. Simoudis, J. Han & U. M. Fayyad (Eds.), *Proceedings of the Second International Conference on Knowledge Discovery and Data Mining (KDD-96), Portland, Oregon, USA* (pp. 189–194). AAAI Press.

Murakami, K. & Uno, T. (2013). Efficient algorithms for dualizing large-scale hypergraphs. In *ALENEX* (pp. 1–13).

Pasquier, N., Bastide, Y., Taouil, R., & Lakhal, L. (1999). Efficient mining of association rules using closed itemset lattices. *Information Systems, 24*(1), 25–46.

Rioult, F., Zanuttini, B., & Crémilleux, B. (2010). Nonredundant generalized rules and their impact in classification. In Z. W. Ras & L.-S. Tsay (Eds.), *Advances in intelligent information systems. Studies in computational intelligence* (Vol. 265, pp. 3–25). Heidelberg: Springer.

Soulet, A., & Crémilleux, B. (2008). Adequate condensed representations of patterns. *Data Mining and Knowledge Discovery, 17*(1), 94–110.

Soulet, A., Crémilleux, B., & Rioult, F. (2004). Condensed representation of EPs and patterns quantified by frequency-based measures. In *Post-proceedings of knowledge discovery in inductive databases, pise*. Heidelberg: Springer.

Soulet, A., & Rioult, F. (2014). Efficiently depth-first minimal pattern mining. In V. S. Tseng., T. B. Ho., Z. Zhou., A. L. P. Chen., & H. Kao (Eds.), *Proceedings 18th Pacific-Asia Conference on Advances in Knowledge Discovery and Data Mining, PAKDD 2014, Part I, Tainan, Taiwan, May 13–16, 2014. Lecture notes in computer science* (Vol. 8443, pp. 28–39). Heidelberg: Springer.

Szathmary, L., Valtchev, P., Napoli, A., & Godin, R. (2009). Efficient vertical mining of frequent closures and generators. In *IDA*. LNCS (Vol. 5772, pp. 393–404). Heidelberg: Springer.

Zaki, M.J. (2000). Generating non-redundant association rules. In *KDD* (pp. 34–43).

Zeng, Z., Wang, J., Zhang, J., & Zhou, L. (2009). FOGGER: an algorithm for graph generator discovery. In *EDBT* (pp. 517–528).

Part II
Quality Measures, Dissimilarities and Ultrametrics

Comparison of Proximity Measures for a Topological Discrimination

Rafik Abdesselam and Fatima-Zahra Aazi

Abstract The results of any operation of clustering or classification of objects strongly depend on the proximity measure chosen. The user has to select one measure among many existing ones. Yet, according to the notion of topological equivalence chosen, some measures are more or less equivalent. In this paper, we propose a new approach to compare and classify proximity measures in a topological structure and in a context of discrimination. The concept of topological equivalence uses the basic notion of local neighborhood. We define the topological equivalence between two proximity measures, in the context of discrimination, through the topological structure induced by each measure. We propose a criterion for choosing the "best" measure, adapted to the data considered, among some of the most used proximity measures for quantitative or qualitative data. The principle of the proposed approach is illustrated using two real datasets with conventional proximity measures of literature for quantitative and qualitative variables. Afterward, we conduct experiments to evaluate the performance of this discriminant topological approach and to test if the proximity measure selected as the "best" discriminant changes in terms of the size or the dimensions of the used data. The "best" discriminating proximity measure will be verified *a posteriori* using a supervised learning method of type Support Vector Machine, discriminant analysis or Logistic regression applied in a topological context.

R. Abdesselam (✉)
COACTIS-ISH, University of Lyon, Lumière Lyon 2,
14/16, Avenue Berthelot, 69363 Lyon Cedex 07, France
e-mail: rafik.abdesselam@univ-lyon2.fr

F.-Z. Aazi
ERIC & LM2CE, Universities Lumière Lyon 2, Lyon, France
e-mail: faazi@mail.univ-lyon2.fr

F.-Z. Aazi
Hassan 1er, Settat, Morocco

F.-Z. Aazi
5, Avenue Pierre Mends-France, 69676 Bron Cedex, France

© Springer International Publishing Switzerland 2017
F. Guillet et al. (eds.), *Advances in Knowledge Discovery and Management*,
Studies in Computational Intelligence 665, DOI 10.1007/978-3-319-45763-5_5

1 Introduction

The comparison of objects, situations or ideas are essential tasks to assess a situation, to rank preferences or to structure a set of tangible or abstract elements, etc. In a word, to understand and act, we have to compare. These comparisons that the brain naturally performs, however, must be clarified if we want them to be done by a machine. For this purpose, we use proximity measures. A proximity measure is a function which measures the similarity or dissimilarity between two objects of a set. These proximity measures have mathematical properties and specific axioms. But are such measures equivalent? Can they be used in practice in a undifferentiated way? Do they produce the same learning database that will serve as input to the estimation of the membership class of a new object? If we know that the answer is negative, then, how to decide which one to use? Of course, the context of the study and the type of the data considered can help to select few proximity measures but which one to choose from this selection?

We find this problematic in the context of a supervised classification or a discrimination. The assignment or the classification of an anonymous object to a class partly depends on the used learning database. According to the selected proximity measure, this database changes and therefore the result of the classification changes too. We are interested here in the degree of topological equivalence of these proximity measures in discrimination. Several studies on topological equivalence of proximity measures have been proposed (Richter 1992; Batagelj and Bren 1992; Rifqi et al. 2003; Batagelj and Bren 1995; Lesot et al. 2009; Zighed et al. 2012) but neither of these propositions has an objective of discrimination.

Table 1 Some proximity measures for continuous data

Measure	Distance—dissimilarity
Euclidean	$u_E(x, y) = \sqrt{\sum_{j=1}^{p} (x_j - y_j)^2}$
Mahalanobis	$u_{Mah}(x, y) = \sqrt{(x - y)^t \sum^{-1}(x - y)}$
Manhattan	$u_{Man}(x, y) = \sum_{j=1}^{p} |x_j - y_j|$
Tchebychev	$u_{Tch}(x, y) = \max_{1 \le j \le p} |x_j - y_j|$
Cosine dissimilarity	$u_{Cos}(x, y) = 1 - \dfrac{\sum_{j=1}^{p} x_j y_j}{\sqrt{\sum_{j=1}^{p} x_j^2} \sqrt{\sum_{j=1}^{p} y_j^2}} = 1 - \dfrac{<x, y>}{\|x\| \|y\|}$
Normalized Euclidean	$u_{NE}(x, y) = \sqrt{\sum_{j=1}^{p} (\dfrac{x_j - y_j}{\sigma_j})^2}$
Minkowski	$u_{Min_\gamma}(x, y) = (\sum_{j=1}^{p} |x_j - y_j|^\gamma)^{\frac{1}{\gamma}}$
Pearson correlation	$u_{Cor}(x, y) = \dfrac{\sum_{j=1}^{p} (x_j - \bar{x})(y_j - \bar{y})}{\sqrt{\sum_{j=1}^{p} (x_j - \bar{x})^2} \sqrt{\sum_{j=1}^{p} (y_j - \bar{y})^2}} = \dfrac{<x - \bar{x}, y - \bar{y}>}{\|x - \bar{x}\| \|y - \bar{y}\|}$

Where, p is the dimension of space, $x = (x_j)_{j=1,...,p}$ and $y = (y_j)_{j=1,...,p}$ two points in R^p, $(\alpha_j)_{j=1,...,p} \ge 0$, $\sum^{-1}$ the inverse of the variance and covariance matrix, σ_j^2 the variance, $\gamma > 0$

Therefore, this article focuses on how to construct the adjacency matrix induced by a proximity measure, taking into account the membership classes of the objects, by juxtaposing the Within-groups and Between-groups adjacency matrices (Abdesselam 2014).

A criterion for selecting the "best" proximity measure is proposed. We check *a posteriori* whether the chosen measure is a good discriminant one using the Multi-class SVM method (MSVM).

This article is organized as follows. In Sect. 2, after recalling the basic notions of structure, graph and topological equivalence, we present how to build the adjacency matrix for discrimination, the choice of a measure of the degree of topological equivalence between two proximity measures and the selection criterion of the "best" discriminant measure. Two illustrative examples, one with continuous data and the other with binary data are discussed in Sect. 3 as well as other experiments to

Table 2 Some proximity measures for binary data

Measure	Similarity	Dissimilarity
Jaccard	$s_{Jac} = \frac{a}{a+b+c}$	$u_{Jac} = 1 - s_{Jac}$
Dice	$s_{Dic} = \frac{2a}{2a+b+c}$	$u_{Dic} = 1 - s_{Dic}$
Kulczynski	$s_{Kul} = \frac{1}{2}(\frac{a}{a+b} + \frac{a}{a+c})$	$u_{Kul} = 1 - s_{Kul}$
Ochiai	$s_{Och} = \frac{a}{\sqrt{(a+b)(a+c)}}$	$u_{Och} = 1 - s_{Och}$
Sokal and Sneath 1	$s_{SS1} = \frac{2(a+d)}{2(a+d)+b+c}$	$u_{SS1} = 1 - s_{SS1}$
Sokal and Sneath 2	$s_{SS2} = \frac{a}{a+2(b+c)}$	$u_{SS2} = 1 - s_{SS2}$
Sokal and Sneath 4	$s_{SS4} = \frac{1}{4}(\frac{a}{a+b} + \frac{a}{a+c} + \frac{d}{d+b} + \frac{d}{d+c})$	$u_{SS4} = 1 - s_{SS4}$
Sokal and Sneath 5	$s_{SS5} = \frac{ad}{\sqrt{(a+b)(a+c)(d+b)(d+c)}}$	$u_{SS5} = 1 - s_{SS5}$
Russel and Rao	$s_{RR} = \frac{a}{a+b+c+d}$	$u_{RR} = 1 - s_{RR}$
Rogers and Tanimoto	$s_{RT} = \frac{a+d}{a+2(b+c)+d}$	$u_{RT} = 1 - s_{RT}$
Hamann	$s_{Hama} = \frac{a+d-b-c}{a+b+c+d}$	$u_{Hama} = \frac{1-s_{Hama}}{2}$
Y-Yule	$s_{YY} = \frac{\sqrt{ad}-\sqrt{bc}}{\sqrt{ad}+\sqrt{bc}}$	$u_{YY} = \frac{1-s_{YY}}{2}$
Q-Yule	$s_{QY} = \frac{ad-bc}{ad+bc}$	$u_{QY} = \frac{1-s_{QY}}{2}$
Hamming distance		$u_{Hamm} = \sum_{j=1}^{p}(x_j - y_j)^2$

Let $x = (x_i)_{i=1,...,p}$ and $y = (y_i)_{i=1,...,p}$ be two points in $\{0, 1\}^p$ representing respectively the attributes of two any objects x and y. Where, $a = |X \cap Y| = \sum_{i=1}^{p} x_i y_i$ is the number of attributes common to both points x and y, $b = |X - Y| = \sum_{i=1}^{p} x_i(1 - y_i)$ is the number of attributes present in x but not in y, $c = |Y - X| = \sum_{i=1}^{p}(1 - x_i)y_i$ is the number of attributes present in y but not in x and $d = |\overline{X} \cap \overline{Y}| = \sum_{i=1}^{p}(1 - x_i)(1 - y_i)$ is the number of attributes in neither x or y $X = \{j/x_j = 1\}$ and $Y = \{j/y_j = 1\}$ are the sets of attributes present in data point x and y respectively, and $|.|$ the cardinality of a set. The cardinals a, b, c and d are linked by the relation $a + b + c + d = p$

evaluate the effects of the dimensions and the size of data on the choice of the "best"
discriminant proximity measure. A general conclusion and some perspectives of this
work are given in Sect. 4.

Table 1 shows some classic proximity measures used for continuous data, defined
on R^p. For binary data, we give in Table 2 the definition of 14 proximity measures
defined on $\{0, 1\}^p$. All the datasets used are from the UCI Machine Learning Repository (UCI 2013).

2 Topological Equivalence

The topological equivalence is based on the concept of topological graph also referred
to as neighborhood graph. The basic idea is actually quite simple: two proximity
measures are equivalent if the corresponding topological graphs induced on the set
of objects remain identical. Measuring the similarity between proximity measures
consists in comparing the neighborhood graphs and measure their similarity. We will
first define more precisely what a topological graph is and how to build it. Then, we
propose a measure of proximity between topological graphs that will subsequently
be used to compare the proximity measures.

2.1 Topological Graph

Consider a set $E = \{x, y, z, \ldots\}$ of $n = |E|$ objects in R^p. We can, by means of a
proximity measure u, define a neighborhood relationship V_u to be a binary relationship on $E \times E$. There are many possibilities for building this neighborhood binary
relationship.

Thus, for a given proximity measure u, we can build a neighborhood graph on a
set of individuals-objects, where the vertices are the individuals and the edges are
defined by a property of neighborhood relationship.

Many definitions are possible to build this Binary neighborhood relationship. One
can choose, the Minimal Spanning Tree (MST) (Kim and Lee 2003), the Gabriel
Graph (GG) (Park et al. 2006) or, which is the case here, the Relative Neighborhood
Graph (RNG) (Toussaint 1980; Jaromczyk and Toussaint 1992), where, all pairs of
neighbour points (x, y) satisfy the following property:

$$\begin{cases} V_u(x, y) = 1 \ if \ u(x, y) \leq \max(u(x, z), u(y, z)) \, ; \ \forall x \in E \, ; \ \forall y \in E \, ; \ \forall z \in E - \{x, y\} \\ V_u(x, y) = 0 \ otherwise \end{cases} \quad (1)$$

That is, if the pairs of points verify or not the ultra-triangular inequality (1), ultrametric condition. Which means geometrically that the hyper-lunula (the intersection
of the two hyperspheres centered on two points) is empty.

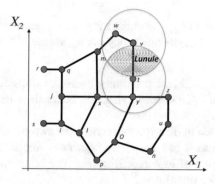

$$\begin{pmatrix} V_u & \cdots & x & y & z & t & u & \cdots \\ \vdots & & \vdots & \vdots & \vdots & \vdots & \vdots & \\ & \cdots & & & & & & \cdots \\ x & \cdots & 1 & 1 & 0 & 0 & 0 & \cdots \\ y & \cdots & 1 & 1 & 1 & 1 & 0 & \cdots \\ z & \cdots & 0 & 1 & 1 & 0 & 1 & \cdots \\ t & \cdots & 0 & 1 & 0 & 1 & 0 & \cdots \\ u & \cdots & 0 & 0 & 1 & 0 & 1 & \cdots \\ \vdots & & \vdots & \vdots & \vdots & \vdots & \vdots & \\ & \cdots & & & & & & \cdots \end{pmatrix}$$

Fig. 1 Topological graph RNG

Figure 1 shows, an example of a topological graph RNG perfectly defined in $\mathbb{R}^2$ by the associated adjacency matrix V_u, containing $0s$ and 1.

In this case, $u(x, y) = u_E(x, y) = \sqrt{(\sum_{i=1}^{p}(x_i - y_i)^2)}$ is the Euclidean distance.

For a given neighborhood property (MST, GG or RNG), each measure u generates a topological structure on the objects in E which are totally described by the adjacency matrix V_u.

2.2 Comparison of Proximity Measures

Let p be the number of explanatory variables (predictors) $\{x^j; j = 1, .., p\}$ and y a target qualitative variable to explain, partition of $n = \sum_{k=1}^{q} n_k$ individuals-objects into q modalities-subgroups $\{G_k; k = 1, .., q\}$.

For any given proximity measure u_i, we construct, according to Property (1), the overall binary adjacency matrix V_{u_i} stands as a juxtaposition of q symmetrical Within-groups adjacency matrices $\{V_{u_i}^k; k = 1, .., q\}$ and $q(q - 1)$ Between-groups adjacency matrices $\{V_{u_i}^{kl}; k \neq l; k, l = 1, .., q\}$:

$$\begin{cases} V_{u_i}^k(x, y) = 1 \ if \ u_i(x, y) \leq \max(u_i(x, z), u_i(y, z)) \ ; \ \forall x, y, z \in G_k, \ z \neq x \ and \ z \neq y \\ V_{u_i}^k(x, y) = 0 \ otherwise \end{cases}$$

$$\begin{cases} V_{u_i}^{kl}(x, y) = 1 \ if \ u_i(x, y) \leq \max(u_i(x, z), u_i(y, z)); \ \forall x \in G_k, \forall y \in G_l, \ \forall z \in G_l, \ z \neq y \\ V_{u_i}^{kl}(x, y) = 0 \ otherwise \end{cases}$$

$$V_{u_i} = \begin{pmatrix} V_{u_i}^1 & \cdots & V_{u_i}^{lk} & \cdots & V_{u_i}^{1q} \\ & \cdots & & & \\ V_{u_i}^{k1} & \cdots & V_{u_i}^k & \cdots & V_{u_i}^{kq} \\ & \cdots & & & \\ V_{u_i}^{q1} & \cdots & V_{u_i}^{qk} & \cdots & V_{u_i}^q \end{pmatrix}$$

Note that the partitioned adjacency matrix V_{u_i} thus constructed, is not symmetrical. Indeed, for two objects $x \in G_k$ and $y \in G_l$, the adjacency binary values $V_{u_i}^{kl}(x, y)$ and $V_{u_i}^{lk}(y, x)$ can be different.

- The first objective is to regroup the different proximity measures considered, according to their topological similarity in order to visualize better their resemblance in a context of discrimination.

  To measure the topological equivalence in discrimination between two proximity measures u_i and u_j, we propose to test if the associated adjacency matrices V_{u_i} and V_{u_j} are different or not. The degree of topological equivalence between two proximity measures is measured by the quantity:

$$S(V_{u_i}, V_{u_j}) = \frac{\sum_{k=1}^{n} \sum_{l=1}^{n} \delta_{kl}}{n^2} \quad with \quad \delta_{kl} = \begin{cases} 1 \text{ if } V_{u_i}(k, l) = V_{u_j}(k, l) \\ 0 \text{ otherwise.} \end{cases}$$

- The second objective is to define a criterion to assist in the selection of the "best" proximity measure, among the considered ones, that discriminates at the best the q groups.

  We note, $V_{u*} = diag(\mathbb{K}_{G_1}, \ldots, \mathbb{K}_{G_k}, \ldots, \mathbb{K}_{G_q})$ the adjacency block diagonal reference matrix, "perfect discrimination of the q groups" according to an unknown proximity measure denoted $u*$. Where $\mathbb{1}_{n_k}$ is the vector of order n_k whose all components are equal to 1 and $\mathbb{K}_{G_k} = \mathbb{1}_{n_k}{}^t\mathbb{1}_{n_k}$, is the symmetric matrix of order n_k whose elements are all equal to 1.

$$V_{u*} = \begin{pmatrix} \mathbb{K}_{G_1} & & & & \\ 0 & \cdots & & & \\ 0 & 0 & \mathbb{K}_{G_k} & & \\ 0 & 0 & 0 & \cdots & \\ 0 & 0 & 0 & 0 & \mathbb{K}_{G_q} \end{pmatrix}$$

Thus, we can establish the degree of topological equivalence of discrimination $S(V_{u_i}, V_{u*})$ between each considered proximity measures u_i and the reference measure u^*.

Finally, in order to evaluate otherwise the choice of the "best" discriminant proximity measure proposed by this approach, we *a posteriori* applied a Multiclass SVM method (MSVM) on the adjacency matrix associated to each considered proximity measure including the reference one u^*.

3   Illustration Examples

To illustrate our approach, we consider here two sets of well-known and relatively simple data, the Iris (Fisher 1936; Anderson 1935) and Animals Zoo, presented in

Table 3 Data sets

Number	Name	Explanatory variables type & $X_{(n \times p)}$	Variable to explain $Y_{(q)}$
1	Iris	Continuous 150 × 4	3
2	Zoo	Binary 74 × 15	3

Table 3. These two sets of respectively continuous and binary explanatory variables are references for discriminant analysis and clustering. The complete data and the dictionary of variables are especially in the UCI Machine Learning Repository (UCI 2013).

Let $X_{(n,p)}$ be a set of data with n objects and p explanatory variables, and $Y_{(q)}$ be a qualitative variable to be explained with q modalities-classes.

3.1 Comparison and Classification of Proximity Measures

The main results of the proposed approach in the case of continuous and binary data, are presented in the following tables and graphs. They allow to visualize the measures that are close to each other in a context of discrimination.

For the continuous data set, Table 4 summarizes the similarities in pairs between the eight proximity measures and shows that, independently of the other measures, the two by two similarity value between the reference measure and each of the proximity measures is most important, $S(V_{u_{Tch}}, V_{u^*}) = 68.10\%$, with the Tchebychev measure u_{Tch}.

A Principal Component Analysis (PCA) followed by Ascendant Hierarchical Classification (AHC) were performed from the similarity matrix between the eight proximity measures considered, to partition them into homogeneous groups and to view their similarities.

Table 4 Continuous data—similarities $S(V_{u_i}, V_{u_j})$ and $S(V_{u_j}, V_{u^*})$

S	u_E	u_{Mah}	u_{Man}	u_{Tch}	u_{Cos}	u_{NE}	$u_{Min_{y=5}}$	u_{Cor}
u_E	1							
u_{Mah}	0.953	1						
u_{Man}	0.977	0.947	1					
u_{Tch}	0.968	0.934	0.949	1				
u_{Cos}	0.955	0.946	0.949	0.939	1			
u_{NE}	0.968	0.956	0.969	0.945	0.950	1		
$u_{Min_{y=5}}$	0.992	0.951	0.971	0.975	0.953	0.965	1	
u_{Cor}	0.949	0.943	0.944	0.930	0.966	0.946	0.948	1
u^*	0.675	0.673	0.678	**0.681**	0.675	0.674	0.675	0.673

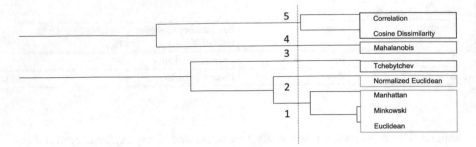

Fig. 2 Hierarchical tree of the continuous proximity measures

Table 5 Continuous measures—assignment of the reference measure

Number	Class 1	Class 2	Class 3	Class 4	Class 5
Frequency	3	1	1	1	2
Active measures	u_E, u_{Min}, u_{Man}	u_{NE}	u_{Tch}	u_{Mah}	u_{Cos}, u_{Cor}
Supplementary measure			u^*		

The AHC algorithm according to the Ward criterion, (Ward Jr 1963), provides the dendrogram of Fig. 2.

The similarity vector $S(V_{u_i}, V_{u^*})$ of the reference measure with the considered proximity measures is positioned as illustrative element in the analysis.

In view of the results presented in Table 5, for the selected partition into 5 classes of proximity measures, the reference measure u^*, projected as additional element, would be closer to the measures of the third class, i.e., the Tchebychev proximity measure u_{Tch} which would be, for these data, the "best" proximity measure among the eight measures considered.

For binary data, the results of pairwise comparisons presented in Table 6, are somewhat different, some are closer than others. We note that pairs of proximity measures of these sub-sets: $(u_{Jac}, u_{Dic}, u_{Kul}, u_{Och}, u_{SS2})$, $(u_{SS1}, u_{RT}, u_{Hama})$, $(u_{RT}, u_{Hama}, u_{Hamm})$ and $(u_{QY}, u_{YY}, u_{Hamm})$ are in perfect topological equivalence of discrimination $S(V_{u_i}, Vu_j) = 1$. The measures u_{QY} and u_{YY} of Yule, independently of the other measures, are those which have a greatest similarity with the reference measure $S(V_{u_{QY}}, V_{u^*}) = S(V_{u_{YY}}, V_{u^*}) = 75.40\%$, followed by the measure u_{RR} of Russel & Rao $S(V_{u_{RR}}, V_{u^*}) = 71.60\%$.

The AHC algorithm according to the Ward criterion, provides the dendrogram of Fig. 3. In view of the results presented in Table 7, for the selected partition into 4 classes of proximity measures, the reference measure u^*, projected as additional element, would be closer to the measures of the fourth class, i.e., the Russel & Rao proximity measure u_{RR} would be, for these data, the "best" proximity measure among the 14 considered.

Table 6 Binary data—similarities $S(V_{u_i}, V_{u_j})$ and $S(V_{u_j}, V_{u*})$

S	u_{Jac}	u_{Dic}	u_{Kul}	u_{Och}	u_{SS1}	u_{SS2}	u_{SS4}	u_{SS5}	u_{RR}	u_{RT}	u_{Hama}	u_{YY}	u_{QY}	u_{Hamm}
u_{Jac}	1													
u_{Dic}	1	1												
u_{Kul}	1	1	1											
u_{Och}	1	1	1	1										
u_{SS1}	0.987	0.987	.987	0.987	1									
u_{SS2}	1	1	1	1	0.987	1								
u_{SS4}	0.997	0.997	0.997	0.997	0.986	0.997	1							
u_{SS5}	0.997	0.997	0.997	0.997	0.986	0.997	1	1						
u_{RR}	0.826	0.826	0.826	0.826	0.814	0.826	0.824	0.824	1					
u_{RT}	0.987	0.987	0.987	0.987	1	0.987	0.986	0.986	0.814	1				
u_{Hama}	0.987	0.987	0.987	0.987	1	0.987	0.986	0.986	0.814	1	1			
u_{YY}	0.938	0.938	0.938	0.938	0.926	0.938	0.940	0.940	0.884	0.926	0.926	1		
u_{QY}	0.938	0.938	0.938	0.938	0.926	0.938	0.940	0.940	0.884	0.926	0.926	1	1	
u_{Hamm}	0.987	0.987	0.987	0.987	1	0.987	0.986	0.986	0.814	1	1	0.926	0.926	1
u^*	0.695	0.695	0.695	0.695	0.683	0.695	0.694	0.694	0.716	0.683	0.683	**0.754**	**0.754**	0.683

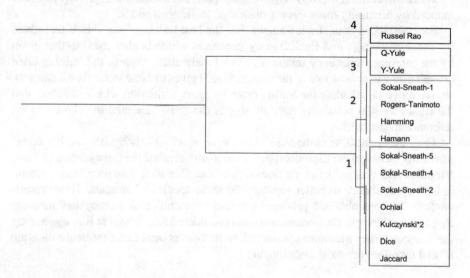

Fig. 3 Hierarchical tree of the binary proximity measures

Table 7 Binary measures—assignment of the reference measure

Number	Class 1	Class 2	Class 3	Class 4
Frequency	7	4	2	1
Active measures	$u_{Jac}, u_{DC}, u_{Kul},$ $u_{DKO} u_{SS2}, u_{SS4}, u_{SS5}$	$u_{SS1}, u_{RT}, u_{Hama},$ u_{Hamm}	u_{YY}, u_{QY}	u_{RR}
Supplementary measure				u^*

3.2 Discriminant Measures According to the MSVM Method

This part consists in validating *a posteriori* the results of choosing the best measure in view of the reference matrix using MSVM. We use the $MSVM_{LLW}$ model (Lee et al. 2004), considered as the most theoretically based of MSVM models as it is the only one that implements asymptotically the Bayes decision rule.

Working with the $MSVM_{LLW}$ model involves the choice of optimal values of its parameters, namely, C, representing the weight of learning errors, and the parameter(s) of the kernel function if we decide to change the data space.

For our two datasets, we choose to work in the original data space and therefore to use a linear kernel. The only parameter to be optimized is C. To do this, we will test several values and choose the one that minimizes the testing error calculated by cross-validation. For both examples, we test 10 values of the parameter C for all databases. After simulations, the chosen value is $C = 1$.

The main results of the $MSVM_{LLW}$ model, applied to each of the adjacency matrices induced by proximity measures are presented in Tables 8 and 9.

For continuous data, Table 8 shows that the best training error rate is that given by Tchebychev u_{Tch} and Euclidean u_E measures which is also equal to that given by the reference adjacency matrix V_{u^*}. For binary data, Table 9, the training error doesn't allow to choose one of the measures as it gives to same value for all datasets, so, we move to calculate the testing error by cross validation which indicates that the Russel & Rao proximity measure u_{RR} is the "best" one and the closest to the reference measure u^*.

Thus, the application of the MSVM model reveals that Tchebychev and Euclidean proximity measures are the most appropriate to differentiate the three species (Setosa, Virginica and Versicolor) of iris flowers, and that Russel & Rao proximity measure is the one to choose to better separate the three species of animals. Those results confirm the ones obtained previously, namely the choice of Tchebychev measure u_{Tch} among the eight continuous measures considered and Russel & Rao u_{RR} among the fourteen binary measures considered as the nearest ones to the reference measure u^* and therefore the most discriminant.

3.3 Experimentations

We conduct experiments on more datasets to evaluate the effect of the data, their size and/or their dimensions on the results of the classification of proximity measures for the purpose of discrimination. For instance, are the proximity measures grouped differently depending on the dataset used? Depending on the sample size and/or the number of explanatory variables considered in the same set of data?

To answer these questions, we have therefore applied the proposed approach on the different datasets presented in Table 10, all from the repository (UCI 2013). The

Table 8 Results of the MSVM model—continuous iris data

Name	Measure	Training error (%)	Confusion matrix	Rank
Euclidean	u_E	**0**	$\begin{pmatrix} 50 & 0 & 0 \\ 0 & 50 & 0 \\ 0 & 0 & 50 \end{pmatrix}$	1
Mahalanobis	u_{Mah}	0.66	$\begin{pmatrix} 50 & 0 & 0 \\ 0 & 49 & 1 \\ 0 & 0 & 50 \end{pmatrix}$	3
Manhattan	u_{Man}	0.66	$\begin{pmatrix} 50 & 0 & 0 \\ 0 & 49 & 1 \\ 0 & 0 & 50 \end{pmatrix}$	3
Tchebychev	u_{Tch}	**0**	$\begin{pmatrix} 50 & 0 & 0 \\ 0 & 50 & 0 \\ 0 & 0 & 50 \end{pmatrix}$	1
Cosine dissimilarity	u_{Cos}	0.66	$\begin{pmatrix} 50 & 0 & 0 \\ 0 & 49 & 1 \\ 0 & 0 & 50 \end{pmatrix}$	3
Normalized Euclidean	u_{NE}	1.33	$\begin{pmatrix} 50 & 0 & 0 \\ 0 & 50 & 0 \\ 0 & 2 & 48 \end{pmatrix}$	6
Minkowski	$u_{Min_\gamma=5}$	1.33	$\begin{pmatrix} 50 & 0 & 0 \\ 0 & 49 & 1 \\ 0 & 1 & 49 \end{pmatrix}$	6
Pearson correlation	u_{Cor}	1.33	$\begin{pmatrix} 50 & 0 & 0 \\ 0 & 49 & 1 \\ 0 & 1 & 49 \end{pmatrix}$	6
Reference measure	u^*	**0**	$\begin{pmatrix} 50 & 0 & 0 \\ 0 & 50 & 0 \\ 0 & 0 & 50 \end{pmatrix}$	

objective is to compare the results of the classification of proximity measures and the choice of the "best" discriminant measure proposed for each of these datasets.

To analyze the effect of the change of dimensions, we consider the continuous data set "Waveform Database Generator" to generate 3 samples (number 4) of size $n = 2000$ objects and p dimension respectively equal to 40, 20 and 10 explanatory variables. Similarly, to evaluate the impact of the change in sample size, we also generated 3 other samples (number 5) of size n, respectively, equal to 3000, 1500 and 500 objects with the same dimension p equal to 30 explanatory variables.

Table 9 Results of the MSVM model—binary zoo data

Name	Measure	Training error (%)	Test error (%)	Confusion matrix	Rank
Jaccard	u_{Jac}	0	4.05	$\begin{pmatrix} 39 & 2 & 0 \\ 0 & 20 & 0 \\ 1 & 0 & 12 \end{pmatrix}$	4
Dice	u_{Dic}	0	4.05	$\begin{pmatrix} 39 & 2 & 0 \\ 0 & 20 & 0 \\ 1 & 0 & 12 \end{pmatrix}$	4
Kulczynski	u_{Kul}	0	4.05	$\begin{pmatrix} 39 & 2 & 0 \\ 0 & 20 & 0 \\ 1 & 0 & 12 \end{pmatrix}$	4
Ochiai	u_{Och}	0	4.05	$\begin{pmatrix} 39 & 2 & 0 \\ 0 & 20 & 0 \\ 1 & 0 & 12 \end{pmatrix}$	4
Sokal and Sneath 1	u_{SS1}	0	5.41	$\begin{pmatrix} 40 & 1 & 0 \\ 2 & 18 & 0 \\ 1 & 0 & 12 \end{pmatrix}$	9
Sokal and Sneath 2	u_{SS2}	0	4.05	$\begin{pmatrix} 39 & 2 & 0 \\ 0 & 20 & 0 \\ 1 & 0 & 12 \end{pmatrix}$	4
Sokal and Sneath 4	u_{SS4}	0	6.76	$\begin{pmatrix} 38 & 3 & 0 \\ 1 & 19 & 0 \\ 1 & 0 & 12 \end{pmatrix}$	13
Sokal and Sneath 5	u_{SS5}	0	6.76	$\begin{pmatrix} 38 & 3 & 0 \\ 1 & 19 & 0 \\ 1 & 0 & 12 \end{pmatrix}$	1
Russel and Rao	u_{RR}	0	**1.35**	$\begin{pmatrix} 41 & 0 & 0 \\ 0 & 20 & 0 \\ 0 & 0 & 13 \end{pmatrix}$	1
Rogers and Tanimoto	u_{RT}	0	5.41	$\begin{pmatrix} 40 & 1 & 0 \\ 2 & 18 & 0 \\ 1 & 0 & 12 \end{pmatrix}$	9
Hamann	u_{Hama}	0	5.41	$\begin{pmatrix} 40 & 1 & 0 \\ 2 & 18 & 0 \\ 1 & 0 & 12 \end{pmatrix}$	9
Y-Yule	u_{YY}	0	2.70	$\begin{pmatrix} 39 & 2 & 0 \\ 0 & 20 & 0 \\ 0 & 0 & 13 \end{pmatrix}$	2
Q-Yule	u_{QY}	0	2.70	$\begin{pmatrix} 39 & 2 & 0 \\ 0 & 20 & 0 \\ 0 & 0 & 13 \end{pmatrix}$	2
Hamming distance	u_{Hamm}	0	5.41	$\begin{pmatrix} 40 & 1 & 0 \\ 2 & 18 & 0 \\ 1 & 0 & 12 \end{pmatrix}$	9
Reference measure	u^*	0	**0**	$\begin{pmatrix} 41 & 0 & 0 \\ 0 & 20 & 0 \\ 0 & 0 & 13 \end{pmatrix}$	

Table 10 Continuous data sets

Number	Name	Explanatory variables $X_{(n \times p)}$	Variable to explain $Y_{(q)}$
1	Iris	150×4	3
2	Wine	178×13	3
3	Wine quality	3000×11	2
4_1	Waveform database generator	2000×40	3
4_2	Waveform database generator	2000×20	3
4_3	Waveform database generator	2000×10	3
5_1	Waveform database generator	3000×30	3
5_2	Waveform database generator	1500×30	3
5_3	Waveform database generator	500×30	3

The main results of these experiments, namely the topological equivalence of proximity measures and the assignment of the reference measure u^* to the nearest class are presented in Table 11.

For each of these experiments, we selected a partition into five classes of proximity measures to compare and well distinguish the measures of the membership class of the reference measure, that is to say the most discriminating ones.

Clusters of proximity measures obtained for the three data sets number 4 are virtually identical, so there's not really dimension effect.

As to clusters of proximity measures of the three data sets number 5, they are almost identical, so there is no sample size effect.

Note that all the samples number 4 and 5, are generated from the same data set "Waveform Generator Database", the ideal reference measure u^* for discrimination is close to the same proximity measure, i.e. here, the Tchebychev measure u_{Tch}. This result shows that there is no size or dimensionality effect on the result of choosing the "best" discriminant measure.

Table 11 Clusters and assignment of the reference measure u^*

Number	Class 1	Class 2	Class 3	Class 4	Class 5
1	u_{Cos}, u_{Cor}	u_E, u_{Min}, u_{Man}	u_{Mah}	u_{NE}	u_{Tch}, u^*
2	u_{Cos}, u_{Cor}	u_E, u_{Min}, u_{Tch}	u_{Mah}, u^*	u_{NE}	u_{Man}
3	u_{Cos}, u_{Cor}	u_E, u_{Min}, u_{Man}	u_{Mah}	u_{NE}, u^*	u_{Tch}
4_1	u_{Cos}, u_{Cor}, u_E	u_{Man}, u_{NE}	u_{Mah}	u_{Min}	u_{Tch}, u^*
4_2	$u_{Cos}, u_{Cor}, u_E, u_{NE}$	u_{Man}	u_{Mah}	u_{Min}	u_{Tch}, u^*
4_3	u_{Cos}, u_{Cor}	u_E, u_{Man}, u_{NE}	u_{Mah}	u_{Min}	u_{Tch}, u^*
5_1	u_{Cos}, u_{Cor}, u_E	u_{Man}, u_{NE}	u_{Mah}	u_{Min}	u_{Tch}, u^*
5_2	u_{Cos}, u_{Cor}, u_E	u_{Man}, u_{NE}	u_{Mah}	u_{Min}	u_{Tch}, u^*
5_3	u_{Cos}, u_{Cor}, u_E	u_{Man}, u_{NE}	u_{Mah}	u_{Min}	u_{Tch}, u^*

With regard to all experiments, we can see a slight change in the clusters of the proximity measures. However, we can also note equivalences between certain measures such as u_{Cos}, u_{Cor}, u_E and u_{NE}, u_{Man}. Others are isolated such as u_{Tch}, u_{Mah} and u_{Min}.

4 Conclusion and Perspectives

The choice of a proximity measure is very subjective, it is often based on habits or on criteria such as the interpretation of the *a posteriori* results. This work proposes a new approach for equivalence between proximity measures in the context of discrimination.

This topological approach is based on the concept of neighborhood graph induced by the proximity measure. From a practical point of view, in this paper, we compared several measures built either on continuous or binary data. But this work may well be extended to mixed data (quantitative and qualitative) by choosing the right topological structure and the adapted adjacency matrix.

We plan to extend this work to other topological structures and to use a comparison criteria (Demsar 2006; Schneider and Borlund 2007), other than classification techniques, in order to validate the degree of equivalence between two proximity measures. For example, evaluate the degree of topological equivalence in discrimination between two proximity measures using the non-parametric Test Kappa coefficient of concordance, calculated from the associated adjacency matrices (Abdesselam and Zighed 2011). This will allow to give a statistical significance of the degree of agreement between two similarity matrices and to validate or not the topological equivalence in discrimination, i.e., whether or not they induce the same neighborhood structure on the groups of objects to be separated.

The experiments conducted on different data sets have shown that there is no effect of samples size and no real effect of dimension on both clusters of proximity measures and the result of the choice of the best discriminant measure.

References

Abdesselam, R. (2014). Proximity measures in topological structure for discrimination. In C. H. Skiadas (Ed.), *SMTDA-2014, 3rd Stochastic Modeling Techniques and Data Analysis, International Conference, Lisbon* (pp. 599–606). ISAST.

Abdesselam, R. & Zighed, D. (2011). Comparaison topologique de mesures de proximite. In *Actes des XVIIIeme Rencontres de la Societe Francophone de Classification* (pp. 79–82).

Anderson, E. (1935). The irises of the gaspe peninsula. *Bulletin of the American Iris Society, 59*, 2–5.

Batagelj, V., & Bren, M. (1992). Comparing resemblance measures. Technical report, Proceedings of International Meeting on Distance Analysis (DISTANCIA'92).

Batagelj, V., & Bren, M. (1995). Comparing resemblance measures. *Journal of classification, 12*, 73–90.

Demsar, J. (2006). Statistical comparisons of classifiers over multiple data sets. *The Journal of Machine Learning Research, 7*, 1–30.

Fisher, R. (1936). The use of multiple measurements in taxonomic problems. *Annals of Eugenics, Part II, 7*, 179–188.

Jaromczyk, J.-W., & Toussaint, G.-T. (1992). Relative neighborhood graphs and their relatives. *Proceedings of IEEE, 80*(9), 1502–1517.

Kim, J., & Lee, S. (2003). Tail bound for the minimal spanning tree of a complete graph. *Statistics & Probability Letters, 64*(4), 425–430.

Lee, Y., Lin, Y., & Wahba, G. (2004). Multicategory support vector machines, theory and application to the classification of microarray data and satellite radiance data. *Journal of the American Statistical Association, 465*, 67–81.

Lesot, M.-J., Rifqi, M., & Benhadda, H. (2009). Similarity measures for binary and numerical data: a survey. *IJKESDP, 1*(1), 63–84.

Park, J., Shin, H., & Choi, B. (2006). Elliptic Gabriel graph for finding neighbors in a point set and its application to normal vector estimation. *Computer-Aided Design, 38*(6), 619–626.

Richter, M. (1992). Classification and learning of similarity measures. In *Proceedings der Jahrestagung der Gesellschaft fur Klassifikation*. Studies in classification, data analysis and knowledge organisation. Berlin: Springer

Rifqi, M., Detyniecki, M., & Bouchon-Meunier, B. (2003). *2003*. In IFSA: Discrimination power of measures of resemblance.

Schneider, J., & Borlund, P. (2007b). Matrix comparison, part 2: Measuring the resemblance between proximity measures or ordination results by use of the mantel and procrustes statistics. *Journal American Society for Information Science and Technology, 58*(11), 1596–1609.

Toussaint, G. (1980). The relative neighbourhood graph of a finite planar set. *Pattern Recognition, 12*(4), 261–268.

UCI. (2013). Machine learning repository. http://archive.ics.uci.edu/ml. Irvine, CA: University of California, School of Information and Computer Science.

Ward, J, Jr. (1963). Hierarchical grouping to optimize an objective function. *Journal of the American Statistical Association, 58*(301), 236–244.

Zighed, D., Abdesselam, R., & Hadgu, A. (2012). Topological comparisons of proximity measures. In P.-N. Tan et al. (Eds.), *The 16th PAKDD 2012 Conference*. Part I, LNAI. (Vol. 7301, pp. 379–391). Berlin: Springer.

Comparison of Linear Modularization Criteria Using the Relational Formalism, an Approach to Easily Identify Resolution Limit

Patricia Conde-Céspedes, Jean-François Marcotorchino
and Emmanuel Viennet

Abstract The modularization of large graphs or community detection in networks
is usually approached as an optimization problem of a quality function or criterion,
for instance, the modularity of Newman-Girvan. There exist other clustering criteria,
with their own properties leading to different solutions. In this paper we present six
linear modularization criteria in relational notation such as the Newman-Girvan modularity, Zahn-Condorcet, Owsiński-Zadrożny, the Deviation to Uniformity index, the
Deviation to Indetermination index and the Balanced-Modularity. We use a generic
version of Louvain algorithm to approach the optimal partition of the criteria with
real networks of different sizes. We have found that those partitions present important differences concerning the number of clusters. The relational formalism allows
us to justify these differences from a theoretical point of view. Moreover, this notation enables to easily identify the criteria having a resolution limit (a phenomenon
which causes the criterion to fail to identify modules smaller than a given scale).
This finding is confirmed in artificial benchmark LFR graphs.

1 Introduction

Networks are studied in numerous contexts such as biology, sociology, online social
networks, marketing, etc. Graphs are mathematical representations of networks,
where the entities are called nodes and the connections are called edges. Very large
graphs are difficult to analyse and it is often profitable to divide them in smaller

P. Conde-Céspedes (✉) · E. Viennet
L2TI, Institut Galilée, Université Paris 13,
99, av. Jean-Baptiste Clément, 93430 Villetaneuse, France
e-mail: patricia.conde-cespedes@isep.fr

E. Viennet
e-mail: emmanuel.viennet@univ-paris13.fr

J.-F. Marcotorchino
Thales Communications et Sécurité, 4 av. des Louvresses,
92230 Gennevilliers, France
e-mail: jfmmarco3@gmail.com

© Springer International Publishing Switzerland 2017
F. Guillet et al. (eds.), *Advances in Knowledge Discovery and Management*,
Studies in Computational Intelligence 665, DOI 10.1007/978-3-319-45763-5_6

homogeneous components easier to handle. The process of decomposing a network has received different names: graph clustering (in data analysis), modularization, community structure identification. The clusters can be called communities or modules; in this paper we use those words as synonyms.

Assessing the quality of a graph partition requires a modularization criterion. This function will be optimized to find the best partition. Various modularization criteria were formulated in the past to address different practical applications. Those criteria differ in the definition given to the notion of community or cluster.

To understand the differences between the optimal partitions obtained by each criterion we show how to represent them using the same basic formalism. In this paper we use the Mathematical Relational Analysis (MRA) to express six linear modularization criteria. Linear criteria are easy to handle, for instance, the Louvain method can be adapted to linear quality functions (see Campigotto et al. (2014)). The six criteria studied are: the Newman-Girvan modularity, the Zahn-Condorcet criterion, the Owsiński-Zadrozny criterion, the Deviation to Uniformity, the Deviation to Indetermination index and the Balanced Modularity (details in Sect. 3). The relational representation makes clear the properties of those modularization criteria. It allows to easily identify the criteria suffering from a resolution limit, first discussed by Fortunato and Barthelemy (2006). We will complete this theoretical study by some experiments on real and synthetic networks, demonstrating the effectiveness of our classification.

In this paper, we deal only with linear criteria. Nevertheless, it is important to mention that thanks to the formalism of the MRA it is also possible to express non-linear criteria in relational notations. For instance, we can mention some very well-known criteria such as the Mancoridis-Gansner criterion (see Mancoridis et al. (1998)) in cluster-programming, the Ratio-Cuts by Wei and Cheng (1989), the Michalski criterion (see Michalski and Stepp (1983) and its relational notation given in Decaestecker (1992)), etc. The interested reader can see Conde-Céspedes and Marcotorchino (2012) and Conde-Céspedes (2013).

This paper is organized as follows: Sect. 2 presents the Mathematical Relational Analysis approach and introduces the property of *balance* for linear criteria and its relation to the property of resolution limit. In Sect. 3, six linear modularization criteria in the relational formalism are formulated. Next, Sect. 4 discusses some experiments on real and artificial graphs to confirm the theoretical properties found previously.

2   Relational Analysis Approach

There is a strong link between the Mathematical Relational Analysis[1] and graph theory: *a graph is a mathematical structure that represents binary relations between objects belonging to the same set*. Therefore, a non-oriented and non-weighted graph $G = (V, E)$, with $N = |V|$ nodes and $M = |E|$ edges, is a binary symmetric relation on its set of nodes V represented by its adjacency matrix $\mathbf{A}$ as follows:

---

[1]For more details about Relational Analysis theory see Marcotorchino and Michaud (1979) and Marcotorchino (1984).

$$a_{ii'} = \begin{cases} 1 & \text{if there exists an edge between } i \text{ and } i' \; \forall (i, i') \in V \times V \\ 0 & \text{otherwise} \end{cases} \tag{1}$$

We denote the *degree* d_i of node i the number of edges incident to i. It can be calculated by summing up the terms of the row (or column) i of the adjacency matrix: $d_i = \sum_{i'} a_{ii'} = \sum_{i'} a_{i'i} = a_{i.} = a_{.i}$. We denote $\delta = \frac{2M}{N^2}$ the density of edges of the whole graph.

Partitioning a graph implies defining an equivalence relation on the set of nodes V, that means a symmetric, reflexive and transitive relation. Mathematically, an equivalence relation is represented by a square matrix X of order $N = |V|$, whose entries are defined as follows:

$$x_{ii'} = \begin{cases} 1 & \text{if } i \text{ and } i' \text{ are in the same cluster } \forall (i, i') \in V \times V \\ 0 & \text{otherwise} \end{cases} \tag{2}$$

Modularizing a graph implies to find X as close as possible to A. A modularization criterion $F(X)$ is a function which measures either a *similarity* or a distance between A and X. Therefore, the problem of modularization can be written as a function to optimize $F(X)$ where the unknown X is subject to the constraints of an equivalence relation. In fact, the problem of modularization can be written in the general form:

$$\underset{X}{Max}(F(X)) \tag{3}$$

subject to the constraints of an equivalence relation:

$$\begin{array}{lll} x_{ii'} \in \{0, 1\} & & \text{Binary} \\ x_{ii} = 1 & \forall i & \text{Reflexivity} \\ x_{ii'} - x_{i'i} = 0 & \forall (i, i') & \text{Symmetry} \\ x_{ii'} + x_{i'i''} - x_{ii''} \leq 1 & \forall (i, i', i'') & \text{Transitivity} \end{array}$$

The exact solving of this $0 - 1$ linear program due to the size of the constraints is impractical for big networks. So, heuristic approaches are the only reasonable way to proceed.

We define as well $\bar{X}$ and $\bar{A}$ as the inverse relation of X and A respectively. Their entries are defined as $\bar{x}_{ii'} = 1 - x_{ii'}$ and $\bar{a}_{ii'} = 1 - a_{ii'}$ respectively. In the following we denote κ the optimal number of clusters, that means the number of clusters of the partition X which maximizes the criterion $F(X)$.

2.1 Linear Balanced Criteria

Every linear criterion is an affine function of X, therefore in relational notation it can be written as:

$$F(X) = \sum_{i=1}^{N} \sum_{i'=1}^{N} \Phi(a_{ii'})x_{ii'} + K, \tag{4}$$

where $\Phi(a_{ii'})$ denotes any function depending only on the original data (for instance the adjacency matrix) and K denotes any *constant* depending only on the original data. Therefore, K does not intervene in the optimization problem.

Definition 1 (*Property of linear balance*). A linear criterion is *balanced* if it can be written in the following general form:

$$F(X) = \sum_{i=1}^{N} \sum_{i'=1}^{N} \phi(a_{ii'})x_{ii'} + \sum_{i=1}^{N} \sum_{i'=1}^{N} \bar{\phi}(a_{ii'})\bar{x}_{ii'} + K. \tag{5}$$

where $\phi(.)$ and $\bar{\phi}(.)$ are non negative functions depending only on the original data and verifying $\sum_{i=1}^{N} \sum_{i'=1}^{N} \phi_{ii'} > 0$ and $\sum_{i=1}^{N} \sum_{i'=1}^{N} \bar{\phi}_{ii'} > 0$.
So, they can not be all null simultaneously.

By replacing $\bar{x}$ by its definition $1 - x_{ii'}$, Eq. (5) can be rewritten as follows:

$$F(X) = \sum_{i=1}^{N} \sum_{i'=1}^{N} (\phi_{ii'} - \bar{\phi}_{ii'})x_{ii'} + K. \tag{6}$$

2.1.1 Interpretation of Functions $\phi(.)$ and $\bar{\phi}(.)$

At this point, we can give the intuition behind functions $\phi(.)$ and $\bar{\phi}(.)$. From expression (6) we deduce the importance of the property of *balance* for linear criteria. If the criterion is a function to maximize, the presence and/or absence of the terms $\phi_{ii'}$ and $\bar{\phi}_{ii'}$ has the following impact on the optimal solution:

- If $\bar{\phi}_{ii'} = 0 \, \forall i, i'$ the solution that maximizes $F(X)$ is the partition where all nodes are clustered together in a single cluster, so $\kappa = 1$ and $x_{ii'} = 1 \;\; \forall (i, i')$ and $F(X) = \sum_{i=1}^{N} \sum_{i'=1}^{N} \phi_{ii'}$.
- If $\phi_{ii'} = 0 \, \forall i, i'$ then the optimal solution that maximizes $F(X)$ is the partition where all nodes are separated, so $\kappa = N$ and $x_{ii'} = 0 \, \forall i \neq i'$ and $x_{ii} = 1 \, \forall i$ therefore $F(X) = \sum_{i=1}^{N} \sum_{i'=1}^{N} \bar{\phi}_{ii}$.

In other words, the optimization of a linear criterion who does not verify the property of *balance* will either *cluster all the nodes in a single cluster* or *isolate each node in its own cluster*, therefore forcing the user to fix the number of clusters in advance.

We can deduce from the previous paragraphs that the values taken by the functions ϕ and $\bar{\phi}$ create a sort of *balance* between the fact of generating as many clusters as possible, $\kappa = N$, and the fact generating only one cluster, $\kappa = 1$.

In the following we will call the quantity $\sum_{i=1}^{N} \sum_{i'=1}^{N} \phi(a_{ii'}) x_{ii'}$ the term of *positive agreements* and the quantity $\sum_{i=1}^{N} \sum_{i'=1}^{N} \bar{\phi}(a_{ii'}) \bar{x}_{ii'}$ the term of *negative agreements*.

2.2 Different Levels of Balance

We define two levels of balance for all linear balanced criterion:

Definition 2 (*Property of local balance*). A balanced linear criterion whose functions $\phi_{ii'}$ and $\bar{\phi}_{ii'}$ depend only upon the pair (i, i') (therefore not depending on global properties of the graph) has the property of local balance.

Some remarks about Definition 2:

- When we talk about global properties we refer to the total number of nodes, the total number of edges or other properties describing the global structure of the graph.
- For the particular case of local balance where $\phi_{ii'} + \bar{\phi}_{ll'} = K$ (that is $\phi_{ii'}$ and $\bar{\phi}_{ii'}$ sum up to a constant), we can conclude that whereas $\phi_{ii'}$ increases $\bar{\phi}_{ii'}$ decreases and vice versa.

Let us consider the special case where $\phi(a_{ii'}) = a_{ii'}$, the general term of the adjacency matrix. A *null model* is a graph with the same total number of edges and nodes and where the edges are randomly distributed. Let us denote the general term of the adjacency matrix of this random graph $\bar{\phi}(a_{ii'})$. A criterion based on a null model considers that a random graph does not have community structure. The goal of such a criterion is to maximize the deviation between the real graph, represented by $\phi(a_{ii'})$ and the null model version of this graph, represented by $\bar{\phi}(a_{ii'})$ as shown in Eq. (6). Since the original graph and the null model have the same number of edges M, we have $\sum_{i=1}^{N} \sum_{i'=1}^{N} \phi_{ii'} = \sum_{i=1}^{N} \sum_{i'=1}^{N} \bar{\phi}_{ii'} = 2M$. If this constraint causes $\bar{\phi}_{ii'}$ to depend upon the total number of edges M, then a criterion based on a null model does not verify the property of local balance. Consequently, it is not scale invariant because it depends on a global characteristic of the graph.

The definition of null model for linear criteria can be generalized as follows:

Definition 3 (*Criterion based on a null model*). A balanced linear criterion that seeks to maximize the deviation between the real graph and a null model is a criterion based on a null model. In its formulation, the real graph is represented by $\phi(a_{ii'})$ whereas the null model is represented by $\bar{\phi}(a_{ii'})$. The functions $\phi_{ii'}$ and $\bar{\phi}_{ii'}$ satisfy the following condition:

$$\sum_{i=1}^{N} \sum_{i'=1}^{N} \phi_{ii'} = \sum_{i=1}^{N} \sum_{i'=1}^{N} \bar{\phi}_{ii'}$$

such that the functions $\phi_{ii'}$ and $\bar{\phi}_{ii'}$ depend on global properties of the graph.

The global properties of the graph can be, for example, the total number of edges or the total number of nodes.

We can deduce from Definitions 2 and 3 that a linear criterion cannot be locally balanced and based on a null model at the same time.

In the particular case where $\bar{\phi}$ decreases with the size of the network, it becomes negligible for large graphs. As explained previously, if this term tends towards zero, the optimization of the criterion will tend to group the nodes more easily. For instance, a single edge between two sub-graphs would be interpreted by the criterion as a sign of a strong correlation between the two clusters, and optimizing the criterion would lead to the merge of the two clusters. Such a criterion is said to have a *resolution limit*.

The resolution limit was introduced by Fortunato and Barthelemy (2006), where the authors studied the resolution limit of the modularity of Newman-Girvan. They demonstrated that modularity optimization may fail to identify modules smaller than a given size which depends on global characteristics of the graph. Even weakly interconnected complete sub-graphs—the best identifiable communities—would be merged by this kind of optimization criteria if the network is sufficiently large. According to Kumpula et al. (2007) the resolution limit is present in any modularization criterion based on global optimization of intra-cluster edges and extra-community links and on a comparison to any null model.

In Sect. 4, we will show how criteria having a resolution limit fail to detect certain groups of densely connected nodes.

3 Modularization Criteria in Relational Notation

Graph clustering criteria depend strongly on the meaning given to the notion of *community*. In this section, we describe six linear modularization criteria and their relational coding in Table 1. We assume that the graphs we want to modularize are scale-free, that means that their degree distribution follows a power law.

1. **The Zahn-Condorcet criterion (1785, 1964)**: C.T. Zahn was the first who studied the problem of finding an equivalence relation **X**, which best approximates a given symmetric relation **A** in the sense of minimizing the distance of the symmetric difference (Zahn 1964). The criterion defined by Zahn corresponds to the dual Condorcet's criterion (Condorcet 1785) introduced in Relational Consensus whose relational coding was given by Marcotorchino and Michaud (1979). This criterion requires that every node in each cluster be connected with at least as half as the total nodes inside the cluster. Consequently, for each cluster the fraction of within cluster edges is at least 50 % (see Conde-Céspedes (2013)) and Appendix for proof).

2. **The Owsiński-Zadrożny criterion (1986)** (see Owsiński and Zadrożny (1986)) it is a generalization of Condorcet's function. It has a parameter α, which allows, according to the context, to define the minimal percentage of required within-cluster edges: α. For $\alpha = 0.5$ this criterion is equivalent to Condorcet's criterion. The parameter α defines the balance between the positive agreements term and the negative agreements term. For each cluster the density of edges is at least α % (see Conde-Céspedes(2013)).

3. **The Newman-Girvan criterion (2004)** (see Newman and Girvan (2004)): It is the best known modularization criterion, called sometimes simply *modularity*. It relies upon a null model. Its definition involves a comparison of the number of within-cluster edges in the real network and the expected number of such edges in a

Table 1 Relational notation of linear modularity functions

Criterion	Relational notation
Zahn-Condorcet (1785, 1964)	$F_{ZC}(X) = \sum_{i=1}^{N} \sum_{i'=1}^{N} (a_{ii'} x_{ii'} + \bar{a}_{ii'} \bar{x}_{ii'})$
Owsiński-Zadrożny (1986)	$F_{Z_{OZ}}(X) = \sum_{i=1}^{N} \sum_{i'=1}^{N} ((1 - \alpha) a_{ii'} x_{ii'} + \alpha \bar{a}_{ii'} \bar{x}_{ii'})$ with $0 < \alpha < 1$
Newman-Girvan (2004)	$F_{NG}(X) = \frac{1}{2M} \sum_{i=1}^{N} \sum_{i'=1}^{N} \left(a_{ii'} - \frac{a_{i.} a_{.i'}}{2M} \right) x_{ii'}$
Deviation to Uniformity (2013)	$F_{UNIF}(X) = \frac{1}{2M} \sum_{i=1}^{N} \sum_{i'=1}^{N} \left(a_{ii'} - \frac{2M}{N^2} \right) x_{ii'}$
Deviation to Indetermination (2013)	$F_{DI}(X) = \frac{1}{2M} \sum_{i=1}^{N} \sum_{i'=1}^{N} \left(a_{ii'} - \frac{a_{i.}}{N} - \frac{a_{.i'}}{N} + \frac{2M}{N^2} \right) x_{ii'}$
The Balanced Modularity (2013)	$F_{BM}(X) = \sum_{i=1}^{N} \sum_{i'=1}^{N} ((a_{ii'} - P_{ii'}) x_{ii'} + (\bar{a}_{ii'} - \bar{P}_{ii'}) \bar{x}_{ii'})$ where $P_{ii'} = \frac{a_{i.} a_{.i'}}{2M}$ and $\bar{P}_{ii'} = \left(\bar{a}_{ii'} - \frac{(N-a_{i.})(N-a_{.i'})}{N^2 - 2M} \right)$

random graph where edges are distributed following the *independence structure* (a network without regard to community structure). In fact, the *modularity* measures the *deviation to independence*.

As mention in the previous section, this criterion, based on a null model and it has a resolution limit (see Fortunato and Barthelemy (2006)). In fact, as the network becomes larger $M \longrightarrow \infty$, the term $\bar{\phi}_{ii'} = \dfrac{a_i.a_{.i'}}{2M}$ tends to zero since the degree distribution follows a power law.

4. **The Deviation to Uniformity (2013)** This criterion maximizes the deviation to the *uniformity structure*, it was proposed in Conde-Céspedes (2013). It compares the number of within-cluster edges in the real graph and the expected number of such edges in a random graph (the null model) where edges are uniformly distributed, thus all the nodes have the same degree equal to the average degree of the graph. This criterion is based on a null model and it has a resolution limit. indeed $\delta \longrightarrow 0$ as $N \longrightarrow \infty$.

5. **The Deviation to Indetermination (2013)** Analogously to Newman-Girvan function, this criterion compares the number of within-cluster edges in the real network and the expected number of such edges in a random graph where edges are distributed following the *indetermination structure*[2] (a graph without regard to community structure) (Marcotorchino and Conde-Céspedes 2013; Marcotorchino 2013). The Deviation to Indetermination is based on a null model, therefore it has a resolution limit.

6. **The Balanced modularity**[3] **(2013)** This criterion, introduced in Conde-Céspedes and Marcotorchino (2013), was constructed by adding to the Newman-Girvan modularity a term taking into account the absence of edges $\bar{A}$. Whereas Newman-Girvan modularity compares the actual value of $a_{ii'}$ to its equivalent in the case of a random graph $\dfrac{a_i.a_{.i'}}{2M}$, the new term compares the value of $\bar{a}_{ii'}$ to its version in case of a random graph $\dfrac{(N - a_{i.})(N - a_{.i'})}{N^2 - 2M}$. It is based on a null model and it has a resolution limit.

The six linear criteria of Table 1 verify the property of *balance*, so it is not necessary to set in advance the number of clusters. Table 2 specifically focuses on the fonctions $\phi_{ii'}$ and $\bar{\phi}_{ii'}$ for each criterion.

From Tables 1 and 2 one can easily deduce that two criteria: Zahn-Condorcet and Owsiński-Zadrożny verify the property of local balance. Furthermore, Table 2 clearly shows that the functions $\phi_{ii'}$ and $\bar{\phi}_{ii'}$ add up to a constant $K_{ii'}$ for these two criteria. The quantity $\bar{\phi}_{ii'}$ decreases with the size of the graph for all criteria that have a resolution limit.

---

[2] There exists a duality between the independence structure and the indetermination structure (Marcotorchino 1984, 1985; Ah-Pine and Marcotorchino 2007).

[3] Although the name of this criterion contains the word *balanced*, its definition is not related to the property of balance given in Definition 1.

Table 2 Balance property for linear criteria

Criterion	General balance		
	Local balance	Null model	Comment
Zahn-Condorcet	X		$\phi_{ii'} + \bar{\phi}_{ii'} = a_{ii'} + \bar{a}_{ii'} = 1.$
Owsiński-Zadrożny	X		$\phi_{ii'} + \bar{\phi}_{ii'} = (1 - \alpha)a_{ii'} + \alpha\bar{a}_{ii'}.$
Newman-Girvan		X	$\sum_{i=1}^{N}\sum_{i'=1}^{N}\bar{\phi}_{ii'} = \sum_{i=1}^{N}\sum_{i'=1}^{N}\dfrac{a_{i.}a_{.i'}}{2M} = 2M.$
Deviation to uniformity		X	$\sum_{i=1}^{N}\sum_{i'=1}^{N}\bar{\phi}_{ii'} = \sum_{i=1}^{N}\sum_{i'=1}^{N}\dfrac{2M}{N^2} = 2M$
Deviation to indetermination		X	$\sum_{i=1}^{N}\sum_{i'=1}^{N}\left(\dfrac{a_{i.}}{N} + \dfrac{a_{.i'}}{N} - \dfrac{2M}{N^2}\right) = 2M$
Balanced modularity		X	$\sum_{i,i'=1}^{N}\sum_{i'=1}^{N}\bar{p}_{ii'} = \sum_{i=1}^{N}\sum_{i'=1}^{N}\bar{a}_{ii'} = N^2 - 2M$

4 The Impact of Merging Two Clusters

We modularized five real networks of different sizes: Jazz (Gleiser and Danon 2003), Internet (Hoerdt and Magoni 2003), Web nd.edu (Albert et al. 1999), Amazon (Yang and Leskovec 2012)[4] and Youtube (Mislove et al. 2007). We ran a generic version of Louvain Algorithm (see Campigotto et al. (2014) and Blondel et al. (2008)) until achievement of a stable value of each criterion. The number of clusters obtained for each network is shown in Table 3.

Table 3 shows that the Zahn-Condorcet and Owsiński-Zadrożny criteria generate many more clusters than the other criteria having a resolution limit, for which the number of clusters is rather comparable. Moreover, this difference increases with the network size. Notice that the number of clusters for the Owsiński-Zadrożny criterion decreases with α, that is the minimal required fraction of within-cluster edges, so the criterion becomes more flexible.

In order to explain these differences we measure the impact of merging two clusters on the value of each criterion. Let us suppose we want to merge two clusters $\mathscr{C}_1$ and $\mathscr{C}_2$ in the network of sizes n_1 and n_2 respectively. Let us suppose as well they are connected by l edges as shown in Fig. 1.

Let us denote C_F the contribution of merging two clusters to the value of a criterion F. The contribution C_F can be easily calculated from (6) (for the proof see Conde-Céspedes (2013)):

$$C_F = \sum_{i \in \mathscr{C}_1}^{n_1} \sum_{i' \in \mathscr{C}_2}^{n_2} (\phi_{ii'} - \bar{\phi}_{ii'}) \tag{7}$$

---

[4]The data was taken from http://snap.stanford.edu/data/com-Amazon.html.

Table 3 Ref: Zahn-Condorcet (ZC), Owsiński-Zadrożny (OZ), Deviation to Uniformity (UNIF), Newman-Girvan (NG), Deviation to Indetermination (DI) and Balanced Modularity (BM)

Network	Jazz	Internet	Web nd.edu	Amazon	Youtube
$N \sim$	198	70 k	325 k	334 k	1 M
$M \sim$	3 k	351 k	1 M	925 k	3 M
δ	0,14	1.44×10^{-04}	2.77×10^{-05}	1.65×10^{-05}	4.64×10^{-06}
Criterion	κ	κ	κ	κ	κ
ZC	38	40,123	201,647	161,439	878,849
OZ $\alpha = 0.4$	34	30,897	220,967	121,370	744,680
OZ $\alpha = 0.2$	23	24,470	184,087	77,700	601,800
UNIF	20	173	711	265	51,584
NG	4	46	511	250	5,567
DI	6	39	324	246	13,985
BM	5	41	333	230	6,410

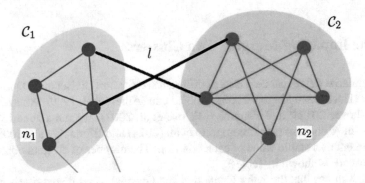

Fig. 1 Two sub graphs of the entire network we want to merge

- If $C > 0$ the merger of the two clusters increases the value of the criterion.
- If $C < 0$ the merger of the two clusters decreases the value of the criterion.

Equation (7) shows that the decision of merging or not the two clusters depends on a comparison between the quantity $\sum\limits_{i \in \mathscr{C}_1} \sum\limits_{i' \in \mathscr{C}_2} \phi_{ii'}$ and the quantity $\sum\limits_{i \in \mathscr{C}_1} \sum\limits_{i' \in \mathscr{C}_2} \bar{\phi}_{ii'}$. Giving the fact that both are positive, it is the one with the highest value that decides to merge or not to merge. Thus, whereas the first one is *for* fusion the second one is *against* the fusion.

Table 4 shows the explicit expression of the contribution for the linear criteria described below.[5]

---

[5]The contribution for the Balanced Modularity will be given later.

Table 4 Contribution of merging two clusters for linear criteria

Criterion: F	$C_F = \sum\limits_{i \in \mathscr{C}_1}^{n_1} \sum\limits_{i' \in \mathscr{C}_2}^{n_2} (\phi_{ii'} - \bar{\phi}_{ii'})$
Zahn-Condorcet	$C_{ZC} = \left(l - \dfrac{n_1 n_2}{2}\right)$
Owsiński-Zadrożny	$C_{OZ} = (l - n_1 n_2 \alpha) \quad 0 < \alpha < 1$
Deviation to uniformity	$C_{UNIF} = (l - n_1 n_2 \delta)$
Newman-Girvan	$C_{NG} = \left(l - n_1 n_2 \dfrac{d_{av}^1 d_{av}^2}{2M}\right)$
Deviation to indetermination	$C_{DI} = \left(l - n_1 n_2 \left(\dfrac{d_{av}^1}{N} + \dfrac{d_{av}^2}{N} - \dfrac{2M}{N^2}\right)\right)$

where $d_{av} = \dfrac{\sum_{i \in V}^{N} a_{i.}}{N}$ is the average degree of the whole graph, $d_{av}^1 = \dfrac{\sum_{i \in \mathscr{C}_1}^{n_1} a_{i.}}{n_1}$ and $d_{av}^2 = \dfrac{\sum_{i' \in \mathscr{C}_2}^{n_2} a_{.i'}}{n_2}$ are the average degrees of $\mathscr{C}_1$ and $\mathscr{C}_2$ respectively.

We can remark from Table 4 that for the five criteria the contribution compares "the number of edges between $\mathscr{C}_1$ and $\mathscr{C}_2$: l" to the quantity in bold. We can see as well that the *contribution* for locally balanced criteria depends only upon local properties: $l, \bar{l}, n_1, n_2$. In fact, locally balanced criteria are scale invariant. In contrast, for the other criteria having a resolution limit the contribution depends and is *decreasing* on the global size of the network. We remark as well that for three criteria: Newman-Girvan, Deviation to Indetermination and Balanced Modularity the contribution depends on the degree distribution of the two clusters. According to Barabasi and Albert (1999) many real networks fall into the class of scale-free networks, meaning that their degree distribution follows a power-law. In a scale-free network a few nodes called hubs have many connexions whereas most nodes have few connexions.

4.1 Impact on the Optimal Number of Clusters

From the previous results we can deduce the main characteristics of the optimal partition found by the optimization of each criterion (see Table 5). In addition, we remark the following facts:

- **The Zahn-Condorcet criterion**: According to Table 4 for merging the two clusters $\mathscr{C}_1$ and $\mathscr{C}_2$, these ones must be connected by at least as many edges as the half of the maximum possible number of edges,[6] that is $l > \dfrac{n_1 n_2}{2}$.

---

[6]This result is a consequence of the rule this criterion relies on: "*The rule of absolute majority of Condorcet*" in voting theory.

Table 5 Summary by criterion

Criterion	Characteristics of the optimal partition
Zahn-Condorcet	• The density of edges of each cluster is at least equal to 50%
	• No resolution limit
	• For real networks the optimal partition contains many small clusters or single nodes
Owsiński-Zadrożny	• It gives the choice to define the minimum required within-cluster density, α
	• For $\alpha = 0.5$ the Owsiński-Zadrożny criterion $\equiv$ the Zahn-Condorcet criterion
	• No resolution limit
	• The optimal partition depends on the parameter α
Deviation to uniformity	• A particular case of Owsiński-Zadrożny criterion with $\alpha = \delta$
	• The density of within cluster edges of each cluster is at least the global density δ
	• It has a resolution limit
Newman-Girvan	• It depends on the degree distribution
	• It has a resolution limit
	• The optimal partition has no single nodes
Deviation to indetermination	• It depends on the degree distribution
	• It has a resolution limit
Balanced modularity	• It depends on the degree distribution
	• It has a resolution limit

- **The Owsiński-Zadrożny criterion**: For merging the two clusters $\mathcal{C}_1$ and $\mathcal{C}_2$, these ones must be connected by at least as $\alpha\%$ as the maximum possible number of edges.
- **The Deviation to Uniformity**: According to Table 4 for the merge to take place the fraction of edges between $\mathcal{C}_1$ and $\mathcal{C}_2$ must be at least equal to the global density of the whole graph.
- **Newman-Girvan criterion**: From Table 4 we can deduce that the optimal partition does not have clusters with a single node (this result was already demonstrated in Brandes et al. (2008)). In fact, if $\mathcal{C}_1$ has only one node with only one connection to $\mathcal{C}_2$, thus $n_1 = 1$, $d_{av}^1 = 1$, $l = 1$ and consequently the contribution is always positive: $C_{NG} = \left(1 - \frac{\sum_{i=1}^{n_2} a_{i.}}{2M}\right) > 0$.
- **Balanced Modularity**: It is easy to understand the behaviour of the contribution of Balanced Modularity when we compare it to those of Newman-Girvan and

Deviation to Indetermination (see Conde-Céspedes (2013) for proof).[7] Indeed, we demostrated in Conde-Céspedes (2013) that:

$$C_{BM} = 2C_{NG} + n_1 n_2 \frac{(d_{av}^1 - d_{av})(d_{av}^2 - d_{av})}{2M(1 - \delta)} \tag{8}$$

and

$$C_{BM} = 2C_{DI} + n_1 n_2 \left(2 - \frac{1}{\delta}\right) \frac{(d_{av}^1 - d_{av})(d_{av}^2 - d_{av})}{N^2(1 - \delta)}. \tag{9}$$

Although the contribution for the Balanced Modularity is increasing in both the contribution of Newman Girvan C_{NG} and in the contribution of Deviation to Indetermination C_{DI}, in both cases C_{BM} has an additional term that we can treat as *regulator*: $\left(n_1 n_2 \frac{(d_{av}^1 - d_{av})(d_{av}^2 - d_{av})}{2M(1-\delta)}\right)$ and $\left(n_1 n_2 \left(2 - \frac{1}{\delta}\right) \frac{(d_{av}^1 - d_{av})(d_{av}^2 - d_{av})}{N^2(1-\delta)}\right)$ respectively. These two regulators have opposite sign for real networks. In fact, the coefficient $\left(2 - \frac{1}{\delta}\right)$ of the second regulator is almost surely negative for real graphs because the density $\delta \ll 0.5$ for scale-free networks. That is why the Balanced Modularity behaves as a regulator between both criteria: Newman-Girvan and Balanced Modularity. However, when the network size increases $N \longrightarrow \infty$ and $M \longrightarrow \infty$ the regulator terms tend to zero.

Only ground-truth overlapping communities are defined on real networks in Table 3. This fact makes difficult to judge the quality of the obtained partitions because we can not directly compare a partition to overlapping communities. That is why in the next section we will consider artificial networks with a predefined community structure.

---

[7]These expressions are deduced from the two following expressions of Balanced Modularity in terms of Newman-Girvan and Deviation to Indetermination criteria:

$$F_{BM} = 2F_{NG} + \sum_{i=1}^{N} \sum_{i'=1}^{N} \left(\frac{(a_{i.} - d_{av})(a_{.i'} - d_{av})}{2M(1 - \delta)}\right) x_{ii'}$$

and

$$F_{BM} = 2F_{DI} + \left(2 - \frac{1}{\delta}\right) \sum_{i=1}^{N} \sum_{i'=1}^{N} \left(\frac{(a_{i.} - d_{av})(a_{.i'} - d_{av})}{N^2(1 - \delta)}\right) x_{ii'}.$$

5 Experiments with Artificial Networks

In order to judge the quality of the partitions obtained by each criterion we generated benchmark LFR graphs[8] (see Lancichinetti et al. (2008)) of different sizes 1000, 5000, 10000, 50000, 100000 and 500000. The input parameters are the same as those considered in Lancichinetti and Fortunato (2009). The average degree is 20, the maximum degree 50, the exponent of the degree distribution is −2 and that of the community size distribution is −1. In order to test the existence of resolution limit we chose small communities sizes, ranging from 10 to 50 nodes, and low values of mixing parameter, 0.10, 0.20 et 0.30. Figure 2 shows the average number of clusters for 100 runs of the generic Louvain algorithm.

In Fig. 2 it is hard to see the curve of the real number of clusters (in black) beacuse it is almost overlapped with those of OZ1 and OZ2.

Figure 2 shows clearly the difference between the behavior of those criteria having a resolution limit (NG, DU, DI and BM) and the behavior of criteria locally defined (ZC and OZ). As the size of the network increases the four criteria suffering from resolution-limit detect fewer clusters than those predefined. The number of clusters is rather comparable for these four functions, one reason can be the fact that the term of negative agreements tends to zero when the network gets bigger. Conversely, the

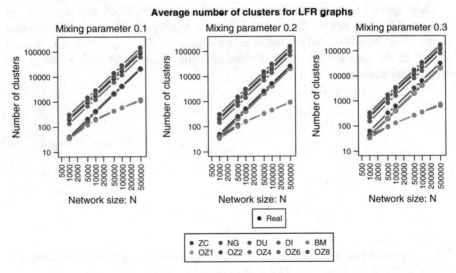

Fig. 2 Average number of cluster for artificial LFR graphs (logarithmic scale). The *curve* of the real number of clusters (in *black*) it is almost overlapped with that of OZ1 and OZ2

---

[8]LFR graphs are benchmark graphs introduced in Lancichinetti et al. (2008) that aim to reproduce as much as possible the structure that reflects the real properties of nodes and communities found in real networks. These artificial graphs have predefined community structure based on the mixing parameter of each node. As stated in Lancichinetti et al. (2008), for each node the mixing parameter is the fraction of its links it shares with the nodes of the network outside its community.

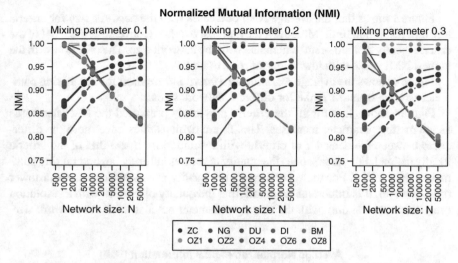

Fig. 3 The average Normalized Mutual Information (NMI) on the graphs in Fig. 2 (logarithmic scale)

number of clusters of criteria locally defined increases nearly at the same rate as the real number of clusters. Whereas OZ with high α identifies more clusters than those predefined, the criterion which best approaches the real number of clusters is OZ with low values of $\alpha = 0.2$ and $\alpha = 0.1$.

Figure 3 shows the *Normalized Mutual Information*[9] (NMI) for the partitions in Fig. 2.

---

[9]The normalized mutual information (NMI) is a measure of similarity of two partitions. It was originated in information theory to measure the departure from independence between two random variables. Given a set of objects V and two partitions P_1 and P_2 defined on V, intuitively, the mutual information measures the information that P_1 and P_2 share. It is normalized between 0 and 1. It is worth 0 if the two partitions are independent and 1 if they are identical. Let p and q be the total number of clusters of partitions P_1 and P_2 respectively. The NMI is calculated as follows:

$$NMI(P_1, P_2) = \frac{2I(P_1, P_2)}{H(P_1) + H(P_2)}$$

where:

- $I(P_1, P_2) = \sum_{u=1}^{p} \sum_{v=1}^{q} p_{uv} \ln \left(\frac{p_{uv}}{p_{u.} p_{.v}} \right)$ is the mutual information of partitions P_1 and P_2. I tells how much we learn about P_1 if we know P_2 and vice versa. The quantity $p_{uv} = \frac{n_{uv}}{N}$ is the fraction of objects who belong simultaneously to cluster u of partition P_1 and to cluster v of partition P_2. Analogously $p_{uv} = \frac{n_{u.}}{N}$ is the fraction of objects who belong to cluster u of partition P_1 and $p_{uv} = \frac{n_{.v}}{N}$ is the fraction of objects who belong to cluster v of partition P_2 and $|V| = N$. In the case $n_{uv} = 0$ we assume $\ln \left(\frac{p_{uv}}{p_{u.} p_{.v}} \right) = 0$.
- $H(P_1) = -\sum_{u=1}^{p} p_{u.} \ln p_{u.}$ represents the Shanon entropy of P_1 and $H(P_2) = -\sum_{v=1}^{q} p_{.v} \ln p_{.v}$ represents the Shanon entropy of P_2 (see Shannon (1948)).

Figure 3 shows that the average NMI decreases with the network size for criteria having a resolution limit. Moreover, they almost overlap. Conversely, the NMI of the criteria locally defined seem to increase with the network size. The criterion with the highest NMI is OZ with low values of α, 0.1 and 0.2.

Figure 4 shows the average *Normalized Mutual Information* for the mixing parameter ranging from 0.1 to 0.8 for different network sizes.

Figure 4 shows that for all the criteria previously presented the NMI decreases as the mixing parameter increases. This figure demonstrates once more the differences between the behavior of criteria with resolution limit and that of the criteria locally defined. For the first ones the quality decreases abruptly beyond mixing parameter equal to 0.6. For the second ones, the quality seems to decrease at a lower rate. However, it is important to remark that the quality of criteria with a resolution limit decreases not only with the mixing parameter but also with the network size.

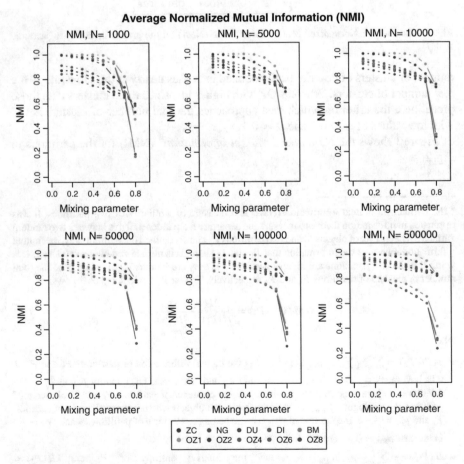

Fig. 4 The average Normalized Mutual Information (NMI) according to mixing parameter for networks of 6 different sizes: 1000, 5000, 10000, 50000, 100000 and 500000

Converserly, the behavior of the NMI of locally defined criteria seem to have the same behavour independtly of the size of the whole network.

Another point to remark is that even when the mixing parameter is high all the criteria find a community structure. In fact, the pre-defined communities in the LFR graphs are based on mixing parameter, whereas all the criteria analysed in this article have their own definition of *graph with no community structure* which is not based on the mixing parameter.

Table 5 presents a summary of the results found by the previous analysis.

6 Conclusions

We have presented six linear modularization criteria in relational notation, Zahn-Condorcet, Owsiński-Zadrożny, the Newman-Girvan modularity, the Deviation to Uniformity index, the Deviation to Indetermination index and the Balanced-Modularity. This notation allowed us to easily identify the criteria suffering from a resolution limit. We found that the first two criteria had a local definition, whereas the others, based on a null model, had a resolution limit. These findings were confirmed by modularizing real and artificial graphs using a generic version of the Louvain algorithm. We compared the number of clusters found by the six criteria and the Normalized Mutual information for artificial graphs. The results showed that criteria based on a local definition had a better performance than those based on a null model when the size of the graph increases, experimentally the crition having the best behavior was Owsiński-Zadrożny with low values of parameter α. However, it is important to remark that these results are based on a particular kind of graphs, more precisely, graphs with a low mixing parameter, small communities,[10] node degrees and community sizes distributed according to a power law.

Acknowledgments This work is supported by REQUEST and Open Food System projects.

Appendix

Theorem 1 (The density of clusters obtained by maximization of Zahn-Condorcet criterion is least 50 %). *Given a connected, non-oriented and unweighted graph $G = (V, E)$, the optimal partition obtained by optimizing the Zahn-Condorcet criterion has the following property: the number of within-cluster edges of each cluster is at least as half as the possible maximum existing within-cluster edges, that is to say the number of existing edges in the case the cluster is a clique. Furthermore, every node in each cluster is connected with at least as half as the total nodes inside the cluster.*

---

[10]What we call *small* are communities ranging from 10 to 50 nodes, that is the same sizes considered by the authors of LFR graphs (see Lancichinetti and Fortunato (2009)).

Proof. Considering the constraints of reflexivity and symmetry of the relational variable $x_{ii'}$ (i.e. $x_{ii} = 1 \forall i$ and $x_{ii'} = x_{i'i}$), the expression of Zahn-Condorcet criterion in Table 2 can be written as follows:

$$F_{ZC}(X) = \sum_{i>i'} (a_{ii'} - \bar{a}_{ii'}) x_{ii'} + N^2 - 2M - N.$$

where:

- $\sum_{i>i'} a_{ii'} x_{ii'}$ is the number of within-cluster edges for all clusters.
- $\sum_{i>i'} \bar{a}_{ii'} x_{ii'}$ is the number of missing within-cluster edges for all clusters.

If we denote E_j the number of within edges of cluster j, the total number of missing edges for the cluster j will be $\left(\frac{n_j(n_j-1)}{2} - E_j \right)$. So, the criterion Zahn-Condorcet will become:

$$F_{ZC}(\mathscr{C}) = \sum_{j=1}^{\kappa} \left(E_j - \left(\frac{n_j(n_j-1)}{2} - Ej \right) \right) + N^2 - 2M - N,$$
or
$$F_{ZC}(\mathscr{C}) = \sum_{j=1}^{\kappa} (2E_j - \frac{n_j(n_j-1)}{2}) + N^2 - 2M - N.$$

the term $(2E_j - \frac{n_j(n_j-1)}{2})$ represents the contribution of cluster j to the value of the criterion. For each cluster of the *optimal partition* this term must be positive or null. Otherwise it would be possible to obtain a better partition by isolating each node in cluster j (the contribution to the value of the criterion by a cluster of an isolated node is null). This implies:

$$(2E_j - \frac{n_j(n_j-1)}{2}) \geq 0, \text{ or } E_j \geq \frac{n_j(n_j-1)}{4}.$$

So, each cluster j has a density of at least 50 %.

This result can be extended to every node of each cluster of the optimal partition. In fact, let us suppose that there is a cluster j containing a node n_0 which is connected with less than half of the total nodes in the cluster. Let us denote E_{j_0} the connexions of n_0 to nodes in C_j. So, $E_{j_0} <= \frac{(n_j-1)}{2}$.

It is always possible to obtain a better partition by isolating n_0. In fact, the contribution of the two resulting clusters after isolation of node n_0 is:

$$2(E_j - E_{j_0}) - \frac{(n_j-1)(n_j-2)}{2}$$

this last expression is greater than the contribution of cluster j, given by $(2E_j - \frac{n_j(n_j-1)}{2})$, if n_0 is connected with less than half of nodes in C_j.

This also proves why the partitions obtaining by optimizing Zahn-Condorcet criterion contain sometimes clusters of isolates nodes. $\square$

References

Ah-Pine, J., & Marcotorchino, F. (2007). Statistical, geometrical and logical independences between categorical variables. In *Proceedings of the ASMDA2007 Symposium*, Chania, Greece.

Albert, R., Jeong, H., & Barabási, A. (1999). Internet: Diameter of the world-wide web. *Nature*, *401*(6749), 130–131.

Barabasi, A. L., & Albert, R. (1999). Emergence of scaling in random networks. *Science*, *286*, 509–512.

Blondel, V., Guillaume, J.-L., Lambiotte, R., & Lefebvre, E. (2008). Fast unfolding of communities in large networks. *Journal of Statistical Mechanics: Theory and Experiment*, *2008*(10), P10008.

Brandes, U., Delling, D., Gaertler, M., Grke, R., Hoefer, M., Nikoloski, Z., et al. (2008). On modularity clustering. *IEEE Transactions on Knowledge and Data Engineering*, *20*(2), 172–188.

Campigotto, R., Conde-Céspedes, P., & Guillaume, J. (2014). A generalized and adaptive method for community detection. CoRR abs/1406.2518.

Conde-Céspedes, P. (2013). *Modélisations et extensions du formalisme de l'Analyse Relationnelle Mathématique à la modularisation des grands graphes*. Thèse de doctorat: Université Pierre et Marie Curie.

Conde-Céspedes, P., & Marcotorchino, J. (2012). Modularisation et recherche de communautés dans les réseaux complexes par unification relationnelle. In *Revue des Nouvelles Technologies de l'Information*, Apprentissage Artificiel et Fouille de Données, RNTI-A-6 (pp. 71–97).

Conde-Céspedes, P., & Marcotorchino, F. (2013). Comparison different modularization criteria using relational metric. In F. Nielsen & F. Barbaresco (Eds.), *Proceedings First International Conference, Geometric Science of Information* (Vol. 1, pp. 180–187). Paris: Springer.

Condorcet, C. A. M. (1785). Essai sur l'application de l'analyse à la probabilité des décisions rendues à la pluralité des voix. *Journal of Mathematical Sociology*, *1*(1), 113–120.

Decaestecker, C. (1992). *Apprentissage en classification conceptuelle incrémentale*. Ph.D. thesis, Université Libre de Bruxelles (Faculté des Sciences).

Fortunato, S., & Barthelemy, M. (2006). Resolution limit in community detection. In *Proceedings of the National Academy of Sciences of the United States of America*.

Gleiser, P., & Danon, L. (2003). Community structure in jazz. *Advances in Complex Systems (ACS)*, *06*(04), 565–573.

Hoerdt, M., & Magoni, D. (2003). *Proceedings of the 11th International Conference on Software, Telecommunications and Computer Networks* (vol. 257).

Kumpula, J., Saramäki, J., Kaski, K., & Kertesz, J. (2007). Limited resolution in complex network community detection with potts model approach. *The European Physical Journal B*, *56*(1), 41–45.

Lancichinetti, A., & Fortunato, S. (2009). Community detection algorithms: A comparative analysis. *Physical Review E*, *80*, 056117.

Lancichinetti, A., Fortunato, S., & Radicchi, F. (2008). Benchmark graphs for testing community detection algorithms. *Physical Review E*, *78*, 046110.

Mancoridis, S., Mitchell, B., Rorres, C., Chen, Y., & Gansner, E. (1998). Using automatic clustering to produce high-level system organizations of source code. In *The IEEE Proceedings of the 1998 International Workshop on Program Understanding (IWPC 1998)* (pp. 45–52). Ischia: IEEE Computer Society.

Marcotorchino, F. (1984). Utilisation des comparaisons par paires en statistique des contingences (partie i). *Publication du Centre Scientifique IBM de Paris, F057, et Cahiers du Séminaire Analyse des Données et Processus Stochastiques Université Libre de Bruxelles* (pp. 1–57).

Marcotorchino, F. (1985). Utilisation des comparaisons par paires en statistique des contingences (partie iii). *Etude F-081 du Centre Scientifique IBM de Paris* (pp. 1–39).

Marcotorchino, F. (2013). Optimal transport, spatial interaction models and related problems, impacts on relational metrics, adaptation to large graphs and networks modularity. Internal Publication of Thales.

Marcotorchino, F., & Conde-Céspedes, P. (2013). Optimal transport and minimal trade problem, impacts on relational metrics and applications to large graphs and networks modularity. In F. Nielsen & F. Barbaresco (Eds.), *Proceedings of First International Conference, Geometric Science of Information* (Vol. 8085, pp. 169–179). Heidelberg: Springer.

Marcotorchino, F., & Michaud, P. (1979). *Optimisation en Analyse ordinale des données*. Paris: Masson.

Michalski, R., & Stepp, R. (1983). Learning from observation: Conceptual clustering. In R. Michalski, J. Carbonell, T. Mitchell, & M. Kaufmann (Eds.), *Machine learning: An artificial intelligence approach, Chap. 11* (Vol. 1, pp. 331–367). Heidelberg: Springer.

Mislove, A., Marcon, M., Gummadi, K., Druschel, P., & Bhattacharjee, B. (2007). Measurement and analysis of online social networks. In *Proceedings of the 5th ACM/Usenix Internet Measurement Conference (IMC 2007)*, San Diego, CA.

Newman, M., & Girvan, M. (2004). Finding and evaluating community structure in networks. *Physical Review E, 69*, 026113.

Owsiński, J., & Zadrożny, S. (1986). Clustering for ordinal data: A linear programming formulation. *Control and Cybernetics, 15*(2), 183–193.

Shannon, C. (1948). A mathematical theory of communication. *Bell System Technical Journal, 27*(379–423), 623–656.

Wei, Y., & Cheng, C. (1989). Towards efficient hierarchical designs by ratio cut partitioning. In *IEEE International Conference on Computer-Aided Design* (pp. 298–301).

Yang, J., & Leskovec, J. (2012). Defining and evaluating network communities based on ground-truth. In *International Conference on Data Mining* (pp. 745–754). IEEE Computer Society. abs/1205.6233.

Zahn, C. (1964). Approximating symmetric relations by equivalence relations. *SIAM Journal on Applied Mathematics, 12*, 840–847.

A Novel Approach to Feature Selection Based on Quality Estimation Metrics

Jean-Charles Lamirel, Pascal Cuxac and Kafil Hajlaoui

Abstract Feature maximization (F-max) is an unbiased quality estimation metric of unsupervised classification (clustering) that favours clusters with a maximal feature F-measure value. In this article we show that an adaptation of this metric within the framework of supervised classification allows efficient feature selection and feature contrasting to be performed. We experiment the method on different types of textual data. In this context, we demonstrate that this technique significantly improves the performance of classification methods as compared with the use of state-of-the art feature selection techniques, notably in the case of the classification of unbalanced, highly multidimensional and noisy textual data gathered in similar classes.

1 Introduction

Since the 1990s, progress in computing and storage capacities has allowed the handling of extremely large volumes of data: it is not rare to deal with space for the description of several thousand, or even tens of thousands, of features. It could be thought that the classification algorithms are more effective with a myriad of features, but the situation is not so simple. The first problem is the increase in the calculation time. Additionally, the fact that many features are redundant for the classification task, or irrelevant, considerably disrupts the functioning of the classifiers. Furthermore, most training algorithms use probabilities whose distribution may be difficult to estimate in the presence of a large number of features. The integration of a process of feature selection into the frame of large dimension data classification has thus become a central issue. In the literature, essentially three types of approach are proposed for feature selection: approaches directly incorporated into the classification methods, known as "embedded", approaches based on techniques of optimisation,

J.-C. Lamirel (✉)
SYNALP Team-LORIA, INRIA Nancy-Grand Est, Vandoeuvre-lès-Nancy, France
e-mail: jean-charles.lamirel@loria.fr

P. Cuxac · K. Hajlaoui
CNRS-Inist, Vandoeuvre-lès-Nancy, France
e-mail: pascal.cuxac@inist.fr

© Springer International Publishing Switzerland 2017
F. Guillet et al. (eds.), *Advances in Knowledge Discovery and Management*,
Studies in Computational Intelligence 665, DOI 10.1007/978-3-319-45763-5_7

or "wrappers", and approaches based on statistical tests, also named filter-based approaches. Thorough states-of-the-art have been described by many authors, such as Ladha and Deepa (2011), Bolón-Canedo et al. (2012), Guyon and Elisseeff (2003) or Daviet (2009). Therefore, below we will simply give a brief overview of the existing approaches.

"Embedded" approaches integrate feature selection into the learning process (Breiman et al. 1984). The most popular methods in this category are those based on SVM (Support Vector Machines) and those founded on neural networks. For example, RFE-SVM (Recursive Feature Elimination for Support Vector Machines) (Guyon et al. 2002) is an integrated process, where feature selection is carried out in an iterative manner using an SVM classifier, and suppressed features are those that are the most distant from the decision boundary.

For their part, "wrapper" methods use a performance criterion to seek out a pertinent sub-group of predictors (Kohavi and John 1997). Most often the performance criterion used is the error rate (but it can be a prediction cost, or the area under the ROC curve). As an example, the WrapperSubsetEval method begins with an empty feature set and continues until the addition of new features no longer improves performance. It uses cross-validation to estimate learning for a given group of features (Witten and Frank 2005). Comparisons between methods, such as that described above with that of Forman (Forman 2003), clearly demonstrate that without taking their effectiveness into account, one of the principal drawbacks of these two classes of methods is that they require long calculation times. This prohibits their use in the case of highly multidimensional data. In this context, a possible alternative is to exploit filter-based methods.

Filter-based approaches are selection methods that are used upstream and independently of the learning algorithm. Based on statistical tests, they require less calculation time than do other approaches. The most classical examples of filter-based methods are the chi-squared method (Ladha and Deepa 2011), mutual information-based methods, like MIFS (Mutual Information Feature Selection) (Hall and Smith 1999), information gain-based methods, like CBF (Consistency-based Filter) (Dash and Liu 2003), correlation-based methods, like MODTREE (Lallich and Rakotomalala 2000), or, nearest-neighbour-based methods, like Relief (Kira and Rendell 1995) or RLF (ReliefF) (Konokenko 1994).

The chi-squared method uses a common statistical test, which measures the difference from an expected distribution, presuming that the features are independent of class labels (Ladha and Deepa 2011). Equally, information gain is one of the most frequent methods of feature selection. This univariate filter supplies an organized classification of all variables. In this approach, the variables retained are those that obtain a positive value of information gain (Hall and Smith 1999).

In the MIFS method, a feature is added to a sub-group of features that have already been selected, provided that its link to the target class surpasses the average connection to predictors that have already been selected. The method takes into account both relevance and redundancy (Hall and Smith 1999).

The CBF method (Consistency-based Filter) evaluates the relevance of a subgroup of features, by estimating the coherence level of classes that results when training samples are projected onto this sub-group (Dash and Liu 2003).

The MODTREE method uses a filtering procedure that relies on the principle of calculation of pairwise correlations. It works in the space of pairs of individuals described by indicators of co-labelling attached to each original feature. To that end, a correlation coefficient for each pair of features is used, which represents the linear correlation between the two elements of a pair. The calculation of partial correlation coefficients thus allows the step-wise selection of features (Lallich and Rakotomalala 2000).

The fundamental hypothesis of the Relief method (Kira and Rendell 1995) is inspired by the nearest neighbour principle. This method considers a variable relevant if it discriminates correctly data in the positive class compared to its nearest neighbour in the negative class. The feature score is cumulative and is calculated using a random draw of data samples. RLF is an extension of Relief, where the ability to solve multi-class problems has been added. In addition, this variant is more robust and can treat incomplete and noisy data (Konokenko 1994). RLF is considered to be one of the most efficient feature selection methods.

As for all statistical tests, filter-based approaches behave erratically in the case of very low frequency features, which are common in text classification (Ladha and Deepa 2011). In this article we show that, despite their diversity, all existing approaches are inoperative, or even detrimental, in the case of extremely imbalanced, highly multidimensional and noisy data, that has a high degree of similitude between classes. As an alternative, we propose a new method of feature selection and contrast, based on the recently developed maximization metric feature. Furthermore, we compare the performance of this method to that of classical techniques in the context of help with patent validation. Then we extend the range of our study to habitually used textual reference data. The rest of this manuscript is structured as follows: Sect. 2 presents our new approach for feature selection; Sect. 3 details the data used; Sect. 4 compares the results of the different data corpora of the classification, with and without the use of the proposed approach; Sect. 5 presents our conclusions and further research directions.

2   Labelling Maximization for the Features Selection

Feature maximization (F-max) is an unbiased metric with which to estimate the quality of an unsupervised classification. It uses the properties (i.e. the features) of data associated with each cluster, without prior examination of the cluster profiles (Lamirel et al. 2004). Its principal advantage is that it is totally independent of the classification method and of its operating mode. When used after learning, it can be exploited to establish global indices of clustering quality (Lamirel et al. 2010), or for cluster labelling (Lamirel and Ta 2008).

Consider a partition C which results from a clustering method applied to a dataset D represented by a group of features F. The feature maximization measure favours clusters with a maximal feature F-measure. The feature F-measure $FF_c(f)$ of a feature f associated with a cluster c is defined as the harmonic mean of the feature recall $FR_c(f)$ and of the feature predominance $FP_c(f)$, which are themselves defined as follows:

$$FR_c(f) = \frac{\Sigma_{d \in c} W_d^f}{\Sigma_{c \in C} \Sigma_{d \in c} W_d^f} \quad FP_c(f) = \frac{\Sigma_{d \in c} W_d^f}{\Sigma_{f' \in F_c, d \in c} W_d^{f'}} \tag{1}$$

with

$$FF_c(f) = 2 \left(\frac{FR_c(f) \times FP_c(f)}{FR_c(f) + FP_c(f)} \right) \tag{2}$$

where W_d^f represents the weight of the feature f for the data d and F_c represents all the features present in the dataset associated with the cluster c.

Former experiments in cluster labelling clearly highlighted that the feature maximization metric has similar discrimination capabilities to the Chi-square metric, whilst having much better generalization capabilities (Lamirel and Ta 2008). Taking into account the basic definition of the labelling maximization metric, its use for the task of variable selection in the context of supervised learning becomes a simple process. Therefore, this generic metric can be applied to data associated with a class, as well as those associated with a cluster. The selection process can thus be defined as parameter-free, based on classes in which a class variable is characterised using both its capacity to discriminate between classes ($FP_c(f)$ index) and its ability to faithfully represent the class data ($FR_c(f)$ index). The set S_c of features that are characteristic of a given class c belonging to a class set C is translated by:

$$S_c = \left\{ f \in F_c \mid FF_c(f) > \overline{FF}(f) \text{ and } FF_c(f) > \overline{FF}_D \right\} \text{ where} \tag{3}$$

$$\overline{FF}(f) = \Sigma_{c' \in C} \frac{FF'_c(f)}{|C_{/f}|} \text{ and } \overline{FF}_D = \Sigma_{f \in F} \frac{\overline{FF}(f)}{|F|} \tag{4}$$

where $C_{/f}$ represents the subset of C in which the feature f occurs.

Finally, the set of all selected features S_C is the subset of F defined by:

$$S_C = \cup_{c \in C} S_c. \tag{5}$$

In other words, the features that are judged relevant for a given class are those whose representations are better than average in this class and better than the average representation of all features in terms of the feature F-measure.

In the specific context of the feature maximization process, a step of improvement by contrast can be exploited as a complement to the first selection step. The role of this is to adapt the description of each piece of data to the specific characteristics of its associated class. This consists of modifying the data weighting schema in a distinct way for each class, taking into account the information gain supplied by the feature F-measure of the local features in this class.

The information gain is proportional to the relation between the F-measure value of a variable in the $FF_c(f)$ class and the average F-measure value of this variable for the whole partition. This concerns only a single piece of data and one feature describing it. Thus, the resulting gain acts as a contrast factor. This factor adjusts the weight of the feature in the data profile, optionally taking into account its prior establishment. For a feature f belonging to the group of selected features S_c from a class c, the gain $G_c(f)$ is expressed as:

$$G_c(f) = (FF_c(f)/\overline{FF}(f))^k \tag{6}$$

where k is an amplification factor that can be optimised according to the precision obtained.

The active features of a class are those for which the information gain is greater than 1. Given that the proposed method is one of selection and of contrast based on the classes, the average number of active features per class is comparable to the total number of features singled out in the case of habitual selection methods.

Below we give an example of the operating mode of the method, on the basis of a toy-dataset encompassing two classes (*Men (M)*, *Women (F)*) described with 3 features: *Nose_Size Hair_Length, Shoes_Size*. Figure 1 shows the source data and how the F-measure calculation of the *Shoes_Size* feature operates in the *Men* class.

As shown in Fig. 2, the second step in the process is to calculate the marginal average F-measure for each feature and the overall average F-measure for the combination of all features and all classes. In this figure, notation $\overline{F(.,.)}$ stands here for overall average $\overline{FF}_D$ presented in (Eq. 3) and notation $\overline{F(x,.)}$ stands for marginal average of class x, which is itself computed as:

$$\overline{F(x,.)} = \Sigma_{f \in S_x} \frac{FF_x(f)}{|S_x|} \tag{7}$$

Shoes_Size	Hair_Length	Nose_Size	Class
9	5	5	M
9	10	5	M
9	20	6	M
5	15	5	W
6	25	6	W
5	25	5	W

$FR(S,M) = 27/43 = 0.62$

$FP(S,M) = 27/78 = 0.35$

$FF(S,M) = \dfrac{2(FR(S,M) \times FP(S,M))}{FR(S,M) + FP(S,M)}$

$= 0.48$

Fig. 1 Principle of feature F-measure computation for sample data

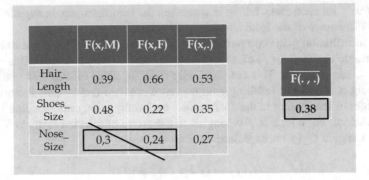

Fig. 2 Principle of computation of overall feature F-measure average and elimination of irrelevant features

Features with F-measures that are systematically lower than the overall average are eliminated. The *Nose_Size* feature is thus removed. Remaining features (i.e. selected features) are considered active in the classes in which their F-measure is above the marginal average:

1. *Shoes_Size* is active in the *Men's* class,
2. *Hair_Length* is active in the *Women's* class.

Contrast ratio highlights the degree of activity and passivity of selected features as regards their F-measure marginal average in different classes. Figure 3 illustrates how the contrast is calculated for the example presented. In the context of this example, the contrast may be considered as a function that will virtually have the following effects:

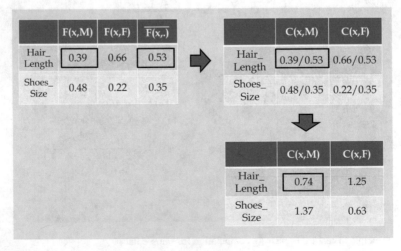

Fig. 3 Principle of computation of contrast for selected features

1. Increase the length of women's hair,
2. Increase the size of the men's shoes,
3. Decrease the length of the men's hair,
4. Reduce the size of women's shoes.

3 Experimental Data

One of the goals of the QUAERO project is to use bibliographic information to help experts to judge patent precedence. Thus, initially it was necessary to prove that it is possible to associate such bibliographic information with the patent classes in a pertinent manner; or in other words, to classify it correctly within such classes. The main experimental data source comprised 6,387 patents from the pharmacological domain (A61K class; medical preparation) in an XML format, grouped into 15 sub-classes. The bibliographic references in the patents were extracted from the Medline database.[1] 25,887 citations were extracted in total. Interrogation of the Medline database with the extracted citations allowed bibliographic notices to be recovered for 7,501 references. Each notice was then labelled with the first classification code of the citing patent (Hajlaoui et al. 2012). Each notice's abstract was treated and transformed into a "bag of words" (Salton 1971) using the TreeTagger tool (Schmid 1994). To reduce the noise generated by this tool, a frequency threshold of 45 (i.e. an average threshold of 3 per class) was applied to the extracted descriptors. The result was a description space limited to the 1,804 dimension. A last TF-IDF weighting step was applied (Salton 1971). The series of labelled notices, which were thus pre-treated, represented the final corpus on which training was carried out. This last corpus was highly unbalanced. The smallest class (A61K41) contained 22 articles, whereas the largest contained 2,500 (A61K31 class). The inter-class similarity was calculated using a cosine correlation. This indicated that more than 70 % of pairs of classes had a similarity of between 0.5 and 0.9. Thus, the ability of a classification model to precisely detect the correct class is strongly reduced. A solution commonly used to contend with an imbalance in classes' data is sub-sampling of the larger classes (Good 2006) and/or over-sampling of the smaller ones (Chawla et al. 2002). However, re-sampling, which introduces redundancy into the data, does not improve the performance of this dataset, as was shown by Hajlaoui et al. (2012). Additionally, these authors showed that exploiting different documentary weighting schemes has little influence on performance. Therefore, we have proposed an alternative solution detailed below, namely to edit out the features that are judged irrelevant and to contrast those considered reliable (Lamirel et al. 2014).

As a complement, 5 other well-known reference text datasets have been exploited for validation of the method:

---

[1] http://www.ncbi.nlm.nih.gov/pubmed/.

- The R8 and R52 corpora were obtained by Cardoso Cachopo[2] from the R10 and R90 datasets, which are derived from the Reuters 21,578 collection[3]. The aim of these adjustments was to retain only data that had a single label. We merely considered monothematic documents and classes that still had at least one example of training and one example of test conditions. R8 is a reduction of the R10 corpus (the 10 most frequent classes to 8 classes) and R52 is a reduction of the R90 corpus (90 classes to 52 classes).
- The Amazon[tm] corpus (AMZ) is a UCI dataset (Bache and Lichman 2013) derived from the recommendations of clients of the Amazon web site that are usable for author identification. To evaluate the robustness of the classification algorithms with respect to a large number of target classes, 50 of the most active users who have frequently posted comments in these newsgroups were identified. Thirty messages were collected for each user. Each message included the author's linguistic style, such as the use of figures, punctuation, frequent words and sentences.
- The 20 Newsgroups dataset (Lang 1995) is a collection of approximately 20,000 documents almost uniformly distributed among 20 different discussion groups. We consider two "bag of words" versions of this dataset in our experiments. In the (20N-AT) version, all words are preserved and non-alphabetic characters are converted into spaces. This resulted in an 11,153 word description space. The (20N-ST) version is obtained after a additional stemming step. Words of less than 2 characters, as well as stopwords (S24 SMART list (Salton 1971)), are eliminated. Stemming was performed using Porter's algorithm (Porter 1980). The description space was thus reduced to 5,473 words (Porter 1980). The description space is thus reduced to 5,473 words.
- The WebKB dataset (WKB) contains 8,282 pages collected from the departments of computer science of various universities in January 1997 by the World Wide-Knowledge Base, a project of the CMU text learning group[4] (Carnegie Mellon University, Pittsburgh). The pages were manually divided into 7 classes: student, faculty, departmental, course, personal, project, other. We operate on the Cardoso Cachopo's reduced version, in which the "departmental" and "staff" classes were rejected due to their low number of pages, and the class "other" had been deleted. Cleaning and stemming methods used for the 20 Newsgroups dataset were then applied to the reduced dataset. This resulted in a 4,158 item dataset with a 1,805 word description space.
- The Chirac-Mitterrand dataset (CHM) is a well-known corpus of talks by presidents Chirac and Mitterrand that issued from the DEFT'05 challenge. The corpus was constituted by (Alphonse et al. 2005) and includes 73,255 of J. Chirac's sentences and 12,320 of F. Mitterrand's. It is thus rather ill-balanced. We focused on the most difficult task of the challenge, which consisted of identifying Mitterrand's sentences in bodies having neither years nor names of people. We used a simple "bag of words" model in which all single words and punctuation signs are retained.

---

[2]http://web.ist.utl.pt/~acardoso/datasets/.

[3]http://www.research.att.com/~lewis/reuters21578.html.

[4]http://www.cs.cmu.edu/afs/cs.cmu.edu/project/theo-20/www/data/.

This resulted in a 55,355 word description space. In the specific context of this dataset, we additionally compared our classification results to the best reported results of the DEFT'05 challenge.

4 Experiments and Results

4.1 Experiments

To carry out our experiments, we first took into consideration different classification algorithms that are implemented in the Weka tool box[5]: decision trees (J48) (Quinlan 1993), random forests (RF) (Breiman 2001), KNN (Aha et al. 1991), habitual Bayesian algorithms, i.e. the Multinomial Naïve Bayes (MNB) and Bayesian Network (NE), and finally, the SMO-SVM algorithm (SMO) (Platt 1999). Default parameters were used during the implementation of these algorithms, apart from KNN, for which the number of neighbours was optimized based on the resulting precision. Secondly, we emphasized particularly tests of the efficacy of feature selection approaches, including our proposed feature maximization contrasting method (FMC). In our test, we included a panel of filter-based approaches applicable to large dimension data, using once again the Weka platform. The methods tested include: chi-squared (Ladha and Deepa 2011), information gain (Hall and Smith 1999), CBF (Dash and Liu 2003), symmetric incertitude (Yu and Liu 2003), RLF (Konokenko 1994), PCA (Principal Component Analysis) (Pearson 1901). Default parameters were used for most of these methods, except for PCA where the explained variance percentage is tuned with respect to the resulting accuracy. Initially, we tested the methods separately. In a second phase, we combined the feature selection supplied by the different methods with the FMC method that we have proposed (Eqs. 3–6). We used a 10-fold cross-validation in all our experiments.

4.2 Results

The different results are presented in Tables 1, 2, 3, 4, 5, 6, 7, 8, 9, 10, 11 and 12. They are based on measurements of standard performance (level of true positives [TP] or recall [R], level of false positives [FP], precision [P], F-measure [F] and ROC) weighted by class size, then averaged for all the classes. For each table and each combination of selection and classification methods, an indicator of performance gain or loss (TP Incr) is calculated using the TP of SMO level on the original data as a reference. Finally, as the results of chi-squared, information gain and symmetric

---

[5]http://www.cs.waikato.ac.nz/ml/weka/.

Table 1 Classification results on initial data

	TP(R)	FP	P	F	ROC	TP Incr
J48	0.42	0.16	0.40	0.40	0.63	−23 %
RandomForest	0.45	0.23	0.46	0.38	0.72	−17 %
SMO	**0.54**	**0.14**	**0.53**	**0.52**	**0.80**	**0 % (Ref)**
BN	0.48	0.14	0.47	0.47	0.78	−10 %
MNB	0.53	0.18	0.54	0.47	0.85	−2 %
KNN (k=3)	0.53	0.16	0.53	0.51	0.77	−2 %

Table 2 Results of classification after feature selection (BN classifier)

	TP(R)	FP	P	F	ROC	Nbr. var.	TP Incr
CHI+	0.52	0.17	0.51	0.47	0.80	282	−4 %
CBF	0.47	0.21	0.44	0.41	0.75	37	−13 %
PCA (50 % vr.)	0.47	0.18	0.47	0.44	0.77	483	−13 %
RLF	0.52	0.16	0.53	0.48	0.81	937	−4 %
FMC	**0.99**	**0.003**	**0.99**	**0.99**	**1**	**262/cl**	**+90 %**

incertitude were identical, they only figure once in the tables, as results of the chi-squared type (and are noted CHI+).

For our main patent collection, Table 1 shows that the performances of all classification methods are weak for the dataset considered, if no feature selection process is carried out. In this context, this table confirms the superiority of SMO, KNN and BN, compared to the other two methods, based on decision trees. Additionally, SMO gave the best global performance in terms of discrimination, as demonstrated by the highest ROC value. However, this method is clearly not usable in an operational context of patent evaluation such as QUAERO, because of the major confusion between classes. This shows its intrinsic in-ability to cope with the attraction effect of the largest classes. Every time that a standard feature selection method is applied in our context, in association with the best classification methods, the quality of the results is slightly altered, as indicated in Table 2. Table 2 also underlines the fact that the reduction in the number of features assessed by the FMC method is similar to that of CHI+ (in terms of active features; see Sect. 2 for more details). However, its use stimulates the performance of classification methods, particularly BN (Table 3), leading to impressive classification results in the context of highly complex classification: 0.987 accuracy i.e. only 94 misclassed data with the BN method, for a total of 7,252.

The results presented in Table 4 illustrate more precisely the efficiency of our feature contrasting method that acts on data description (Eq. 6). In experiments relating to this table, the contrast is applied individually to the features extracted by each selection method. Then in a second step, a BN classifier is applied to the contrasted data. The results show that, irrespective of the type of method used for feature

Table 3 Results of classification after FMC selection

	TP(R)	FP	P	F	ROC	TP Incr
J48	0.80	0.05	0.79	0.79	0.92	+48%
RandomForest	0.76	0.09	0.79	0.73	0.96	+40%
SMO	0.92	0.03	0.92	0.91	0.98	+70%
BN	**0.99**	**0.003**	**0.99**	**0.99**	**1**	**+90%**
MNB	0.92	0.03	0.92	0.92	0.99	+71%
KNN (k=3)	0.66	0.14	0.71	0.63	0.85	+22%

Table 4 Results of classification with different feature selection methods coupled with our additional feature contrasting method (BN classifier)

	TP(R)	FP	P	F	ROC	Nbr. var.	TP Incr
CHI+	0.79	0.08	0.82	0.78	0.98	282	+46%
CBF	0.63	0.15	0.69	0.59	0.90	37	+16%
PCA (50% vr.)	0.71	0.11	0.73	0.67	0.53	483	+31%
RLF	0.79	0.08	0.81	0.78	0.98	937	+46%
FMC	**0.99**	**0.003**	**0.99**	**0.99**	**1**	**262/cl**	**+90%**

selection, the performances of the resulting classification are re-enforced each time that our feature contrasting is applied downstream of the selection. The average performance increase is 44%. Finally, Table 5 illustrates the ability of the FMC approach to confront efficiently the problems of imbalance and class similitude. The examination of TP level variations (especially in the small classes) seen in this table shows that the attraction effect of data from the largest classes, produced at a high level in the case of the use of original data, is practically systematically overcome each time the FMC approach is exploited. The ability of this approach to correct class imbalance is equally clearly demonstrated by the homogeneous distribution of active features in the different classes, despite the extremely heterogeneous class size (Fig. 4).

The summary of the results of the 5 complementary datasets is presented in Tables 6, 7, 8, 9, 10 and 11. These tables highlight the fact that the FMC method can significantly improve the performance of the classifiers in different types of cases. As in the context of our previous experience (patents), the best performances were obtained with the use of the FMC method in combination with the MNB and BN Bayesian classifiers. Table 7 presents the comparative results of such a combination. It demonstrates that the FMC method is particularly effective in increasing the performance of the classifiers when the complexity of the classification task rises because of an increasing number of classes (AMZ corpus). Tables 8 and 9 supplies general information about the data and behaviour upon use of the FMC selection method. These parameters illustrate a significant reduction in classification complexity obtained with FMC, because of the drop in the number of features to manage, as

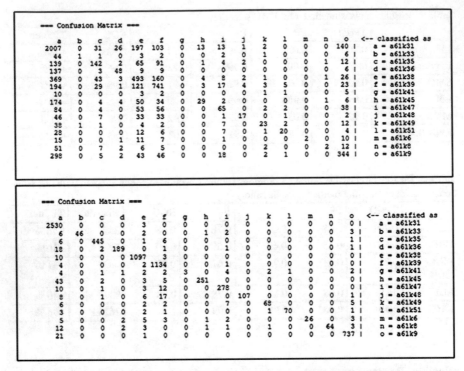

Fig. 4 Confusion matrix of the optimal results before and after feature selection on the PAT-QUAERO dataset (SMO classification)

well as a concomitant decrease in badly classed data. Table 9 also stresses the calculation time, which is highly curbed for this method (the calculation is carried out on Linux using a laptop computer equipped with an Intel® Pentium® B970 2.3 Ghz processor and 8 GB of memory).

For these datasets, similar remarks to those mentioned for the patent dataset can be made on the subject of low efficiency of common feature selection methods and re-sampling methods. Tables 8 and 9 also show that the value of the contrast magnification factor (Eq. 6) used to obtain the best performances can vary throughout the experiments (from 1 to 4 in this last context). However, it can be seen that if a fixed value is taken for this factor, for example the highest (here 4), the results are not down-graded. This choice thus represents a good alternative to confront the problem of configuration.

The 5 most contrasted features (lemmes) of the 8 classes issued from the R8 corpus are shown in Table 6. The main lines of the themes covered by the classes can be clearly demonstrated in this way. This illustrates the extraction capacities of subjects by the FMC method. Finally, the acquisition of very good performances by combining the FMC selection approach with a classification method such as MNB

Table 5 Characteristics of classes before and after FMC selection (BN classifier)

Class label	Size	Feat. Select.	% TP FMC	% TP before
a61k31	2, 533	223	**1**	0.79
a61k33	60	276	**0.95**	0.02
a61k35	459	262	**0.99**	0.31
a61k36	212	278	**0.95**	0.23
a61k38	1, 110	237	**1**	0.44
a61k39	1, 141	240	**0.99**	0.65
a61k41	22	225	**0.24**	0
a61k45	304	275	**0.98**	0.09
a61k47	304	278	**0.99**	0.21
a61k48	140	265	**0.98**	0.12
a61k49	90	302	**0.93**	0.26
a61k51	78	251	**0.98**	0.26
a61k6	47	270	**0.82**	0.04
a61k8	87	292	**0.98**	0.02
a61k9	759	250	**1**	0.45

Table 6 List of high contrast features (lemmes) for the 8 classes of the R8 corpus

Trade	Grain	Ship	Acq
6.35 tariff	5.60 agricultur	6.59 ship	5.11 common
5.49 trade	5.44 farmer	6.51 strike	4.97 complet
5.04 practic	5.33 winter	6.41 worker	4.83 file
4.86 impos	5.15 certif	5.79 handl	4.65 subject
4.78 sanction	4.99 land	5.16 flag	4.61 tender
Learn	Money-fx	Interest	Crude
7.57 net	6.13 currenc	5.95 rate	6.99 oil
7.24 loss	5.55 dollar	5.85 prime	5.20 ceil
6.78 profit	5.52 germani	5.12 point	4.94 post
6.19 prior	5.49 shortag	5.10 percentag	4.86 quota
5.97 split	5.16 stabil	4.95 surpris	4.83 crude

is a real advantage for large-scale use, given that the MNB method has incremental abilities and that the two methods have low calculation times.

Concerning the specific case of the CHM corpus, the best results were obtained by the combination of our FMC method with a BN classifier. This approach led to an accuracy value of 99.999 %. As shown in the confusion matrix presented in Table 10, we only observed 12 errors, as compared to approx. 16,850 for the best antecedent approach (El-Bèze et al. 2005), which additionally exploits a complex linguistic model. Furthermore, unlike previous approaches, errors are not bilateral:

Table 7 Results of classification after FMC selection (MNB or BN classifiers)

		TP(R)	FP	P	F	ROC	TP Incr.
R8	-	0.937	0.02	0.942	0.938	0.984	
	FMC	**0.998**	**0.001**	**0.998**	**0.998**	**1**	+6 %
R52	-	0.91	0.01	0.909	0.903	0.985	
	FMC	**0.99**	**0.001**	**0.99**	**0.99**	**0.999**	+10 %
AMZ	-	0.748	0.05	0.782	0.748	0.981	
	FMC	**0.998**	**0.001**	**0.998**	**0.998**	**1**	+33 %
20N-AT	-	0.882	0.006	0.884	0.881	0.988	
	FMC	**0.992**	**0**	**0.992**	**0.1**	**1**	+13 %
20N-ST	-	0.865	0.007	0.866	0.864	0.987	
	FMC	**0.991**	**0.001**	**0.991**	**1**	**1**	+15 %
WKB	-	0.842	0.068	0.841	0.841	0.946	
	FMC	**0.996**	**0.002**	**0.996**	**0.996**	**0.996**	+18 %

Table 8 Dataset information and complementary results after FMC selection (5 reference datasets and MNB or BN classifiers)

	R8	R52	AMZ	20N-AT	20N-ST	WKB
Nb. class	8	52	50	20	20	4
Nb. data	7,674	9,100	1,500	18,820	18,820	4,158
Nb. feat.	3,497	7,369	10,000	11,153	5,473	1,805
Nb. sel. feat.	1,186	2,617	3,318	3,768	4,372	725
Act. feat./class (av.)	268.5	156.05	761.32	616.15	525.95	261
Magnification factor	4	2	1	4	4	4
Misclassed (Std)	373	816	378	2,230	2,544	660
Misclassed (FMC)	**19**	**91**	**3**	**157**	**184**	**17**
Comp. time (s)	1	3	1.6	10.2	4.6	0.8

Mitterrand was confused 12 times with Chirac, but Chirac was never confused with Mitterrand. To achieve these results we did not apply any linguistic processing: there was not even any stemming operation and "empty words" were preserved and proved useful for analysis. Table 11 highlights words with the strongest contrast in the talks of each protagonist and in such a way shows that our approach can be additionally exploited to accurately characterize speakers' profiles. In the case of Mitterrand, many types of linguistic features already highlighted by Habert et al. (2000) obtain high FMC values, but the dominant features panel is much more comprehensive with our approach. Mitterrand talks seem marked by humanistic connotations, as illustrated by the high contrast sequence "gens, assez, capables, penser (people, enough, capable, thinking)" and contain a lot of empty words in relation with the interrogative mode and philosophical discourse. The contrast of dominant features

Table 9 Dataset information an complementary results after FMC selection (Chirac-Mitterrand (CHM) dataset and BN classifier)

	CHM
Nb. class	2
Nb. data	84, 575
Nb. feat.	55, 355
Nb. sel. feat.	5, 321
Act. feat./class (av.)	2, 900
Magnification factor	4
Misclassed (Reference: (El-Bèze et al. 2005))	16, 650
Misclassed (FMC)	**12**
Comp. time (s)	10

Table 10 Confusion matrix for the Chirac-Mitterrand (CHM) dataset (BN classifier)

a	b	
73,255	0	a = Chirac
12	11,308	b = Mitterrand

Table 11 10 most contrasted features in the talks of Mitterrand and Chirac

Mitterrand		Chirac	
Contrast	Feature	Contrast	Feature
1.88	douze	1.93	Partenariat
1.85	est-ce	1.86	Dynamisme
1.80	eh	1.81	Exigence
1.79	quoi	1.78	Compatriotes
1.78	-	1.77	Vision
1.76	gens	1.77	Honneur
1.75	assez	1.76	Asie
1.74	capables	1.76	Efficacité
1.72	penser	1.75	Saluer
1.70	bref	1.74	Soutien

is much more pronounced in the case of Chirac and these features represent mostly nouns, demonstrating a more clearly established language, based on stable values.

Interestingly, complementary results obtained with the numerical UCI Wine dataset show that, with the help of FMC, NB and BN methods can only exploit two features (among 13) for classification as a decision tree classifier like J48 (i.e. C4.5 (Quinlan 1993)) would do on standard data. The difference is that a perfect result is obtained with NB or BN and FMC, whereas this is not the case with J48 (Table 12). Some explanations are provided by looking at the distribution of the class samples on the alternative decision plans of the two methods. In the "Proline-Colour Intensity" decision plan exploited by J48, the different classes cannot clearly be

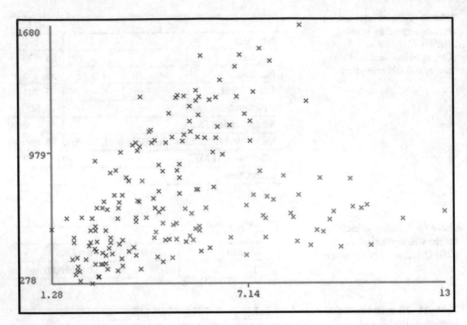

Fig. 5 WINE dataset: "Proline-Color intensity" decision plan generated by J48—Proline is on Y axis on this and next figures

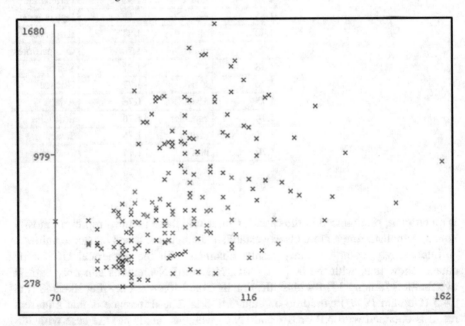

Fig. 6 WINE dataset: "Proline-Magnesium" decision plan generated by FMC (before data contrasting)

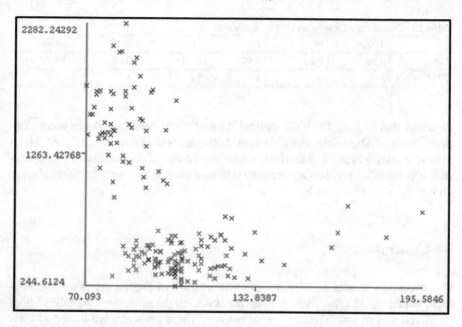

Fig. 7 WINE dataset: "Proline-Magnesium" decision plan generated by FMC (after data contrasting with a magnification factor k = 1)

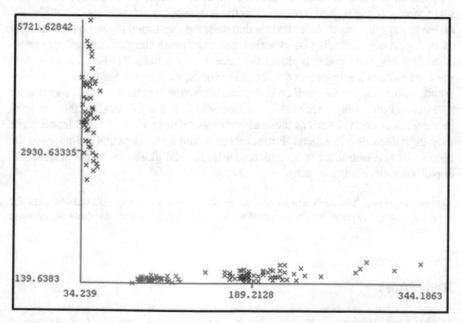

Fig. 8 WINE dataset: "Proline-Magnesium" decision plan generated by FMC (after data contrasting with a magnification factor k = 4)

Table 12 Classification results on UCI Wine dataset

	TP R	FP	P	F	ROC	TP Incr
J48	0.94	0.04	0.94	0.94	0.95	0 % (Ref)
FMC + BN	1	0	1	1	1	**+6 %**

discriminated (Fig. 5). The FMC method "apparently" generates an even more complex "Proline-Magnesium" decision plan if contrast is not considered (Fig. 6). However, as shown in Figs. 7 and 8, with the combined effect of contrast and high magnification factor ($k = 4$) on data features, the different classes can easily be discriminated on that decision plan (Fig. 8).

5 Conclusion

Our main aim was to develop an efficient method of feature selection and contrast, which would allow routine problems linked to the supervised classification of large volumes of textual data to be overcome. These problems are linked to class imbalance, with a high degree of similarity between them, as they house highly multidimensional and noisy data. To achieve our aim, we adapted a recently developed metric in the unsupervised framework to the context of supervised classification. By different experiments on large textual datasets, we illustrated numerous advantages of our approach, including its effectiveness to improve the performance of classifiers. Notably, this approach places the accent on the most flexible classifiers and the least demanding in terms of calculation times, such as the Bayesian classifiers. Another advantage of this method is that it concerns an approach without parameters that depend on a simple variable extraction schema. The method can thus be used in numerous contexts, such as those of incremental or semi-supervised learning, or even digital learning in general. Further work would be to adapt this technique to the domain of text examination, to enrich ontologies and glossaries by the large-scale exploitation of existing corpora.

Acknowledgments This work was carried out in the context of the QUAERO program (http://www.quaero.org) supported by OSEO (http://www.oseo.fr/), Agence française de développement de la recherche.

References

Aha, D., Kibler, D., & Albert, M. (1991). Instance-based learning algorithms. *Machine Learning*, 6, 37–66.

Alphonse, E. E., et al. (2005). Préparation des donnés et analyse des résultats de DEFT'05. In *TALN 2005 - Atelier DEFT 2005* (pp. 99–111).

Bache, K., & Lichman, M. (2013). Uci machine learning repository. University of California, School of Information and Computer Science, Irvine, CA, USA. http://archive.ics.uci.edu/ml.

Bolón-Canedo, V., Sánchez-Maroño, N., & Alonso-Betanzos, A. (2012). A review of feature selection methods on synthetic data. *Knowledge and Information Systems*, 1–37.

Breiman, L. (2001). Random forests. *Machine Learning*, 45(1), 5–32.

Breiman, L., Friedman, J., Olshen, R., & Stone, C. (1984). *Classification and regression trees*. Technical report. Wadsworth International Group, Belmont, CA, USA.

Chawla, N., Bowyer, K., Hall, L., & Kegelmeyer, W. (2002). Synthetic minority oversampling technique. *Journal of Artificial Intelligence Research*, 16, 321–357.

Dash, M., & Liu, H. (2003). Consistency-based search in feature selection. *Artificial Intelligence*, 151(1), 155–176.

Daviet, H. (2009). *Class-Add, une procédure de sélection de variables basée sur une troncature k-additive de l'information mutuelle et sur une classification ascendante hiérarchique en pré-traitement*. Thèse de doctorat, Université de Nantes.

El-Bèze, M., Torres-Moreno, J.-M., & Béchet, F. (2005). Peut-on rendre automatiquement à César ce qui lui appartient. Application au jeu du Chirand-Mitterrac. In *TALN 2005 - Atelier DEFT 2005* (pp. 125–134).

Forman, G. (2003). An extensive empirical study of feature selection metrics for text classification. *Journal of Machine Learning Research*, 3, 1289–1305.

Good, P. (2006). *Resampling methods*. Ed. Birkhauser.

Guyon, I., & Elisseeff, A. (2003). An introduction to variable and feature selection. *Journal of Machine Learning Research*, 3, 1157–1182.

Guyon, I., Weston, J., Barnhill, S., & Vapnik, V. (2002). Gene selection for cancer classification using support vector machines. *Machine Learning*, 46(1), 389–422.

Habert, B., et al. (2000). Profilage de textes: cadre de travail et expérience. In *Proceedings of JADT'2000 (5ièmes journées internationales d'Analyse Statistique des Données Textuelles)*.

Hajlaoui, K., Cuxac, P., Lamirel, J.-C., & Francois, C. (2012). Enhancing patent expertise through automatic matching with scientific papers. In J.-G. Ganascia, P. Lenca & J.-M. Petit (Eds.), *Discovery science*. (Vol. 7569, pp. 299–312), Lecture notes in computer science. Berlin Heidelberg: Springer.

Hall, M., & Smith, L. (1999). Feature selection for machine learning: Comparing a correlation-based filter approach to the wrapper. In *Proceedings of the Twelfth International Florida Artificial Intelligence Research Society Conference* (pp. 235–239).

Kira, K., & Rendell, L. (1995). The feature selection problem: Traditional methods and a new algorithm. In *Proceedings of the Tenth National Conference on Artificial Intelligence* (pp. 129–134).

Kohavi, R., & John, G. (1997). Wrappers for feature subset selection. *Artificial Intelligence*, 97(1–2), 273–324.

Konokenko, I. (1994). Estimating attributes: Analysis and extensions of relief. In *Proceedings of European Conference on Machine Learning* (pp. 171–182).

Ladha, L., & Deepa, T. (2011). Feature selection methods and algorithms. *International Journal on Computer Science and Engineering*, 3(5), 1787–1797.

Lallich, S., & Rakotomalala, R. (2000). Fast feature selection using partial correlation for multi-valued attributes. In D. A. Zighed, J. Komorowski & J. Żytkow (Eds.), *Principles of data mining and knowledge discovery* (Vol. 1910, pp. 221–231), Lecture notes in computer science. Berlin Heidelberg: Springer.

Lamirel, J., Al Shehabi, S., François, C., & Hoffmann, M. (2004). New classification quality estimators for analysis of documentary information: Application to patent analysis and web mapping. *Scientometrics*, 60(3), 445–562.

Lamirel, J., Ghribi, M., & Cuxac, P. (2010). Unsupervised recall and precision measures: a step towards new efficient clustering quality indexes. In *Proceedings of the 19th International Conference on Computational Statistics (COMPSTAT'2010, Paris, France)*.

Lamirel, J., Cuxac, P., Chivukula, A.S., & Hajlaoui, K. (2014). Optimizing text classification through efficient feature selection based on quality metric. *Journal of Intelligent Information Systems, Special issue on PAKDD-QIMIE 2013* (pp. 1–18).

Lamirel, J., & Ta, A. (2008). Combination of hyperbolic visualization and graph-based approach for organizing data analysis results: An application to social network analysis. In *Proceedings of the 4th International Conference on Webometrics, Informetrics and Scientometrics and 9th COLLNET Meetings, Berlin, Germany*.

Lang, K. (1995). Learning to filter netnews. In *Proceedings of the Twelfth International Conference on Machine Learning* (pp. 331–339).

Pearson, K. (1901). On lines an planes of closetst fit to systems of points in space. *Philosophical Magazine, 2*(11), 559–572.

Platt, J. (1999). Fast training of support vector machines using sequential minimal optimization. In: *Advances in kernel methods* (pp. 185–208). Cambridge, MA, USA: MIT Press.

Porter, M. (1980). An algorithm for suffix stripping. *Program, 14*(3), 130–137.

Quinlan, J. R. (1993). *C4.5: programs for machine learning*. San Francisco, CA, USA: Morgan Kaufmann Publishers Inc.

Salton, G. (1971). *Automatic processing of foreign language documents*. Englewood Clifs, NJ, USA: Prentice-Hill.

Schmid, H. (1994). Probabilistic part-of-speech tagging using decision trees. In *Proceedings of International Conference on New Methods in Language Processing*.

Witten, I., & Frank, E. (2005). *Data mining: Practical machine learning tools and techniques*. San Francisco: Morgan Kaufmann.

Yu, L., & Liu, H. (2003). Feature selection for high-dimensional data: A fast correlation-based filter solution. In *Proceedings of ICML 03, Washington DC, USA* (pp. 856–863).

Ultrametricity of Dissimilarity Spaces and Its Significance for Data Mining

Dan A. Simovici, Rosanne Vetro and Kaixun Hua

Abstract We introduce a measure of ultrametricity for dissimilarity spaces and examine transformations of dissimilarities that impact this measure. Then, we study the influence of ultrametricity on the behavior of two classes of data mining algorithms (kNN classification and PAM clustering) applied on dissimilarity spaces. We show that there is an inverse variation between ultrametricity and performance of classifiers. For clustering, increased ultrametricity generate clusterings with better separation. Lowering ultrametricity produces more compact clusters.

1 Introduction

Ultrametrics are dissimilarities that satisfy a stronger version of the triangular inequality (usually associated with metrics) and they occur in many data mining applications such as agglomerative hierarchical clustering algorithms (Leclerc 1985; Gordon 1981, 1987; Contreras and Murtagh 2012; Jardine and Sibson 1971; Diatta and Fichet 1998; Bertrand and Janowitz 2002; Barthélemy and Brucker 2008; Brucker 2006; Barthélemy et al. 2004), and have applications in the study of phylogenetic trees in biology (Ninio 1983; Kimura 1983; Di Summa et al. 2015), p-adic numbers in mathematics (Schikhof 1984; Amice 1975), and certain physical systems (Rammal et al. 1986), etc. Our goal is to evaluate the degree of ultrametricity of dissimilarity spaces and to study the impact of the degree of ultrametricity on performance of classification and clustering algorithms.

Measuring ultrametricity of metric spaces has preoccupied a number of researchers (for example, in Rammal et al. (1985)); however, the proposed measures are usable for the special case of metrics and are linked to the subdominant ultrametric attached

D.A. Simovici (✉) · R. Vetro · K. Hua
University of Massachusetts Boston, Boston, USA
e-mail: dsim@cs.umb.edu

R. Vetro
e-mail: rvetro@cs.umb.edu

K. Hua
e-mail: kingsley@cs.umb.edu

© Springer International Publishing Switzerland 2017
F. Guillet et al. (eds.), *Advances in Knowledge Discovery and Management*,
Studies in Computational Intelligence 665, DOI 10.1007/978-3-319-45763-5_8

to a metric which requires computing a single-link clustering or a minimal spanning tree. We propose an alternative measure referred to as the weak ultrametricity that can be applied to the more general case of dissimilarity spaces.

The set of reals is denoted by $\mathbb{R}$; the set of non-negative reals is denoted by $\mathbb{R}_{\geqslant 0}$.

A *dissimilarity space* is a pair (S, d), where S is a set and $d : S \times S \longrightarrow \mathbb{R}$ is a function such that $d(x, y) \geqslant 0$, $d(x, x) = 0$, and $d(x, y) = d(y, x)$ for $x, y \in S$. We assume that all dissimilarity spaces considered are finite. The set of dissimilarities defined on a set S is denoted by $\mathsf{DISS}(S)$.

A *triangle* in (S, d) is a triple $(x, y, z) \in S^3$. To simplify the notation, we denote $t = (x, y, z)$ by xyz.

The mapping d is a *quasi-metric* if it is a dissimilarity and it satisfies the triangular inequality $d(x, y) \leqslant d(x, z) + d(z, y)$ for $x, y, z \in S$. In addition, if $d(x, y) = 0$ implies $x = y$, then d is a *metric*.

An *quasi-ultrametric* is a dissimilarity $d : S \times S \longrightarrow \mathbb{R}_{\geqslant 0}$ that satisfies the inequality $d(x, y) \leqslant \max\{d(x, z), d(z, y)\}$ for $x, y, z \in S$. If, in addition, $d(x, y) = 0$ implies $x = y$, then d is an *ultrametric*.

In Sect. 2 we introduce a measure of ultrametricity for dissimilarity spaces. A weaker variant of this measure that is less influenced by outliers and therefore is better from a computational point of view is discussed in Sect. 3. Then, we examine transformations of dissimilarities that affect ultrametricity. The influence of ultrametricity of dissimilarities on the performance of classifiers is examined in Sect. 2 using the k-nearest neighbors classifiers. Section 5 is dedicated to the study of the influence of ultrametricity on cluster compactness and separation.

2 Ultrametricity of Dissimilarities

Let r be a non-negative number and let $\mathscr{D}_r(S)$ be the set of dissimilarities defined on a set S that satisfy the inequality $d(x, y)^r \leqslant d(x, z)^r + d(z, y)^r$ for $x, y, z \in S$. Note that every dissimilarity belongs to the set $\mathscr{D}_0$; a dissimilarity in $\mathscr{D}_1$ is a quasi-metric.

Theorem 2.1 *Let (S, d) be a dissimilarity space and let $\mathscr{D}_\infty(S) = \bigcap_{r \geqslant 0} \mathscr{D}_r(S)$. If $d \in \mathscr{D}_\infty(S)$, then d is an ultrametric on S.*

Proof Let $d \in \mathscr{D}_\infty$ and let $t = xyz$ be a triangle in the dissimilarity space (S, d). Assume that $d(x, y) \geqslant d(x, z) \geqslant d(z, y)$.

Suppose intially that $d(x, z) = d(y, z)$. Then, $d \in \mathscr{D}_r(S)$ implies that $d(x, y)^r \leqslant 2d(x, z)^r$, so

$$\left(\frac{d(x, y)}{d(x, z)} \right)^r \leqslant 2$$

for every $r \geqslant 0$. By taking $r \to \infty$ it is clear that this is possible only if $d(x, y) \leqslant d(x, z)$, which implies $d(x, y) = d(x, z) = d(y, z)$; in other words, t is an equilateral triangle.

The alternative supposition is that $d(x, z) > d(y, z)$. Again, since $d \in \mathscr{D}_r(S)$, it follows that

$$d(x, y) \leqslant \left(d(x, z)^r + d(z, y)^r\right)^{\frac{1}{r}}$$

$$= d(x, z)\left(1 + \left(\frac{d(z, y)}{d(x, z)}\right)^r\right)^{\frac{1}{r}}$$

for every $r > 0$. Since $\lim_{r \to \infty} d(x, z)\left(1 + \left(\frac{d(y,z)}{d(x,z)}\right)^r\right)^{\frac{1}{r}} = d(x, z)$, it follows that $d(x, y) \leqslant d(x, z)$ for $x, y, z \in S$. This inequality implies $d(x, y) = d(x, z)$, so the largest two sides of the triangle xyz are equal. This allows us to conclude that d is an ultrametric. □

It is easy to verify that if r and s are positive numbers, then $r \leqslant s$ implies $(d(x, z)^r + d(z, y)^r)^{\frac{1}{r}} \geqslant (d(x, z)^s + d(z, y)^s)^{\frac{1}{s}}$ (see Simovici and Djeraba (2014), Lemma 6.15). Thus, if $r \leqslant s$ we have the inclusion $\mathscr{D}_s \subseteq \mathscr{D}_r$.

Let d and d' be two dissimilarities defined on a set S. We say that d' dominates d if $d(x, y) \leqslant d'(x, y)$ for every $x, y \in S$. The pair $(\mathsf{DISS}(S), \leqslant)$ is a partially ordered set.

Let r, s be two positive numbers such that $r < s$, and let $d \in \mathscr{D}_r(S)$. The family $\mathscr{D}_{s,d}(S)$ of s-dissimilarities on S that are dominated by d has a largest element.

Indeed, since every element of $\mathscr{D}_{s,d}(S)$ is dominated by d, we can define the mapping $\tilde{e} : S \times S \longrightarrow \mathbb{R}_{\geqslant 0}$ as $\tilde{e}(x, y) = \sup\{e(x, y) \mid e \in \mathscr{D}_{s,d}(S)\}$. It is immediate that e is a dissimilarity on S and that $\tilde{e} \leqslant d$. Moreover, we have $e(x, y)^s \leqslant e(x, z)^s + e(z, y)^s \leqslant \tilde{e}(x, z)^s + \tilde{e}(z, y)^s$ for every $x, y, z \in S$, which implies

$$\tilde{e}(x, y)^s \leqslant \tilde{e}(x, z)^s + < \tilde{e}(z, y)^s.$$

Thus, $\tilde{e} \in \mathscr{D}_{s,d}(S)$, which justified our claim.

For $r > 0$ define the function $F_r : \mathbb{R}^2_{\geqslant 0} \longrightarrow \mathbb{R}_{\geqslant 0}$ as $F_r(a, b) = (a^r + b^r)^{\frac{1}{r}}$. It is straightforward to see that $p \geqslant q$ implies $F_p(a, b) \leqslant F_q(a, b)$ for $a, b \in \mathbb{R}_{\geqslant 0}$. Furthermore for $r > 0$ we have $d \in \mathscr{D}_r(S)$ if and only if $d(x, y) \leqslant F_r(d(x, z), d(z, y))$.

Definition 2.2 Let r, s be two positive numbers. An (r, s)-transformation is a function $g : \mathbb{R}_{\geqslant 0} \longrightarrow \mathbb{R}_{\geqslant 0}$ such that

(i) $g(x) = 0$ if and only if $x = 0$;
(ii) g is continuous and strictly monotonic on $\mathbb{R}_{\geqslant 0}$;
(iii) $g(F_r(a, b)) \leqslant F_s(g(a), g(b))$ for $a, b \in \mathbb{R}_{\geqslant 0}$. □

Note that if $d \in \mathscr{D}_r(S)$ and g is an (r, s)-transformation, then $gd \in \mathscr{D}_s(S)$.

3   A Weaker Dissimilarity Measure

The notion of weak ultrametricity that we are about to introduce has some computational advantages over the notion of ultrametricity, especially from the point of view of handling transformations of metrics.

Let (S, d) be a dissimilarity space and let $t = xyz$ be a triangle. Following Lerman's notation (Lerman 1981), we write $S_d(t) = d(x, y)$, $M_d(t) = d(x, z)$, and $L_d(t) = d(y, z)$, if $d(x, y) \geqslant d(x, z) \geqslant d(y, z)$.

Definition 3.1   Let (S, d) be a dissimilarity space and let $t = xyz \in S^3$ be a triangle. The *ultrametricity* of t is the number $u_d(t)$ defined by

$$u_d(t) = \max\{r > 0 \mid S_d(t)^r \leqslant M_d(t)^r + L_d(t)^r\},$$

which is the ultrametricity of the subspace $(\{x, y, z\}, d)$ of (S, d). If $d \in \mathscr{D}_p$, we have $p \leqslant u_d(t)$ for every $t \in S^3$.

The *weak ultrametricity* of the triangle t, $w_d(t)$, is given by

$$w_d(t) = \begin{cases} \dfrac{1}{\log_2 \frac{S_d(t)}{M_d(t)}} & \text{if } S_d(t) > M_d(t) \\ \infty & \text{if } S_d(t) = M_d(t). \end{cases}$$

If $w_d(t) = \infty$, then t is an *ultrametric triple*.

The *weak ultrametricity* of the dissimilarity space (S, d) is the number $w(S, d)$ defined by

$$w(S, d) = \mathtt{median}\{w_d(t) \mid t \in S^3\}. \qquad \square$$

The definition of $w(S, d)$ eliminates the influence of triangles whose ultrametricity is an outlier, and gives a better picture of the global ultrametric property of (S, d).

For a triangle t we have

$$0 \leqslant S_d(t) - M_d(t) = \left(2^{\frac{1}{w_d(t)}} - 1\right) M_d(t) \leqslant \left(2^{\frac{1}{w(S,d)}} - 1\right) M_d(t)$$

Thus, if $w_d(t)$ is sufficiently large, the triangle t is almost isosceles. For example, if $w_d(t) = 5$, the difference between the length of longest side $S_d(t)$ and the median side $M_d(t)$ is less than 15 %.

For every triangle $t \in S^3$ in a dissimilarity space we have $u_d(t) \leqslant w_d(t)$. Indeed, since $S_d(t)^{u_d(t)} \leqslant M_d(t)^{u_d(t)} + L_d(t)^{u_d(t)}$ we have $S_d(t)^{u_d(t)} \leqslant 2M_d(t)^{u_d(t)}$, which is equivalent to $u_d(t) \leqslant w_d(t)$.

Next we discuss dissimilarity transformations that impact the ultrametricity of dissimilarities.

Theorem 3.2   *Let (S, d) be a dissimilarity space and let $f : \mathbb{R}_{\geqslant 0} \longrightarrow \mathbb{R}_{\geqslant 0}$ be a function that satisfies the following conditions:*

(i) $f(0) = 0$;

(ii) f is increasing;

(iii) the function $g : \mathbb{R}_{\geq 0} \longrightarrow \mathbb{R}_{\geq 0}$ given by

$$g(a) = \begin{cases} \frac{f(a)}{a} & \text{if } a > 0, \\ 0 & \text{if } a = 0 \end{cases}$$

is decreasing.

Then the function $e : S \times S \longrightarrow \mathbb{R}_{\geq 0}$ defined by $e(x, y) = f(d(x, y))$ for $x, y \in S$ is a dissimilarity and $w_d(t) \leq w_e(t)$ for every triangle $t \in S^3$.

Proof Let $t = xyz \in S^3$ be a triangle. It is immediate that $e(x, y) = e(y, x)$ and $e(x, x) = 0$.

Since f is an increasing function we have $f(S_d(t)) \geq f(M_d(t)) \geq f(L_d(t))$, so the ordering of the sides of the tranformed triangle is preserved.

Since g is a decreasing function, we have $g(S_d(t)) \leq g(M_d(t))$, that is, $\frac{f(S_d(t))}{S_d(t)} \leq \frac{f(M_d(t))}{M_d(t)}$, or

$$\frac{S_d(t)}{M_d(t)} \geq \frac{f(S_d(t))}{f(M_d(t))}.$$

Therefore,

$$w_d(t) = \frac{1}{\log_2 \frac{S_d(t)}{M_d(t)}} \leq \frac{1}{\log_2 \frac{S_e(t)}{M_e(t)}} = w_e(t).$$

Example 3.3 Let (S, d) be a dissimilarity space and let e be the dissimilarity defined by $e(x, y) = d(x, y)^r$, where $0 < r < 1$. If $f(a) = a^r$, then f is increasing and $f(0) = 0$. Furthermore the function $g : \mathbb{R}_{\geq 0} \longrightarrow \mathbb{R}_{\geq 0}$ given by

$$g(a) = \begin{cases} \frac{f(a)}{a} & \text{if } a > 0, \\ 0 & \text{if } a = 0 \end{cases} = \begin{cases} a^{r-1} & \text{if } a > 0, \\ 0 & \text{if } a = 0 \end{cases}$$

is decreasing. Therefore, we have $w_e(t) \geq w_d(t)$. $\quad\square$

Example 3.4 Let $f : \mathbb{R}_{\geq 0} \longrightarrow \mathbb{R}_{\geq 0}$ be defined by $f(a) = \frac{a}{a+1}$. It is easy to see that f is increasing on $\mathbb{R}_{\geq 0}$, $f(0) = 0$, and

$$g(a) = \begin{cases} \frac{1}{1+a} & \text{if } a > 0, \\ 0 & \text{if } a = 0 \end{cases}$$

is decreasing on the same set. Therefore, the weak ultrametricity of a triangle increases when d is replaced by e given by

$$e(x, y) = \frac{d(x, y)}{1 + d(x, y)}$$

for $x, y \in S$.   □

Example 3.5 For a dissimilarity space (S, d), the Schoenberg transform of d described in Deza and Laurent (1997) is the dissimilarity $e : S^2 \longrightarrow \mathbb{R}_{\geqslant 0}$ defined by

$$e(x, y) = 1 - e^{-kd(x,y)}$$

for $x, y \in S$. Let $f : \mathbb{R}_{\geqslant 0} \longrightarrow \mathbb{R}_{\geqslant}$ be the function $f(a) = 1 - e^{-ka}$ that is used in this transformation. It is immediate that f is a increasing function and $f(0) = 0$. For $a > 0$ we have $g(a) = \frac{1 - e^{-ka}}{a}$, which allows us to write

$$g'(a) = \frac{e^{-ka}(ka + 1) - 1}{a^2}$$

for $a > 0$. Taking into account the obvious inequality $ka + 1 < e^{ka}$ for $k > 0$, it follows that the function g is decreasing. Thus, the weak ultrametricity of a triangle relative to the Schoenberg transform is greater than the weak ultrametricity under the original dissimilarity.   □

4 Classification and Ultrametricity

The k-nearest neighbors algorithm (kNN) is a classification method that is memory-based and does not require a model to fit. The classification is decided according to a simple majority decision among the most similar training set samples.

We show that the performance of kNN applied to a dissimilarity space (S, d) degrades with the increase of the ultrametricity of d. This happens because the increase of ultrametricity among the elements of S promotes the equalization of distances.

We begin with a dissimilarity space (S, d) and we obtain a new dissimilarity $d' = f(d)$, where f is one of the transformations examined in Sect. 2. Algorithm 9 encapsulates the above process. It runs kNN with t-fold cross-validation and computes the confusion matrix generated for each fold as well as the cumulative classification error of the transformed space.

We limit the precision of the transformed dissimilarity d' taking into account, as observed in Murtagh et al. (2008) that ultrametricity can decrease with the increase in precision. Limiting the precision of d' to a few decimal digits promotes the equalization of those distances. We used in our experiments the data sets *Iris* and *ionosphere* available from https://archive.ics.uci.edu/ml/datasets/ and data set *ovarian cancer* obtained from the FDA-NCI Clinical Proteomics Program Databank (https://home. ccr.cancer.gov/ncifdaproteomics).

Algorithm 9: Runs kNN with transformed distance function

Input: A metric or dissimilarity space $S = (M, d)$, the number of nearest neighbors k, the number of folds t and a function f, such that $f(d) = d'$ and $u <= u'$ where u and u' are the ultrametricities of S and $S' = (M, d')$, respectively.

Output: The cumulative classification error of the transformed space S'

1 $d' \leftarrow f(d)$, limited to some decimal precision
2 *partition M in t subsamples*
3 **for** *i=1* **to** *t* **do**
4 | *training = partition(i).training*
5 | *test = partition(i).test*
6 | *testSize(i) = size(test)*
7 | kNN(*training, test, k, d'*)
8 | *err(i) = # misclassified objects*
9 **return** *cerr = sum(err)/sum(testsSize)*

Table 1 Average of 10 computations of the classification error produced by kNN using stratified t-fold cross-validation, for different values of k and t = 10

Diss.	Iris			Ionosphere			Ovarian cancer		
	$k = 3$	$k = 5$	$k = 7$	$k = 3$	$k = 5$	$k = 7$	$k = 3$	$k = 5$	$k = 7$
d	0.1033	0.0467	0.0427	0.3860	0.3701	0.3852	0.1403	0.1394	0.1431
$d^{0.1}$	0.1187	0.0753	0.0567	0.3875	0.4097	0.3897	0.1454	0.1431	0.1477
$d^{0.01}$	0.2700	0.2900	0.3000	0.5211	0.5239	0.5365	0.3574	0.3181	0.3000

Our experiments considered a initial Euclidean space (S, d) where S corresponds to one of the data sets described above and d to the Euclidean distance. We first tested our method on the original space and compared the results to the results generated by the increase of ultrametricity of dissimilarity $d' = f(d)$, where $f(a) = a^r$ for $a \geqslant 0$. We used kNN with both t-fold cross-validation and with stratified t-fold cross-validation (where each fold has roughly equal size and roughly the same class proportions as in the entire data set). The transformed distances were limited to 2 decimal digit precision.

The classification error obtained is consistently higher for the case of the transformed space (S, d'), in both validation scenarios. In Table 1 we show the results for three values of k (the number of neighbors) in stratified 10-fold validation. Similar results are obtained for 5 folds in both validation scenarios.

5 The Impact of Ultrametricity on Cluster Compactness and Separation

Clustering validation evaluates the goodness of the results of a clustering algorithm (Maulik and Bandyopadhyay 2002). We used internal validation measures that rely on information in the data (Tang et al. 2005), namely and compactness and separation (Tang et al. 2005; Zhao and Karypis 2002).

Separation is a measure of distinctiveness between a cluster and the rest of the world. The pairwise distances between cluster centers or the pairwise minimum distances between objects in different clusters are often used as measures of separation.

The compactness of each cluster was evaluated using the average dissimilarity between the observations in the cluster and the medoid of the cluster. Separation was computed using the minimal dissimilarity between an observation of the cluster and an observation of another cluster.

We investigated the impact of ultrametricity on compactness and separation of clusters by using the Partition Around Medoids (PAM) algorithm (Kaufman and Rousseeuw 1990) to cluster objects originally in the Euclidean Space and later in a transformed dissimilarity space with lower or higher ultrametricity.

Experiments show that a transformation on the distance matrix that decreases the ultrametricity of the original Euclidean space can actually improve compactness but also decrease separation of the clusters generated by PAM. However, the compactness improves at a faster ratio than the decrease in separation. We also observed that the increase of ultrametricity produces the reverse effect, degrading compactness and increasing separation, at different ratios. In this case, compactness decreases in a faster ratio than the increase in separation.

Let (S, d) be a dissimilarity space, (S, d') be the transformed dissimilarity space, where $d' = f(d)$ is obtained by applying one of the transformations described in Sect. 2.

The increase of ultrametricity from (S, d) to (S, d') promotes the equalization of dissimilarity values. In the extreme case, we have an ultrametric space where the pairwise distances involved in all triplets of points form an equilateral or isosceles triangle. To explore how the equalization (or the reverse process) may affect clustering quality, a better study of the effects of increased (or decreased) ultrametricity on the results generated by a widely known and robust clustering algorithm was performed.

In order to study the impact of ultrametricity on cluster compactness and separation, we have implemented an algorithm that runs PAM on the original and transformed spaces, and computes those measure for each cluster from S and S'.

Our experiments considered a initial Euclidean space (S, d) where S corresponds to a set of objects. To obtain a valid comparison of compactness and separation, the clusters obtained from a specific data set S must contain the same elements in the original and transformed spaces.

Dissimilarities d^x where $x > 1$ tend to decrease the ultrametricity of the original space, whereas dissimilarities where $0 < x < 1$ tend to increase ultrametricity.

Current existing clustering validation measures and criteria can be affected by various data characteristics (Liu et al. 2010). For instance, data with variable density is challenging for several clustering algorithms. It is known that k-means suffers from an uniformizing effect which tends to divide objects into clusters with relatively equal sizes (Xiong et al. 2009). Likewise, k-means and PAM do not have a good performance when dealing with skewed distribution data sets where clusters have unequal sizes. To determine the impact of ultrametricity in the presence of any of those characteristics, experiments were carried considering 3 different data aspects:

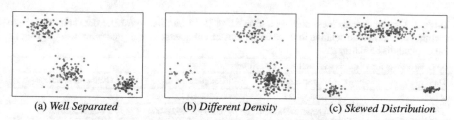

(a) *Well Separated*   (b) *Different Density*   (c) *Skewed Distribution*

Fig. 1 Synthetic data illustrating 3 different data aspects: **a** good separation, **b** different density and **c** skewed distributions

good separation, density, and skewed distributions in three synthetic data sets named *WellSeparated*, *DifferentDensity* and *SkewDistribution*, respectively.

Figure 1 shows the synthetic data that was generated for each aspect. Each data set contains 300 objects.

Table 2 show the results for data sets *WellSeparated*, *DifferentDensity* and *SkewDistribution*, respectively. Table 3 show results for the data set Iris. The measure (compactness or separation) ratio is computed dividing the transformed space measure by the original space measure. The average measure ratio computed for the 3 clusters is presented in each table.

Note that the average measure ratio is less than one for spaces with lower ultrametricity (obtained with dissimilarities d^5 and d^{10}). In this case, the average compactness ratio is also lower than the average separation ratio, showing that the transformations generated intra-cluster dissimilarities that shrunk more than the inter-cluster ones, relatively to the original dissimilarities. In spaces with higher ultrametricity (obtained with dissimilarities $d^{0.1}$ and $d^{0.01}$), the average measure ratio is higher than one. The average compactness ratio is also higher than the average separation ratio, showing that the transformations generated intra-cluster dissimilarities that expanded more than the inter-cluster ones. This explains the equalization effect obtained with the increase in ultrametricity.

In Fig. 2, we show the relationship between compactness and separation ratio for the three synthetic data sets and for the *Iris* data set which exhibit similar variation patterns.

As previously mentioned, data with characteristics such as different density and different cluster sizes might impose a challenge for several clustering algorithms.

We show a scenario where PAM, when applied to the original Euclidean space, does not perform well. Nevertheless, we are able to improve the PAM's results by applying a transformation that decreases the ultrametricity of the original space and running PAM on the transformed space.

Consider the data set presented in Fig. 3a which was synthetically generated in an Euclidean Space with pairwise metric d by three normal distributions with similar standard deviation but different densities. It has 300 points in total, with the densest group including 200 points and the other two containing 75 and 25 points.

Note that the somewhat sparse groups are also located very close to each other. Different symbols ($+$, $\triangle$, $\circ$) are used to identify the three distinct distributions. PAM's

Table 2 Cluster compactness and separation using PAM on three synthetic data sets. Both ratio averages are computed relative to the data set cluster compactness and separation values given by the original dissimilarity d

(a) Compactness for a data set with well-separated clusters

Diss.	Compactness avg.	Compactness std.	Compactness ratio avg.
d	0.1298267265	0.0364421138	1
d^{10}	7.4595950908E-009	9.0835007432E-009	5.7458085055E-008
d^5	0.000048905	4.3815641482E-005	0.0003766941
$d^{0.1}$	0.8231766265	0.0254415565	6.3405790859
$d^{0.01}$	0.9722292326	0.0030358862	7.4886678515

(b) Separation for a data set with well-separated clusters

Diss.	Separation avg.	Separation std.	Separation ratio avg.
d	0.5904521462	0.339733487	1
d^{10}	0.0020607914	0.0035682378	0.0034901921
d^5	0.0473640032	0.0795298042	0.0802164976
$d^{0.1}$	0.9752251248	0.0521762794	1.6516581929
$d^{0.01}$	0.9979573861	0.0052696787	1.6901579451

(c) Compactness for a data set with well-separated clusters

Diss.	Compactness avg.	Compactness std.	Compactness ratio avg.
d	0.2599331876	0.0225831458	1
d^{10}	1.7193980983E-009	8.1299728150E-010	6.6147694106E-009
d^5	4.4663622551E-005	7.7685178838E-006	0.0001718273
$d^{0.1}$	0.8942911252	0.0073467836	3.4404653496
$d^{0.01}$	0.9729198463	0.0174965529	3.7429612403

(d) Separation for a data set with well-separated clusters

Diss.	Separation avg.	Separation std.	Separation ratio avg.
d	0.8716430647	1.4832867815	1
d^{10}	0.0244453421	0.0423405745	0.0280451288
d^5	0.2484825264	0.4303843596	0.2850737147
$d^{0.1}$	0.8400992968	0.2757718021	0.9638111411
$d^{0.01}$	0.9777162094	0.0325513479	1.1216933272

(e) Compactness for a data set with clusters with varied densities

Diss.	Compactness avg.	Compactness std.	Compactness ratio avg.
d	0.1072664803	0.098564337	1
d^{10}	0.000000449	7.7773337902E-007	4.1860674698E-006
d^5	0.0002096486	0.0003626508	0.0019544651
$d^{0.1}$	0.7880494471	0.0792970382	7.3466514879
$d^{0.01}$	0.9633479044	0.0171811278	8.9808848178

(f) Separation for a data set with clusters with varied densities

Diss.	Separation avg.	Separation std.	Separation ratio avg.
d	0.971795701	0.0185685451	1
d^{10}	0.0029611253	0.0005897832	0.0030470656
d^5	0.0932204575	0.0090867253	0.0959259826
$d^{0.1}$	1.0416448857	0.001980664	1.0718764083
$d^{0.01}$	1.0047158048	0.0001909503	1.0338755396

Table 3 Cluster compactness and separation using PAM on the data set *Iris*. Both ratio averages are computed relative to the data set cluster compactness and separation values given by the original dissimilarity d

(a) Compactness results for the *Iris* data set

Diss.	Compactness avg.	Compactness std.	Compactness ratio avg.
d	0.2564313287	0.0572997859	1
d^{10}	4.495583902E-007	3.0731794825E-007	1.7531336456E-006
d^5	0.0007628527	0.0004963497	0.0029748809
$d^{0.1}$	0.8664974196	0.0223773478	3.379062238
$d^{0.01}$	0.9630194558	0.0029079036	3.7554672456

(b) Separation results for the *Iris* data set

Diss.	Separation avg.	Separation std.	Separation ratio avg.
d	0.2841621289	0.3120959612	1
d^{10}	1.1716078298E-005	2.0292841461E-005	4.1230259434E-005
d^5	0.0045832156	0.0079357613	0.0161288754
$d^{0.1}$	0.8715968561	0.0944160231	3.0672519923
$d^{0.01}$	0.9858840558	0.0108572902	3.4694421086

objective function tries to minimize the sum of the dissimilarities of all objects to their nearest medoid. However, it may fail to partition the data according to the original distributions when dealing with cluster with different densities. In this case, the split of the densest cluster may occur. In our example, PAM not only divides the heaviest cluster, but also combines the two sparse clusters that are not well separated. Note that unlike k-means (which also does not perform well in these scenarios but eventually can find the right partition due to the randomness on the selection of the centroids), PAM will most likely fail due to the determinism of its BUILD and SWAP steps combined and the choice of the objective function.

In order to explore the positive effect of increased intra-cluster compactness produced by lower degrees of ultrametricity, we applied the same transformations $f(d) = d^r$ with positive integer exponents ($r > 1$), to the original Euclidean distance matrix obtained from d. Results show significant improvement of the clustering.

Figure 3b shows the result of applying PAM to cluster the synthetic data with dissimilarity d. Note that the clustering result does not correspond to a partition resembling the distributions that were used to generate the data. Figure 3d, c show that PAM also fails to provide a good partition with dissimilarities $d^{0.1}$ and $d^{0.01}$ since the increase in ultrametricity promotes equalization of dissimilarities which may degrade even more the results. Note however that the partitions obtained by PAM using the dissimilarities d^5 and d^{10} form similar clusters to the ones generated by the original distributions. Indeed, the increase in compactness helps PAM to create boundaries that are compliant with the original normal distributions.

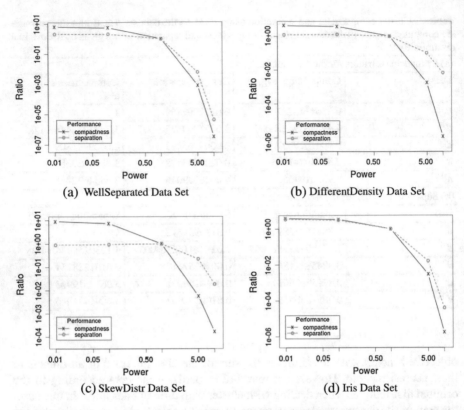

Fig. 2 Relation between Compactness and Separation Ratio for three synthetic data set and for the Iris data set

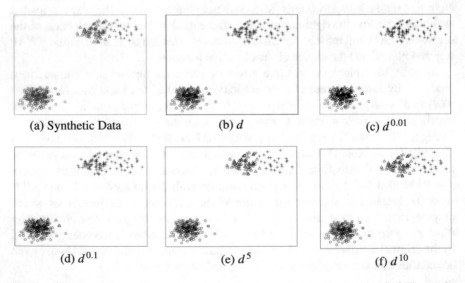

Fig. 3 **a** The synthetic data generated from distributions with different densities. **b–f** The results of PAM using Euclidean distance d and other dissimilarities obtained by transformations on d

Table 4 Cluster compactness and separation using PAM on a synthetic data set comprising clusters with different density

(a) Compactness and clustering quality results for synthetic data set

Diss.	Compactness avg.	Compactness std.	Compactness ratio avg.	Normalized mutual info.
d	0.1386920089	0.0558906421	1	0.6690207374
d^{10}	1.2953679952E-009	6.3343701540E-010	9.3398891934E-009	0.9365672372
d^5	2.8689799313E-005	1.0529323158E-005	0.0002068598	0.9365672372
$d^{0.1}$	0.8428018314	0.0308261682	6.0767872501	0.6702445506
$d^{0.01}$	0.9745718848	0.0037669287	7.026878423	0.6702445506

(b) Separation results for synthetic data set

Diss.	Separation avg.	Separation std.	Separation ratio avg.	
d	0.4604866874	0.7771228672	1	
d^{10}	0.0114269071	0.0197919837	0.0248148479	
d^5	0.104837087	0.1815831588	0.2276658368	
$d^{0.1}$	0.8160827216	0.2379010818	1.7722178381	
$d^{0.01}$	0.978284428	0.0270049282	2.1244575681	

Fig. 4 Relation between compactness and separation ratio for the test data set

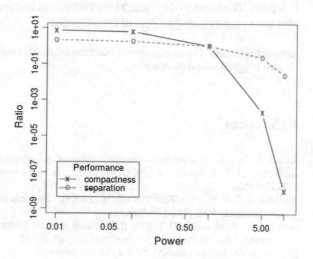

Cluster quality evaluated using the normalized mutual information (Manning et al. 2008) for yet another synthetic data set with diverse densities of clusters is shown in Table 4 together with the compactness and separation ratios. The relationship between compactness and separation ratio for this data set is presented in Fig. 4.

6   Conclusions and Further Work

We examined the influence of ultrametricity of dissimilarity spaces in classification and clustering.

We have shown that there is an inverse variation between ultrametricity and the performance of classifiers. The increase of ultrametricity promotes the equalization of distances between objects. This equalization raises the level of uncertainty during the classification process and degrades the quality of the results generated by classifiers.

For clustering, increased ultrametricity generates clusterings with better separation. However, it also decreases compactness faster than the increase in separation. Lowering ultrametricity produces clusters that are more compact but not as well separated as in the original space. In this case, compactness grows at a faster ratio than the decrease in separation. Finally, we present an example where we are able to improve PAM's results by applying a transformation on the dissimilarity space that reduces its ultrametricity.

There are numerous applications that can benefit from this study. For example, changing the ultrametricity of the original space may help finding patterns in data that do not conform to the expected behavior, in a classical example of anomaly detection. The impact of ultrametricity on various hierarchical clustering algorithms also seems a promising subject of investigation.

Acknowledgments The authors wish to thank the referees for the careful reading of the paper and for their suggestions and observations.

References

Amice, Y. (1975). *Les nombres p-adiques*. Paris: Presses Universitaires de France.

Barthélemy, J.-P., & Brucker, F. (2008). Binary clustering. *Discrete Applied Mathematics, 156*(8), 1237–1250.

Barthélemy, J.-P., Brucker, F., & Osswald, C. (2004). Combinatorial optimization and hierarchical classifications. *4OR, 2*(3), 179–219.

Bertrand, P., & Janowitz, M. F. (2002). Pyramids and weak hierarchies in the ordinal model for clustering. *Discrete Applied Mathematics, 122*(1–3), 55–81.

Brucker, F. (2006). Sub-dominant theory in numerical taxonomy. *Discrete Applied Mathematics, 154*(7), 1085–1099.

Contreras, P., & Murtagh, F. (2012). Fast, linear time hierarchical clustering using the baire metric. *Journal of Classification, 29*(2), 118–143.

Deza, M. M., & Laurent, M. (1997). *Geometry of cuts and metrics*. Heidelberg: Springer.

Di Summa, M., Pritchard, D., & Sanità, L. (2015). Finding the closest ultrametric. *Discrete Applied Mathematics, 180*, 70–80.

Diatta, J., & Fichet, B. (1998). Quasi-ultrametrics and their 2-ball hypergraphs. *Discrete Mathematics, 192*(1–3), 87–102.

Gordon, A. D. (1981). *Classification*. London: Chapman and Hall.

Gordon, A. D. (1987). A review of hierarchical classification. *Journal of the Royal Statistical Society, Series (A), 150*(2), 119–137.

Jardine, N., & Sibson, R. (1971). *Mathematical taxonomy*. New York: Wiley.

Kaufman, L., & Rousseeuw, P. J. (1990). *Finding groups in data: An introduction to cluster analysis*. New York: Wiley.

Kimura, M. (1983). *The neutral theory of molecular evolution*. Cambridge: Cambridge University Press.

Leclerc, B. (1985). La comparaison des hiérarchies: indices et métriques. *Mathématiques et sciences humaines, 92*, 5–40.

Lerman, I. C. (1981). *Classification et Analyse Ordinale des Données*. Paris: Dunod.

Liu, Y., Li, Z., Xiong, H., Gao, X., & Wu, J. (2010). Understanding of internal clustering validation measures. In *2010 IEEE 10th international conference on data mining* (pp. 911–916). IEEE.

Manning, C. D., Raghwan, P., & Schütze, H. (2008). *Introduction to information retrieval*. Cambridge: Cambridge University Press.

Maulik, U., & Bandyopadhyay, S. (2002). Performance evaluation of some clustering algorithms and validity indices. *IEEE Transactions on Pattern Analysis and Machine Intelligence, 24*(12), 1650–1654.

Murtagh, F., Downs, G., & Contreras, P. (2008). Hierarchical clustering of massive, high dimensional data sets by exploiting ultrametric embedding. *SIAM Journal on Scientific Computing, 30*(2), 707–730.

Ninio, J. (1983). *Molecular approaches to evolution*. Princeton: Princeton University.

Rammal, R., Angles d'Auriac, C., & Doucot, D. (1985). On the degree of ultrametricity. *Le Journal de Physique - Letteres, 45*, 945–952.

Rammal, R., Toulouse, G., & Virasoro, M. A. (1986). Ultrametricity for physicists. *Reviews of Modern Physics, 58*, 765–788.

Schikhof, W. H. (1984). *Ultrametric calculus*. Cambridge: Cambridge University Press.

Simovici, D. A., & Djeraba, C. (2014). *Mathematical tools for data mining* (2nd ed.). London: Springer.

Tang, P. N., Steinbach, M., & Kumar, V. (2005). *Introduction to data mining*. Reading: Addison-Wesley.

Xiong, H., Wu, J., & Chen, J. (2009). K-means clustering versus validation measures: a data-distribution perspective. *IEEE Transactions on Systems, Man, and Cybernetics, Part B: Cybernetics, 39*(2), 318–331.

Zhao, Y., & Karypis, G. (2002). Evaluation of hierarchical clustering algorithms for document datasets. In *Proceedings of the Eleventh International Conference on Information and Knowledge Management* (pp. 515–524). ACM.

Part III
Semantics, Ontologies, and Social Networks

SMERA: Semantic Mixed Approach for Web Query Expansion and Reformulation

Bissan Audeh, Philippe Beaune and Michel Beigbeder

Abstract Matching users' information needs and relevant documents is the basic goal of information retrieval systems. However, relevant documents do not necessarily contain the same terms as the ones in users' queries. In this paper, we use semantics to better express users' queries. Furthermore, we distinguish between two types of concepts: those extracted from a set of pseudo relevance documents, and those extracted from a semantic resource such as an ontology. With this distinction in mind we propose a Semantic Mixed query Expansion and Reformulation Approach (SMERA) that uses these two types of concepts to improve web queries. This approach considers several challenges such as the selective choice of expansion terms, the treatment of named entities, and the reformulation of the query in a user-friendly way. We evaluate SMERA on four standard web collections from INEX and TREC evaluation campaigns. Our experiments show that SMERA improves the performance of an information retrieval system compared to non-modified original queries. In addition, our approach provides a statistically significant improvement in precision over a competitive query expansion method while generating concept-based queries that are more comprehensive and easy to interpret.

1 Introduction

Once the domain of librarians and specialists, today the practice of searching for information is open to users from different profiles and backgrounds, all of whom use queries composed of keywords to look for information on the web. The challenge of this online search for content is that retrieval systems need to provide relevant documents for all the users who express the need for a particular piece of

B. Audeh (✉) · P. Beaune · M. Beigbeder
École Nationale Supérieure des Mines de Saint-Étienne,
158 Cours Fauriel, 42023 Saint-Étienne, France
e-mail: audeh@emse.fr

P. Beaune
e-mail: beaune@emse.fr

M. Beigbeder
e-mail: mbeig@emse.fr

© Springer International Publishing Switzerland 2017
F. Guillet et al. (eds.), *Advances in Knowledge Discovery and Management*,
Studies in Computational Intelligence 665, DOI 10.1007/978-3-319-45763-5_9

information using many different queries. In addition, the length of web queries is a major challenge for most query modification approaches.

The issue we are tackling is how to improve the precision of short ambiguous web queries. To achieve this goal, our paper explores semantic related techniques for automatic query reformulation.

Since most web users employ two to three terms in a query to express their information needs (Jansen et al. 2000), it is not easy for a system to retrieve relevant documents at early ranks in the result list. To address this challenge, a number of approaches propose to consider the semantics during the indexing step. In this case, concepts, instead of terms (or stems), are used to index documents and queries. The relevance between a document and a query is then evaluated on the basis of this conceptual indexation. Another option is to keep a keyword-based index and to use semantic approaches to expand and reformulate users' queries. While both of these solutions have been explored in the literature of information retrieval, in general, it is not possible to confirm the advantage of one option over the other one. Many elements could affect the choice of how to use the semantics within an information retrieval system, such as the nature of the document collection (web, closed collection), the context of use (professional, general), the motivation (creating a new retrieval system or improving an existing one), and the cost. In this paper, we are interested in the case where documents and queries are indexed using classical term-based techniques. Thus, we focus on semantically modifying users' queries while preserving the keyword-based retrieval mechanism.

Techniques that automatically modify users' queries have existed since the early years of information retrieval. As a result, the literature is wealthy of terms like "query expansion", "query refinement", "query reformulation", "query enrichment", "local and global analysis" and "relevance feedback". All these techniques intend to improve keyword-based queries even though the number of terms used to describe how this is achieved is confusing. For our work, we employ two commonly used terms: *query expansion* and *query reformulation*. We define *query expansion* as assigning new terms to users' queries, whereas we consider *query reformulation* as the way in which these new terms are integrated within the original query. The literature does not always make a difference between *query expansion* and *query reformulation*, this is because in most cases the query is considered as a bag of words. In general, approaches try to add new terms with eventually optimized weights; hence, reformulating the query is not considered as a separate process.

In this paper, we study the effect of different semantic aspects to automatically improve web queries. To do this, we associate query terms with implicit concepts that we obtain with a pseudo relevance feedback approach, and explicit concepts that we extract from an ontology. Once detected, explicit and implicit concepts are used to obtain sets of expansion terms (Sect. 3.1) and to construct a new query (Sect. 3.2). The new query is still composed of keywords, but it is structured so as to represent the concepts. This allows a straightforward understanding of the relationships between the original user keywords and the detected concepts. In Sect. 4 we compare our proposition versus no query expansion as well as versus a state-of-the-art expansion approach. We begin our paper with a brief state of the art of existing query expansion and query reformulation approaches.

2 Query Expansion and Reformulation in Information Retrieval

Associating new terms to a query requires the use of a data source other than the query itself. This resource can be a collection of documents (Qiu and Frei 1993), a subset of the collection via a relevance or pseudo relevance feedback process (Rocchio and Salton 1965), a completely independent resource that is also a collection of documents (Deveaud et al. 2013), or a semantic resource (Voorhees 1994). All of these approaches have been the subject of many comparisons and surveys that as a whole reveal three common points: an expanded query is often not structured, named entities are processed in the same way as common terms, and no specific consideration is taken regarding the advantage (or disadvantage) of adding a candidate term to the query. In the following subsections, we will focus on query expansion or reformulation approaches that consider these three aspects in the state of the art.

2.1 Concept-Based Query Reformulation

Representing a query completely depends on the query language that the retrieval system can interpret. A bag-of-words representation is the most common way to reformulate an expanded query. With this representation, the query is composed of weighted terms with no explicit operators.

In the literature, several approaches explored the advantages of structured queries, whether by using only original query terms (the case of studies on long queries) (Metzler and Croft 2005; Bendersky and Croft 2008; Maxwell and Croft 2013), or by integrating new terms from different resources with the original query terms (Bendersky et al. 2011, 2012; Deveaud et al. 2013). Query expansion approaches, in the latter case, propose to introduce the notion of concepts into the expanded query, which we call "concept based query representation". For (Bendersky et al. 2011), a concept is one or more terms that must belong to one of the following types: an original query term, a composition of multiple original terms, or term obtained from the pseudo relevance feedback of the original query on different expansion collections. The obtained concepts are then combined to construct a new query using Eq. 1:

$$Score(Q, D) = \sum_{T \in \tau} \sum_{\kappa \in T} \lambda_\kappa f(\kappa, D) \tag{1}$$

where τ is the set of concept types, $f(\kappa, D)$ is the query likelihood retrieval function that matches the concept κ in the document D, and λ_κ is the weight of the concept κ. The weight in this equation takes a set of features into account, especially the frequency of the concept in the expansion collections. Similarly, (Deveaud et al. 2013) work on detecting query concepts but without considering possible associations among original terms. So, a concept in this case is either an original query term or a set of terms from pseudo relevance feedback documents. (Deveaud et al. 2013) use Latent Dirichlet Allocation (LDA) (Blei et al. 2003) on the document sets obtained by pseudo relevance feedback on different collections. The score of a document is computed as shown by Eq. 2:

$$Score(Q, D) = \lambda \cdot P(Q|D) + (1 - \lambda) \cdot \prod_{k \in T_{\hat{k}}} \hat{\delta}_k \prod_{w \in W_k} \hat{\phi}_{k,w} \cdot P(w|D) \qquad (2)$$

where W_k is the set of terms of the concept k, $\hat{\phi}_{k,w}$ is the weight of the term w in the concept k, $\hat{\delta}_k$ is the normalized weight of the concept k, and $T_{\hat{k}}$ is the set of concepts assigned to the query. The authors show that combining four different collections for concept extraction is more effective in precision than the use of any single resource.

All of these approaches did generate structured queries based on the notion of concepts, but they didn't explore the advantage of using formal semantic relationships from a structured resource like an ontology. They also did not consider the specificity of named entities.

2.2 Query Expansion and Named Entities

The approaches in the previous subsection focused on pseudo relevance feedback techniques, where named entities are not considered as special terms. For this reason, it is possible that the expanded query doesn't contain any reference to these important objects. Other approaches focus on the importance of named entities in a query; for example, studies on long queries consider a sub-query containing a named entity as a valuable reformulation candidate (Huston and Croft 2010; Kumaran and Carvalho 2009). Bendersky and Croft (2008) classify noun phrases (eventually named entities) in order to use them in the reformulated query. Recent approaches are becoming increasingly interested in Wikipedia, which is a rich resource of named entities. Xu et al. (2008) extracted terms from Wikipedia, that are semantically close to named entities in the query, while Brandao et al. (2011) proposed an approach based on the infoboxes of Wikipedia to expand named entities.

These approaches explicitly handle named entities differently from other terms. Nevertheless, they rely on a bag-of-words representation for the modified query instead of concept-based representation. In addition, no specific treatment is done to control the quality of expansion terms.

2.3 Quality of Expansion Terms

For most query expansion approaches, new terms are systematically added to all queries, even if in some instances, better results can be obtained without expansion. These approaches do not consider the advantage of adding (or not adding) each term to the query. Though, several methods exist to measure the quality of a query or query terms that could be used in query expansion, (Cronen-Townsend et al. 2002) proposed the clarity measure, which is based on computing the entropy between the query model and the document model. They confirmed the relationship between this measure and query ambiguity. Nevertheless, the effectiveness of using this measure to choose new terms for query expansion was not confirmed (Zobel 2004; Shah and Croft 2004). Other studies focused on measuring the importance of the query terms. (Zhao and Callan 2010) used a technique based on pseudo relevance feedback and latent semantic analysis (Deerwester et al. 1990) to classify terms according to their importance within the query. This approach was only used to evaluate original query terms, not to choose new terms for query expansion purpose.

From this brief presentation of query modification approaches, it can be seen that structured queries, named entities and terms quality aspects are the subject of several studies in the dedicated literature. We consider that an approach that gathers all of these aspects could be effective to improve web queries. For this reason, we propose the semantic mixed expansion and reformulation approach that we thoroughly discuss in the following section.

3   Semantic Mixed Expansion and Reformulation Approach (SMERA)

As discussed in Sect. 2, query expansion and reformulation approaches are not new to information retrieval. In light of the weaknesses revealed by these approaches, we propose SMERA that uses semantics to improve web queries. Our approach utilizes both, but well distinguished, query expansion and query reformulation techniques (Fig. 1). The consideration of concept-based query representation, named entities and the quality of expansion terms allows our approach to generate queries that are comprehensible and easy to interpret. We believe that generating user-friendly queries is important to understand and analyze the relationships between a well-expressed query (from a human point of view) and an effective query[1].

As Fig. 1 shows, our approach is composed of two steps. The first step (expansion) includes detecting query concepts, and choosing the most appropriate and representative terms of these concepts. The second step (reformulation) will use the concepts detected during the first step to reformulate the expanded query in a concept-based representation.

---

[1] In this paper we define an effective query is the one that obtains good results with standard measures used in evaluation campaigns, in particular, precision measures for the case of web queries.

Fig. 1 The main steps of
SMERA

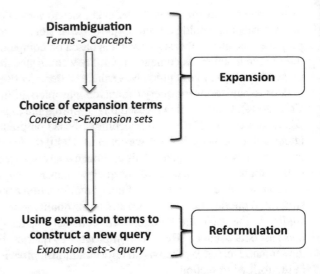

3.1 Expansion

Defining query expansion as a separate step that assigns new terms to the query allows
it to be independent from the matching function of the retrieval model. Our approach
depends on the assumption that each original query term (except stop words) belongs
to a concept (in its abstract meaning). This strong assumption is justified by the fact
that a user doesn't use one term to express two different concepts; on the contrary,
he may use multiple terms to express one concept. Thus, we consider that a query
of k terms corresponds to at most k concepts. This allows us to initialize the number
of concepts and to keep a clear relationship between original terms and represented
concepts in the reformulated query. For each query term, we define an expansion set
that contains semantically similar terms. The nature of web queries (short, ambiguous, and rich with named entities) and the literature of query expansion, oriented
our approach towards mixing two types of expansion resources: the collection of
documents through pseudo relevance feedback and an ontology. We use pseudo relevance feedback documents to detect what we call "implicit concepts", while we
consider named entities in the query as the "explicit concepts" that we identify using
an ontology. In both cases, a concept in our approach is a set of semantically similar
terms.

Figure 2 shows the main steps of our expansion approaches for both named entities
and other terms. If we consider the example in Table 1, SMERA will first detect the
named entity "Jack Robinson". This named entity will be disambiguated and linked
to an explicit concept which is then expanded with the ontology-based approach (cf.
Sect. 3.1.1). The other terms of the query, except stop words, will be expanded based
on implicit concepts extracted by an LSI-based[2] method (cf. Sect. 3.1.2).

---

[2]LSI: Latent Semantic Indexing (Deerwester et al. 1990).

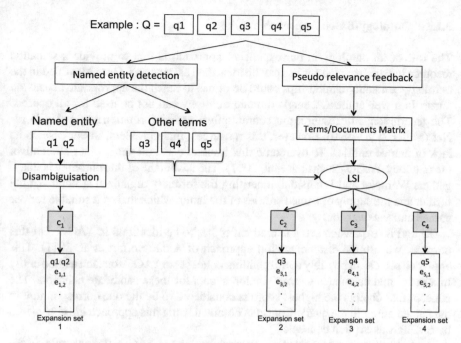

Fig. 2 Generating expansion sets in SMERA from explicit and implicit concepts

Table 1 Term categories and SMERA actions demonstrated on one TREC query (#455)

Original query: When did Jack Robinson appear at his first game?

Category	Values	SMERA action
Named entities	"Jack Robinson"	Expand using explicit concept approach
Non named entity term	Appear, first, game	Expand using implicit concept approach
Stop words	When, did, at, his	Do not expand

Expansion sets do not necessarily have the same size. This is because the number of available terms in the corresponding concepts is not necessarily the same. In addition, we use quality filters that measure the utility of adding a term to the query and eliminate less useful terms. As a result, our approach does not always expand all queries with the same number of expansion terms. A detailed explanation of these steps is in the following subsections.

3.1.1 Ontology-Based Approach

The role of an ontology in our expansion approach is first to provide a semantic resource in which named entities can be identified as concepts. Once identified in the ontology, semantic relationships could be of use to reach the appropriate expansion terms. In a web context, a single domain ontology can not be used for all queries. The generic semantic resource most commonly used in information retrieval is Word-Net (Miller et al. 1990). However, this resource's main problem, in our case, is its lack of named entities. To overcome this issue, we sought yet another alternative: the ontology YAGO (Suchanek et al. 2007). The advantage of this ontology is that it gathers WordNet and Wikipedia, inheriting the formally organized structure of the former and the supply of named entities of the latter, which makes it suitable for our named entity expansion.

To find its expansion set, a named entity has to be identified in YAGO. For this purpose, we use the disambiguation approach of Aida (Hoffart et al. 2011). This approach selects all possibly corresponding concepts in YAGO for each named entity in a query and calculates disambiguation scores for these candidate concepts. The concept that obtains the highest score is considered to be the one corresponding to the named entity in the query. Concepts obtained using this approach are considered by SMERA as explicit concepts.

A wealthy number of semantic relationships exist in YAGO. For example, in the case of concepts corresponding to a named entity, we can find relationships like "lives in" for person entities, or "has the surface" if the named entity is a city. On the other hand, all named entity concepts in YAGO have the semantic relationship "rdf:label". This relationship corresponds to the "redirect" link in Wikipedia, it links the named entity to all its possible appellations. These appellations can be simply orthographic alternative names (e.g., Baltimore-Baltamore), syntactically different names (e.g., Baltimore-Mobtown), or even nominal phrases (e.g., "Aleck Bell"-"The father of the deaf"). In this work, we choose the relationship (rdf:label) to expand named entities. This choice assumes that using alternative appellations to expand named entities leads to less query drift risk than using other semantic relationships in YAGO. In our previous example of Table 1, after disambiguation, the named entity "Jack Robinson" obtains two expansion terms: "Jackie Robinson" and "Jack Roosevelt Robinson".

3.1.2 Pseudo Feedback Approach

The idea of this approach is to detect implicit concepts from a set of pseudo feedback documents related to users' initial query. Several methods exist to extract concepts from a set of documents, such as LDA (Blei et al. 2003), ESA (Gabrilovich and Markovitch 2007) or LSI (Deerwester et al. 1990). We chose to use LSI because of its ability to detect high-level co-occurrence relationships between terms. In other words, two terms that do not occur together in a studied set of documents, but do frequently co-occur with a third term, are considered by LSI as semantically related. To achieve this, LSI (Deerwester et al. 1990) starts by applying singular value decomposition on a matrix A of m lines (m terms) and n columns (n feedback documents), which contains frequencies tf of the terms in document collection (in our case, pseudo feedback documents). The results of this step are the three matrices presented in Eq. 3,

$$A_{\{m,n\}} = U_{\{m,m\}} S_{\{m,n\}} V_{\{n,n\}}^T \tag{3}$$

where S is the diagonal matrix that contains singular values of A. The theory of LSI is that reducing the dimension of the three resulting matrices gives an approximation of the original matrix A and reduces the noise (Eq. 4).

$$A'_{\{m,n\}} = U_{\{m,k\}} S_{\{k,k\}} V_{\{k,n\}}^T \tag{4}$$

In our pseudo feedback expansion approach, we are interested in the matrix $U_{\{m,k\}}$. This matrix contains the m vectors of terms appearing in pseudo relevance feedback documents. These vectors belong to the semantic space of k dimensions created by LSI (Fig. 3).

To find the expansion set of a query term q, we measure its similarity with a term that appears in the feedback documents by calculating its distance with this term[3]. We then suppose that the terms that are the most similar to q belong to the same implicit concept, as presented in Fig. 3. In some cases, an expansion term q' of a term q is also a query term; in this case, we consider that both terms q and q' belong to the same implicit concept (c2 in Fig. 4) and they will both correspond to one expansion set in the reformulated query.

In our example of Table 1, two expansion sets were found for three non-named entity terms: {appear}, {first, play, team, season, game, ball}. From these two sets, we can see that the implicit concepts related to the query terms "first" and "game" were merged resulting in one expansion set for both of these terms.

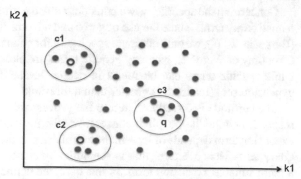

Fig. 3 Terms of feedback documents in the semantic space of LSI (example for the case of 2 dimensions k1 et k2)

● : A term that appears in feedback documents
○ : A term that appears in feedback documents and in the query

---

[3]Our experiments showed no significant difference between using euclidian and cosine distances, in this paper we used euclidian distance because it is more clear for our graphical demonstration in Figs. 3 and 4.

Fig. 4 The fusion of
expansion sets in the case of
query terms that are
semantically close in LSI
semantic space

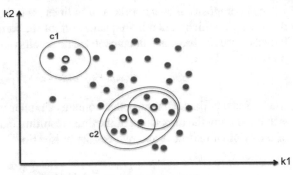

● : A term that appears in feedback documents
○ : A term that appears in feedback documents and in the query

3.1.3 Quality of Expansion Terms

Our ontology-based and feedback-based approaches presented in Sects. 3.1.1 and
3.1.2 respectively generate expansion sets for, at most, each original query term.
In this work, we consider the quality of terms, which means their usefulness in
obtaining relevant documents to the user's information need. From this point of view,
we consider original query terms as the most valuable terms in the query because
they were chosen by the user to express his own information need. An expansion
term, on the other hand, is considered to be useful if it is not too generic, and as far
as we are sure it belongs to a valid query concept. To express these two subjective
conditions, we define *specificity* and *certitude* qualities. *Specificity* is a boolean value.
Certitude is measure with values between 0 and 1. Expansion terms that do not satisfy
a minimum threshold for this measure are rejected from their expansion sets and will
not appear in the reformulated query.

Concerning the specificity, we consider named entities as specific terms. For non
named entity terms, since the use of verbs and adverbs in web queries is not frequent
(Barr et al. 2008), we only compute specificity for nouns. For this purpose, we use the
taxonomy of WordNet, whereby generic terms are placed in the top of the hierarchy
while specific terms can be found in deeper levels. Thus a noun is added to an
expansion set if its depth is greater than a threshold.

The certitude is directly related to the process that links a query term to its corresponding implicit or explicit concept. For the feedback approach, the choice of an
expansion term depends of its semantic similarity, in the LSI space, with the original
query term. Hence, a term that is semantically closer to the original term is likely
a more suitable expansion term. In this case, we define the certitude score between
a term t and a query term q as the euclidean distance between their corresponding
vectors ($\vec{t}$, $\vec{q}$) in the LSI space as defined in Eq. 5.

$$Cert(t, q) = Dist_{euclide}(\vec{t}, \vec{q}) \tag{5}$$

As mentioned in Sect. 3.1.1, an explicit concept is chosen in YAGO for a named entity in the query if it obtains the maximum disambiguation score. For each expansion term that we obtain by the relation "rdf:label", the certitude value is the disambiguation score S_{dis} of the concept to which the query term belongs (Eq. 6)

$$Cert(t, q) = S_{dis}(q, c) \tag{6}$$

where q is a query term, c is the disambiguated YAGO concept associated to q, t is a possible expansion term associated to the concept c, and $S_{dis}(q, c)$ is the disambiguation score, obtained by Aida (Hoffart et al. 2011), for the query term q and the concept c.

3.2 Concept-Based Query Reformulation

Up to this point, the expansion approaches we proposed are independent of the retrieval model. However, reformulating a query depends on the retrieval model and its query language. To achieve a concept-based query representation, we need a structured query language that supports three main elements: proximity between terms, synonymy and term weighting. The model proposed by (Metzler and Croft 2004) is a good environment to apply our idea of semantic reformulation. In the next subsections, we present an overview of this model and how we use it query language to reformulate users' query.

3.2.1   The Retrieval Model of Metzler and Croft (2004)

This information retrieval model is a combination of inference networks and query likelihood models. Like in inference network models, it is possible to handle structured queries, but estimating the probabilities is achieved using a query likelihood language model. The model is implemented within the framework Indri (Strohman et al. 2004), which is part of the Lemur[4] project. Indri proposes a query language model that allows expressing the different functionality of the retrieval model. Table 2 represents some demonstrative examples cited in the Lemur wikipage[5] and shows how the implementation of (Metzler and Croft 2004) in Indri handles the different query language operators.

---

[4]http://sourceforge.net/p/lemur/wiki/The%20Indri%20Query%20Language.
[5]http://sourceforge.net/p/lemur/wiki/Belief%20Operations/.

Table 2 Demonstrative examples of the functionality of Indris operators

Syntax	Interpretation
#combine(dog train)	$0.5log(b(dog)) + 0.5log(b(train))$
#weight(1.0 dog 0.5 train)	$0.67log(b(dog)) + 0.33log(b(train))$
#wsum(1.0 dog 0.5 dog)	$log(0.67b(dog) + 0.33b(dog))$
#syn(car automobile)	one occurrence of "car" or "automobile"
#wsyn(1.0 car 0.5 automobile)	like #syn, but the occurrence of "car" counts as twice the occurrence of "blue"
#n(blue car)	"blue" appears before "car" in a window of maximum n words
#uwn(blue car)	"blue" appears before or after "car" in a window of maximum n words

3.2.2 Representing Concepts in Keyword Query

Our reformulation approach considers the final query to be a linear combination of the user's original query and the combination of the different expansion sets according to three aspects: proximity, synonymy and weighting. The score of this reformulated query is calculated with Eq. 7

$$p(Q|d) = \lambda \prod_q p(q|d) + (1 - \lambda) \prod_{i=1}^k b(r_i)^{w_i} \tag{7}$$

where $p(q|d)$ is the query likelihood probability for the original query term q and a document d, r_i is the combination of terms of an expansion set with an Indri operator (#combine, #weight or #syn), and $b(r_i)$ is the belief calculated according to (Metzler and Croft 2004) as illustrated in Table 2. Finally w_i is the weight of the estimated belief of the representation r_i. In this current study, expansion sets are considered to be equally important to the query ($w_i = 1$, for all i).

For example, the reformulation of the query presented previously in Table 1 is demonstrated in Fig. 5. This figure shows how we combine expansion sets using

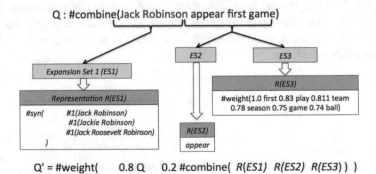

Fig. 5 Example of query reformulation using SMERA

Eq. 7 and the corresponding operators in Indri model (cf. Table 2). The following subsections explain why and how each operator is used to reformulate a query in our approach.

Proximity

Expansion terms that come from the ontology YAGO can be expressions or names composed of one or multiple terms, as we have seen in Sect. 3.1.1. When an element of a named entity expansion set is composed of multiple terms, the proximity between these terms should be highly respected while representing this entity in a query. Expressing the proximity between terms in the query implies defining the maximum distance within which these terms could be considered as related to the entity. In addition, we have to precise if the order in which these terms appear in a text is important. In our work, we suppose that the coverage of semantic alternatives of named entities is the responsibility of the resource, which in our case is the YAGO ontology. For this reason, we consider expansion elements obtained from the semantic resource as blocks that should appear verbatim in a relevant document. Thus, our approach requires that terms that belong to the same expression should be within a window of width 1 and appear in the exact order as in the semantic resource. To represent these types of expansion terms we use the operator #1 (cf. Table 2 and Fig. 5).

Synonymy

Our expansion approaches extract terms that are semantically related to query terms. This semantic similarity is not the direct synonymy in the case of our feedback approach, and we do not use this functionality for feedback expansion sets. In the case of named entities expansion, the semantic similarity is defined by the explicit relation "rdf:label", which will give possible, semantically equal, alternatives of the named entity. When evaluating a document that contains one of these alternatives, we want the matching function of the retrieval model to consider it as any of its other alternatives. For this reason, expansion sets that are obtained from an explicit concept are represented by the operator #syn in the Indri's query language. It should be noted that expansion terms of a named entity are not weighted in our current approach, we consider them as equally important synonyms, though exploring weighting possibility based on popularity or corpus statistics is an interesting area for future work.

Weighting

In our reformulation approach, weighting a term means defining its importance in its expansion set. We consider original query terms as important (weight = 1). The more an expansion term is close to an original query term, the more its weight is close to 1. As we mentioned in the synonymy section, this notion is not defined when the expansion set is obtained from YAGO because we consider all its terms as equal. On the other hand, expansion terms obtained by the feedback approach are terms that are statistically close to the original term in the LSI space, but they cannot be considered as synonyms. In this paper we explore the effect of using the similarity distance from the query term as a weight in the reformulated query. Expansion terms that are obtained from the feedback approach are combined with the operator #weight in the Indri's query language. The euclidean distance between an expansion term and its original term (cf. Eq. 5) is considered as its weight in the #weight expression. In the example of Fig. 5, the original query term "first" has the weight 1, while expansion terms have decreasing weights according to their semantic similarity with this term.

4 Experiments and Evaluation

4.1 Framework

To evaluate our semantic mixed expansion and reformulation approach (SMERA), we used four web collections from TREC and INEX evaluation campaigns, as displayed in Table 3. All of these collections were indexed with the same parameters using Indri: standard stop words were removed and a Krovetz Stemmer was used.

As a baseline, we used the query likelihood language model (Ponte and Croft 1998) to run the users' queries without expansion; we called this the QL model. We also used the relevance model approach (RM3) (Lavrenko and Croft 2001) as a reference model for query expansion. Both QL and RM3 are implemented in the Indri's framework. In addition to these reference approaches, we compared SMERA to the use of only one method for query expansion: the use of LSI via pseudo relevance feedback to expand query terms (both common terms and named entities), we called this the LSI approach, and the use of YAGO to disambiguate and expand named entities (the YAGO approach). The evaluation measures that we used in this experience are precision measures (MRR, P@10 and MAP), which are the most important in our

Table 3 Information about the queries used in our experiments

	# documents	queries	year (track)	nb. judged queries	nb. named entities
Inex 2006	659, 388	544–677	2008 (ad hoc)	70	23
Inex 2009	2, 666, 190	1–115	2009 (ad hoc)	68	21
WT10g	1, 692, 096	451–550	2000–2001 (Web ad hoc)	98	25
Gov2	25, 205, 179	701–850	2004–2006 (Terrabyte)	148	47

Table 4 Free parameters for all the approaches of our experiments

Parameter	Description	Approach
μ	Dirichlet smoothing	QL, SMERA, RM3
n_{Smera}, n_{Rm3}	Number of feedback documents	SMERA, RM3
t	Number of expansion terms	RM3
m	Number of expansion terms per concept	SMERA
k	Number of LSI dimensions	SMERA
$\alpha 1$	The threshold of the certitude filter	SMERA
$\alpha 2$	The depth threshold of the specificity filter	SMERA
$\lambda_{Smera}, \lambda_{Rm3}$	The weight of the original part against the expanded part of the reformulated query (Eq. 7)	SMERA, RM3

web context, in addition to ROM (Audeh et al 2013), which is a Recall Oriented Measure that also takes precision into account.

An interesting aspect of our approach is the scalability. In fact, SMERA applies LSI to a small number of documents retrieved by the initial query. The complexity of LSI is thus independent from the size of the document collection. The approach, on the other hand, uses only the query and the ontology to expand named entities. As a result, the complexity of our approach does not depend on to the number of documents in the collection, except for retrieving feedback documents (which depends on the retrieval model).

Comparing all of the approaches (QL, RM3 and SMERA) in our study depended on many parameters (cf. Table 4). The values of these parameters were chosen by optimizing the average performance of the measure MAP for each collection. These values were obtained after a tuning step. The experience presented in this paper corresponds to the values presented for each collection in Table 5.

4.2 Results

Table 6 presents the values obtained for the evaluation measures on the four collections and for the compared approaches. Statistically significant improvements or degradations for each couple of approaches are presented in Table 7.

In Table 7 we see that SMERA achieves statistically significant improvement in MAP compared to the use of non-expanded queries for INEX 2006, WT10g and Gov2 collections. Analyzing the test case INEX 2009 showed that 57 % of INEX 2009 queries contained at least four useful terms, larger than the average

Table 5 Selected values of the free parameters for our four test cases

	Inex 2006	Inex 2009	WT10g	Gov2
μ	**2500**	**2500**	**2500**	**2500**
n_{Smera}	20	10	30	10
n_{Rm3}	10	10	10	10
m	5	7	3	7
t	20	20	20	20
λ_{Smera}	0.8	0.8	0.5	0.8
λ_{Rm3}	0.5	0.8	0.8	0.8
k	10	5	10	5
$\alpha 1$	**0.4**	**0.4**	**0.4**	**0.4**
$\alpha 2$	7	7	7	7

Table 6 Evaluation results in MAP, P@10, MRR and ROM on the four test collections

		MAP	P@10	MRR	ROM
Inex2006	QL	33.00	53.00	81.97	83.19
	RM3	**35.96**	**55.00**	80.37	**84.61**
	SMERA	34.78	53.71	**84.81**	83.71
Inex2009	QL	34.17	**97.50**	97.79	45.89
	RM3	34.06	96.76	97.43	45.87
	SMERA	**34.41**	97.21	**98.53**	**46.18**
WT10g	QL	20.16	29.18	58.54	70.74
	RM3	20.49	29.08	56.10	71.06
	SMERA	**21.69**	**29.80**	**59.42**	**71.40**
Gov2	QL	29.41	53.51	72.36	70.57
	RM3	29.97	52.97	68.86	71.15
	SMERA	**30.82**	**56.22**	**75.84**	**71.70**

Bold values are the highest in their column

length of web queries. The MAP of the baseline (QL) obtained in this test case was the highest compared to the one obtained for INEX 2006, WT10g and GOV2. In fact, our approach is designed to improve the precision of short ambiguous queries. Expanding long queries that already have good precision has less chance to improve the performance, as it could change the order of relevant documents already retrieved by the original query. Nevertheless, SMERA obtained statistically better MAP than RM3 on this collection. This can be explained by the use of the quality filters defined in Sect. 3.1.3. Because of these filters, SMERA does not systematically expand all queries with the same number of terms; unlike RM3, which systematically adds 20 expansion terms. Most queries of the other three test cases contain from two to three useful terms (which corresponds to the general case of web queries). For these collections, SMERA had between 4.79 and 7.59 % better MAP than QL. The only

Table 7 Improvement or degradation percentage in *MAP*, *P@10*, MRR et *ROM* for each couple of approaches on the four test collections

		MAP	P@10	MRR	ROM
Inex2006	RM3/QL	+8.97*	+3.77*	−1.95	+1.71
	SMERA/QL	+5.39*	+1.40	+3.46	+0.63*
	SMERA/RM3	−3.28	−2.35	+5.52*	−1.06
Inex2009	RM3/QL	−0.32	−0.76	−0.37	+0.04
	SMERA/QL	+0.70	−0.30	+0.76	+0.63
	SMERA/RM3	+1.03	+0.47	+1.13	+0.68
WT10g	RM3/QL	+1.64	+0.34	−4.16	+0.45
	SMERA/QL	+7.59*	+2.12	+1.50	+0.93
	SMERA/RM3	+5.86*	+2.48	+5.92	+0.48
Gov2	RM3/QL	+1.90*	−1.00	−4.84	+0.82*
	SMERA/QL	+4.79*	+5.06*	+4.91*	+1.60*
	SMERA/RM3	+2.84*	+6.13*	+10.14*	+0.77

* indicates statistical significance ($p < 0,05$) for both t-test and randomization test

case in which RM3 obtained better MAP than SMERA was on INEX 2006 test case, which had the particularity of having the smallest document collection. On the other hand, this better performance in MAP of RM3 over SMERA for the case of INEX 2006 was not statistically significant.

The behavior of RM3 and SMERA in P@10 and MRR was similar to their behavior in MAP on the four test cases. Again, the expansion approaches could not obtain significant improvement in P@10 and MRR on the collection INEX 2009. But SMERA achieved significant improvement over RM3 in MRR on the collection INEX 2006, even though RM3 is better (without statistical significance) on the other measures for this collection. The positive results of SMERA on MRR for the four test cases means that it was able to find the first relevant documents in higher ranks than RM3, which is a very appreciable behavior in a web context.

Another interesting observation is the good performance of SMERA on the largest test case, Gov2. It significantly outperformed QL and RM3 in all precision measures. To better understand this observation, we explored the effect of the collection size on the behavior of RM3 and SMERA. In Fig. 6 we plotted the improvements obtained by RM3 and SMERA in P@10 over the use of non-expanded queries on the four collections.

From this figure we note the decreasing relation between the precision at rank 10 of RM3 and the collection size: the larger the collection of documents is, the less improvement RM3 achieves in P@10. Conversely, SMERA reports better improvement in precision at rank 10 with larger collections, which is also a beneficial behavior in the case of the web. The only exception for SMERA is the case of INEX 2009 because of its long queries, which is not the common case of web queries.

Even though in a web context the recall is not a priority, we think that the study of an approach's behavior from different perspectives to helps better use it in the aimed

Fig. 6 Percentage of improvement in *P@10* for RM3 and SMERA on the four test collections in ascending order according to their size (in number of documents)

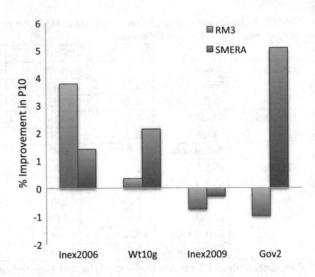

context. The ROM measure shows that both expansion approaches (SMERA and RM3) did not have large neither significant improvements over the baseline. This means that these approaches were not able to find more relevant documents than an approach that uses the basic non-expanded queries. This behavior is due to two main reasons: the first reason is the already high recall of the baseline on all our test cases, as can be seen in Table 8.

The second reason could be the high percentage of non judged documents among the sets of retrieved documents in our test cases (Fig. 7), which is a common but important problem with evaluation campaigns.

This means that even if expansion approaches find new relevant documents, there is a high probability that the documents found were not judged (positively or not) by an assessor.

Finally, we present the advantage of mixing two different approaches of query expansion over the use of each approach separately. While comparing SMERA to the feedback approach, we also analyzed the effect of the number of feedback documents and the number of LSI dimensions, two main parameters that are usually fixed experimentally in similar approaches. In Fig. 8, we fixed the number of feedback documents to 100 and varied the number of dimensions for the collections WT10g and INEX 2006. This performance is compared to SMERA and RM3 with the configurations mentioned in Table 5.

Table 8 The recall at 1000 for the model QL on our four test cases

	Inex2006	Inex2009	WT10g	Gov2
Recall@1000 of QL	83.85	45.95	72.03	71.05

Fig. 7 The average percentage of non judged document per query for our test collections

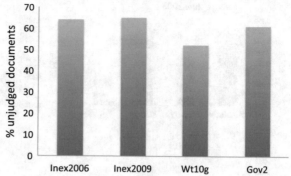

Fig. 8 Mean average precision sensibility to the number of LSI dimensions (*k*) for 100 feedback documents

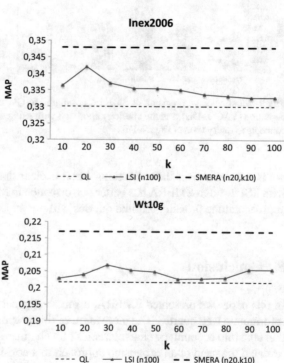

In Fig. 8 we see that using 100 feedback documents with the feedback approach alone could enhance the recall and the precision with 30 and 20 dimensions for the WT10g and INEX 2006 collections respectively, but it was not as good as using the mixed approach of SMERA with 20 to 30 feedback documents for these two collections.

In addition to comparing SMERA to the feedback approach alone, we compared it to the use of the YAGO approach alone. For the later approach, we also considered the effect of disambiguation against the use of the most common concept corresponding to a term in the query. Fig. 9 shows that the effect of using the disambiguation or

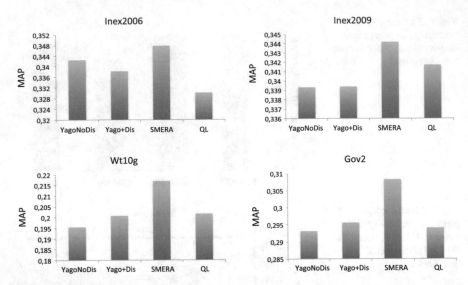

Fig. 9 Mean average precision of SMERA compared to QL, the ontology approach with disambiguation (YAGO+Dis), and the ontology approach with using the most common sense to associate concepts to query terms (YagoNoDis)

not is not stable over the collections–but it is clear that mixing the YAGO approach with LSI through SMERA has better performance in MAP than the use of the YAGO approach alone or using original queries without expansion.

5 Conclusion

In this paper we presented SMERA, a mixed approach to semantically expand and reformulate web queries. The motivation of this proposal was the lack of approaches that take into account the characteristics of web queries. More specifically, our study revealed the need of an expansion approach that considers the importance of named entities and allows an efficient, yet comprehensive, semantic representation of expanded queries. Representing concepts in a keyword query revealed the need to carefully handle the selection of expansion sets and the importance of the way in which these sets should be represented in the final query.

Evaluating our approach on four standard test collections showed the advantage of using SMERA over the use of non-expanded queries and the use of a state-of-the-art expansion method (RM3). Although not very powerful in improving the recall, our approach showed scalability and statistically significant improvements in several precision measures. The analysis of these results, and the comparison to the use of one of the proposed expansion methods in our expansion approach, suggests that SMERA is a well adapted approach for web queries' reformulation.

As a next step, we plan to investigate semantic relationships other than "rdf:label" in YAGO. The idea is to see if a sophisticated choice of the semantic relationship according to the entity type could be of interest. On the other hand, in this work we relied on the assumption that all query concepts (that we discover through our expansion approaches) have the same importance to the query. As we have seen, the approach achieved good performance even with the above assumption. We would like to explore possible solutions to weight concepts' representation, which we would obtain from resources of a different nature: a set of documents (via LSI) and an ontology (via YAGO). Finally, we are convinced of the importance of selective query expansion, which means considering the quality of added terms and not systematically expanding all query terms in the same manner. We saw this aspect investigated in information retrieval, but not explored much by query expansion approaches. Thus, testing existing quality prediction approaches and comparing them to our proposed specificity and certitude filter is an important future step to our work.

References

Audeh, B., Beaune, P., & Beigbeder, M. (2013). Recall-oriented evaluation for information retrieval systems. In: *Information Retrieval Facility Conference (IRFC), Limassol, Chypre*.

Barr, C., Jones, R., & Regelson, M. (2008). The linguistic structure of english web-search queries. In *Proceedings of the Conference on Empirical Methods in Natural Language Processing* (pp. 1021–1030). Association for Computational Linguistics.

Bendersky, M., & Croft, W. B. (2008). Discovering key concepts in verbose queries. In *Proceedings of the 31st Annual International ACM SIGIR Conference on Research and Development in Information Retrieval* (pp. 491–498). ACM.

Bendersky, M., Metzler, D., & Croft, W. B. (2012). Effective query formulation with multiple information sources. In *Proceedings of the Fifth ACM International Conference on Web Search and Data Mining* (pp. 443–452). ACM.

Bendersky, M., Rey, M., & Croft, W. B. (2011). Parameterized concept weighting in verbose queries. In *SIGIR*. ACM Press.

Blei, D., Ng, A., & Jordan, M. (2003). Latent dirichlet allocation. *The Journal of Machine Learning Research, 3*, 993–1022.

Brandao, W., Silva, A., Moura, E., & Ziviani, N. (2011). Exploiting entity semantics for query expansion. In *IADIS International Conference WWW/Internet, Rio de Janeiro*.

Cronen-Townsend, S., Zhou, Y., & Croft, W. B. (2002). Predicting query performance. In *Proceedings of the 25th Annual International ACM SIGIR Conference on Research and Development in Information Retrieval* (p. 299).

Deerwester, S., Dumais, S. T., Furnas, G. W., & Landauer, T. K. (1990). Indexing by latent semantic analysis. *Society, 41*, 391–407.

Deveaud, R., Bonnefoy, L., & Bellot, P. (2013). Quantification et identification des concepts implicites d'une requête. In *CORIA 2013, La dixième édition de la COnférence en Recherche d'Information et Applications, Neuchâtel*.

Gabrilovich, E., & Markovitch, S. (2007). Computing semantic relatedness using wikipedia-based explicit semantic analysis. In *IJCAI*.

Hoffart, J., Yosef, M. A., Bordino, I., Furstenau, H., Pinkal, M., Spaniol, M., Taneva, B., Thater, S., & Weikum, G. (2011). Robust disambiguation of named entities in text. In *EMNLP 2011 Proceedings of the Conference on Empirical Methods in Natural Language Processing* (pp. 782–792).

Huston, S., & Croft, W. B. (2010). Evaluating verbose query processing techniques. In *Proceedings of the 33rd International ACM SIGIR Conference on Research and Development in Information Retrieval* (pp. 291–298). ACM.

Jansen, B. J., Spink, A., & Saracevic, T. (2000). Real life, real users, and real needs: A study and analysis of user queries on the web. *Information Processing and Management, 36*, 207–227.

Kumaran, G., & Carvalho, V. R. (2009). Reducing long queries using query quality predictors. In *Proceedings of the 32nd International ACM SIGIR Conference on Research and Development in Information Retrieval* (p. 564). NY, USA: ACM Press.

Lavrenko, V., & Croft, W. B. (2001). Relevance based language models. In *Proceedings of the 24th Annual International ACM SIGIR Conference on Research and Development in Information Retrieval* (pp. 120–127). NY, USA: ACM Press.

Maxwell, K. T., & Croft, W. B. (2013). Compact query term selection using topically related text. In *Proceedings of the 36th International ACM SIGIR* (pp. 583–592).

Metzler, D., & Croft, W. B. (2004). Combining the language model and inference network approaches to retrieval. *Information Processing and Management, 40*, 735–750.

Metzler, D., & Croft, W. B. (2005). A Markov random field model for term dependencies. In *Proceedings of the 28th Annual International ACM SIGIR Conference on Research and Development in Information Retrieval* (p. 472). NY, USA: ACM Press.

Miller, G. A., Beckwith, R., Fellbaum, C., Gross, D., & Miller, K. J. (1990). Introduction to wordnet: An on-line lexical database. *International Journal of Lexicography, 3*(4), 235–244.

Ponte, J. M., & Croft, W. B. (1998). A language modeling approach to information retrieval. In *Proceedings of the 21st Annual International ACM SIGIR Conference on Research and Development in Information Retrieval* (pp. 275–281). ACM.

Qiu, Y., & Frei, H. (1993). Concept based query expansion. In *Proceedings of the International ACM SIGIR Conference on Research and Development in Informaion Retrieval* (Vol. 11, p. 212). NY: ACM.

Rocchio, J. J., & Salton, G. (1965). Information search optimization and iterative retrieval techniques. In *Fall Joint Computer Conference* (pp. 293–305).

Shah, C., & Croft, W. B. (2004). Evaluating high accuracy retrieval techniques chirag shah. In *SIGIR*. ACM Press.

Strohman, T., Metzler, D., Turtle, H., & Croft, W. (2004). Indri: A language-model based search engine for complex queries. In *Proceedings of the International Conference on Intelligence Analysis*.

Suchanek, F. M., Kasneci, G., & Weikum, G. (2007). Yago: A core of semantic knowledge. In *Proceedings of the 16th International Conference on World Wide Web* (pp. 697–706). ACM.

Voorhees, E. M. (1994). Query expansion using lexical-semantic relations. In *SIGIR 1994*. ACM Press.

Xu, Y., Ding, F., & Wang, B. (2008). Entity-based query reformulation using wikipedia. In *Proceeding of the 17th ACM Conference on Information and Knowledge Mining - CIKM 2008* (p. 1441). NY, USA: ACM Press.

Zhao, L., & Callan, J. (2010). Term necessity prediction. In *Proceedings of the 19th ACM International Conference on Information and Knowledge Management* (pp. 259–268). ACM.

Zobel, J. (2004). Questioning query expansion: An examination of behaviour and parameters. In *SIGIR*. ACM Press.

Multi-layer Ontologies for Integrated 3D Shape Segmentation and Annotation

Thomas Dietenbeck, Fakhri Torkhani, Ahlem Othmani, Marco Attene
and Jean-Marie Favreau

Abstract Mesh segmentation and semantic annotation are used as preprocessing steps for many applications, including shape retrieval, mesh abstraction, and adaptive simplification. In current practice, these two steps are done sequentially: a purely geometrical analysis is employed to extract the relevant parts, and then these parts are annotated. We introduce an original framework where annotation and segmentation are performed simultaneously, so that each of the two steps can take advantage of the other. Inspired by existing methods used in image processing, we employ an expert's knowledge of the context to drive the process while minimizing the use of geometric analysis. For each specific context a multi-layer ontology can be designed on top of a basic knowledge layer which conceptualizes 3D object features from the point of view of their geometry, topology, and possible attributes. Each feature is associated with an elementary algorithm for its detection. An expert can define the upper layers of the ontology to conceptualize a specific domain without the need to reconsider the elementary algorithms. This approach has a twofold advantage: on one hand it allows

T. Dietenbeck (✉)
Sorbonne Université, UPMC Univ Paris 06, INSERM UMRS 1146,
CNRS UMR 7371, Laboratoire d'Imagerie Biomédicale, 75013 Paris, France
e-mail: Thomas.Dietenbeck@upmc.fr

F. Torkhani · A. Othmani · J.-M. Favreau
Clermont Université, Université d'Auvergne, ISIT, BP10448,
63000 Clermont-Ferrand, France
e-mail: Fakhri.Torkhani@udamail.fr

A. Othmani
e-mail: Ahlem.Othmani@udamail.fr

J.-M. Favreau
e-mail: J-Marie.Favreau@udamail.fr

F. Torkhani · A. Othmani · J.-M. Favreau
CNRS, UMR6284, BP10448, 63000 Clermont-Ferrand, France

M. Attene
CNR-IMATI, Via De Marini, 6, 16149 Genova, Italy
e-mail: jaiko@ge.imati.cnr.it

© Springer International Publishing Switzerland 2017   181
F. Guillet et al. (eds.), *Advances in Knowledge Discovery and Management*,
Studies in Computational Intelligence 665, DOI 10.1007/978-3-319-45763-5_10

to leverage domain knowledge from experts even if they have limited or no skills in geometry processing and computer programming; on the other hand, it provides a solid ground to be easily extended in different contexts with a limited effort.

1 Introduction

During the last two decades, adding a semantic description to a scene has been an emerging problematic in the data mining and mesh processing communities. One of the main goals is to be able to extract an abstract description of the manipulated data, using some semantic descriptors. Bridging the gap between raw data and semantic concepts is a very complicated task which usually requires a good knowledge of the specific applicative domain the systems are working on. This link between the expert knowledge and the raw data can be achieved either by using learning techniques or by designing a deterministic system which expresses the knowledge of the expert in a language of computer science.

By assuming that an object belongs to a specific semantic class, advanced mesh segmentation techniques can tag both the whole object and its subparts with semantic terms that describe the shape, the structure and sometimes even the functionality of the object (Laga et al. 2013). When the overall context in known, the identification of an object from its shape may rely on a two-step approach: the object's relevant parts are first recognized through shape segmentation and then labelled using the concepts available in a specific formalization of the context (e.g. an ontology). Some of these approaches (Hudelot et al. 2008; Hassan et al. 2010; Fouquier et al. 2012) are able to exploit the partial semantic description of the scene and to adjust their behaviour to the context. Unfortunately, the algorithms and procedures used in the aforementioned approaches are very specific to their applicative domain and are hardcoded within their implementation.

In this article, we describe an original framework for mesh segmentation that pushes up the semantic approach, thus creating a bridge between an expert knowledge description and the segmentation algorithms. This framework allows an expert in a specific area to formally describe his own ontology, without the need to have any particular skill in geometry processing or shape analysis. The formalized domain description is then used by the system to automatically recognize objects and their features within that domain.

The genericity of our approach is enabled by a multi-layer ontology which formally describes the expert knowledge. The first layer corresponds to the basic properties of any object, such as shapes and structures. The upper layers are specific to each application domain, and describe the functionalities and possible configurations of the objects in this domain. The segmentation and the annotation mechanism is hidden behind the concepts of the first layer, which are associated to specific geometric algorithms.

In Sect. 2 an overview of the existing methods for part-based annotation of 3D shapes is presented. Section 3 introduces our framework for semantics-driven mesh segmentation and annotation; here we describe how to model the expert's knowledge, how our synthetic catalog of segmentation algorithms is composed, and how our expert system works. The elementary algorithms that compute the basic shape features are described in Sect. 4. Section 5 presents some experimental results to illustrate the relevance and feasibility of our approach. Section 6 is reserved for the discussion of possible extensions of this work and future improvements to overcome its current limitations.

2 Related Work

The design of enhanced vision systems can take a significant advantage from semantic formalizations such as ontologies. The ontology paradigm in information science represents one of the most diffused tools to make people from various backgrounds work together (Seifert et al. 2011). Ontology-based interfaces are a key component of ergonomic adaptive computer systems (Maillot and Thonnat 2008), especially in biological/clinical fields for which concepts and standards are constantly shifting (Hassan et al. 2010; Fouquier et al. 2012; Othmani et al. 2010). Furthermore, reasoning capabilities embedded in the logical framework on which ontology softwares are built up can be exploited to improve the performances of existing computer vision systems. Even though still brittle and limited, reasoning inferences out of visual data may enhance the vision system experience (Othmani et al. 2010) as well. In the Computer Graphics community the use of semantics has been a key for significant achievements in applications such as anatomy (Hassan et al. 2010), product design in e-manufacturing (Attene et al. 2009) and robotics (Albrecht et al. 2011; Gurau and Nüchter 2013).

In Camossi et al. (2007), a system to assist a user in the retrieval and the semantic annotation of 3D models of objects in different applications is presented. The ontology provides a representation of the knowledge needed to encode an object's shape, its functionality and its behavior. Then, the annotation and the retrieval are performed based on the functional and the behavior characteristics of the 3D model.

In Attene et al. (2009), the ontology is used to characterize and to annotate the segmented parts of a mesh using a system called "ShapeAnnotator". To this aim, the ontology is loaded according to the input mesh type and the user can link segments to relevant concepts expressed by the ontology. While the annotation of the mesh parts is done by a simple link within the "ShapeAnnotator", Gurau and Nüchter (2013) and Shi et al. (2012) proposed to feed an ontology with a set of user-defined rules (e.g. geometric properties of objects, spatial relationships) and the final annotation is created according to them.

In Hassan et al. (2010), an ontology including an approximation of the geometric shape of some anatomical organs is used to guide the mesh segmentation. The parameters needed to segment the input mesh are provided by the ontology. For the case of semantic classification, an ontology is also used in Albrecht et al. (2011) to "idealize" SLAM-generated 3D point cloud maps. The ontology is used to generate hypotheses of possible object locations and initial poses estimation, and the final result is a hybrid semantic map where all the identified objects are replaced by their corresponding CAD models. Recently, Feng and Pan (2013) proposed a unified framework which bridges semantics and mesh processing. The mesh is divided into a fixed number of parts corresponding to the number of concepts in the ontology. The parts are then annotated based on some rules defined in the ontology and specific to the context (e.g. in a human body the head and a limb are very dissimilar).

In Symonova et al. (2006), the authors focus their work on mesh annotation using an already segmented object. The regions to annotate are defined by connected components in the shape, and the low level labelling is done by using the distribution of local geometric features. The high level annotation is then produced by using connectivity rules and geometrical primitive constraints. Our present work can be seen as a generalization of this original work for the annotation part, while we introduce specific and robust algorithms for the low level labelling, and a complete framework that extends the possibilities introduced by Symonova et al. In our proposal, each new relation or low level label with the corresponding elementary algorithm will be handled automatically by our expert system.

3 Proposed Method

Segmenting an object based on its geometry and associating semantic concepts to each part are non trivial problems, both requiring a very specific processing. Previous works (Hudelot et al. 2008; Attene et al. 2009; Hassan et al. 2010; Fouquier et al. 2012) on this topic are more and more going in the direction of mixing the two problems, in order to help the segmentation using the already extracted semantics, and by extracting the semantics from the partial segmentations.

The framework we present in this section addresses the same goal by focusing on the separation between the segmentation algorithms and semantic reasoning. Thanks to such a separation, addressing a new applicative domain will only require the user to design the corresponding ontology, without code modification. All the necessary geometric machinery and the low-level semantic rules, indeed, are already encoded in the original core of the framework. In the next section, we propose complementary benefits of this approach, and we discuss in the last section several possible extensions.

In a first part, we describe our multi-layer ontology paradigm and how the expert knowledge on a specific domain is implemented on top of elementary concepts. Then we introduce our framework to tackle the problem of semantics-driven mesh segmentation using the defined ontology paradigm. Finally, we give the specifications of the elementary algorithms involved in the segmentation process.

3.1   Terminology

Our work focuses on mesh processing, with the aim to achieve both segmentation and semantic labelling. In the remainder, we call *object* a 3D shape defined by a triangular mesh. A *region* of the mesh is a connected subset of its triangles. A segmentation is a subdivision of the mesh into a collection of regions called *segments*. A region is not necessarily a segment. Finally, we use the term *concept* in the usual way to describe an abstract semantic class.

3.2   Expert Knowledge Description

The idea behind our expert knowledge system is to mimic our visual mechanism of object recognition. Indeed the expert knowledge description follows a "whole-to-part" analysis, where the object of interest is modelled in the top layer of our ontology and is first decomposed into regions according to a criterion (shape, texture, …). A region can potentially be further divided until elementary subparts are obtained corresponding to the intermediate layers of our ontology. Features of these parts as well as relations between them are described using semantic concepts gathered in the last layer of the ontology. Putting these informations together along with user-defined rules allow us to label each part ultimately leading to the recognition of the object.

A segmentation and semantic labelling process implies that the algorithmic part is able to identify and label regions with specific properties, such as geometrical properties (e.g. sphere, cube, vertical region), color or texture properties (e.g. color uniformity, reflectance, texture patterns), but also properties linked to the relative position and configuration of subparts (e.g. parallel regions, $\mathscr{A}$ is up *wrt* $\mathscr{B}$, $\mathscr{A}$ is between $\mathscr{B}$ and $\mathscr{C}$). We note however that all these properties cannot be formalized in the same way. Indeed, some of them extend the knowledge on *a given* part of the mesh by detailing its features (e.g. geometry, color, orientation), while other link *several* regions together (e.g. through topology, distance) thus describing the overall structure of the object.

3.2.1   Features and Unary Properties

A first group of properties characterizing an object are based on the aspect of its parts (e.g. geometry, color, etc.). These properties extend the knowledge on a given part and thus link it to a feature concept (e.g. $\mathscr{A}$ is a cube, is red, etc.). Since each property only involve one part of an object, we will refer them as *unary property*.

More specifically, in the ontology, each unary property corresponds to an object property where the domain is the object parts. The *range concept* corresponds to the features of interest (e.g. shape, orientation, color) which are further specialized

(a)   **(b)**

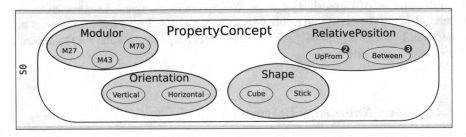

Fig. 1 Examples of unary and *n*-ary property. $\mathscr{A}$ and $\mathscr{B}$ correspond to regions of the object

Fig. 2 A subpart of the elementary semantic concepts of the core ontology. The number beside some concepts indicate that they link several regions together (e.g. UpFrom relates 2 regions $\mathscr{A}$ and $\mathscr{B}$)

into *elementary semantic concepts* corresponding to the possible value of the object property. An example of unary property is given in Fig. 1a and range concepts and their associated elementary semantic concepts are illustrated in Fig. 2 (green ellipses).

3.2.2 Topology and N-Ary Properties

The second group of properties allows to link object's parts together thus giving the overall structure of the object. This kind of properties combines 2 or more parts along with a topological concept (e.g. $\mathscr{A}$ and $\mathscr{B}$ are close, $\mathscr{B}$ is above $\mathscr{A}$, etc.) and will thus be referred to as *n-ary* property.

However, ontologies cannot directly model n-ary properties and we thus model them following the "N-ary Relation" pattern described in Dodds and Davis (2012).[1] More specifically, a concept (e.g. position, distance) is created for the relation and is also specialized into *elementary semantic concepts* (e.g. position → UpFrom, Between, etc.). Each object part corresponds to a ressource of the relation and is related to it through two distinct object properties. The first one "isTargetRegion" allows to define the region we wish to position in space, while the other "isReferenceRegion" states the fixed region used to orientate the space (i.e. to define the semantic).

Considering the example given in Fig. 1b, the n-ary property can be read as $\mathscr{B}$ (the target) is above $\mathscr{A}$ (the reference). The violet ellipse in Fig. 2 illustrates a range of two n-ary relations where the number on the right of each elementary semantic concepts indicates the number of object parts related through this concepts (e.g. 2 parts for UpFrom).

---

[1] See also http://patterns.dataincubator.org/book/nary-relation.html for notations.

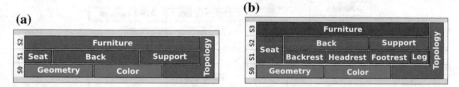

Fig. 3 Example of multi-layer ontologies for furnitures. Note how a more detailed expert knowledge description of the same domain can be achieved by simply adding a layer, e.g. for the concepts of back or support in the case of furnitures

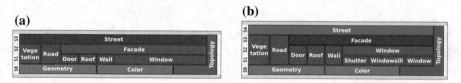

Fig. 4 Example of multi-layer ontologies for streets

3.2.3 Multi-layer Ontology

In order to model our "whole-to-part" recognition mechanism, we structured our ontology in layers. The first layer (called S_0, blue blocks in Figs. 3 and 4) is part of the core of our framework. It contains the feature and topological concepts and associated unary and n-ary concepts described in the previous sections.

The S_0 layer is enriched for each applicative context with specific semantic concepts which are grouped into two (or more) layers: one layer called S_1 (yellow blocks in Figs. 3 and 4), using only references to concepts from the S_0 layer, describes all the object configurations that can be combinatorially built, *i.e* the result of a cartesian product between *elementary semantic concepts*. Figure 5 illustrates an example of such a cartesian product for 2 unary properties (Shape and Orientation).

The supplementary layers ($S_2, \ldots, S_n$, green, red and purple blocks in Figs. 3 and 4) describe the combination rules to populate a complete scene. The top layer of the ontology corresponds to the list of objects one wish to label. The advantage of this representation is that it is very easy to fit the level of details in the description of objects required by the application. Indeed if supplementary informations are required for an object part, it only requires to add a layer in the ontology. Similarly if only a rough description of the part is sufficient, one may remove the unnecessary layer.

To illustrate this motivation, consider the example given in Fig. 3. In Fig. 3a, a furniture is obtained by combining a seat, a back and a support, all of which being described using S_0 concepts. For instance, the back of a furniture is modelled in a rough manner as being a vertical board possibly linked to vertical sticks (e.g. for a chair) and above the seat. However this may not be sufficient to distinguish between all the furnitures: namely a difference between a chair and an armchair resides in

Shape
Board
Cube
Stick

×

Orientation
Horizontal
Vertical

=

Shape × Orientation		Eq. Concept
Board	Horizontal	Seat
Board	Vertical	Foot
Cube	Horizontal	Seat
Cube	Vertical	Impossible
Stick	Horizontal	Impossible
Stick	Vertical	Foot

Fig. 5 Cartesian product between two range of elementary semantic concepts, and the corresponding equivalent concepts

the fact that armchair may have a headrest. To add this supplementary knowledge, a supplementary layer is introduced (see Fig. 3b) describing the back of a furniture as a backrest and a possible headrest above it.

3.2.4 Linking Two Layers: Equivalent Concepts

In addition to the semantic concepts of the application domain, another important expert knowledge consists in how the concepts are linked with each other. This can be expressed as a set of equivalent concepts of the S_n layer describing possible or impossible combinations of concepts of the S_{n-1} layer. The work of the expert is thus strongly simplified since he can describe not only positive rules (i.e. valid/plausible configurations), but also negative rules (i.e. impossible/incompatible combinations). For example, in the furniture ontology a chair leg can be described as a stick shape with a vertical orientation; on the other hand, the combination of a headrest without backrest is an incompatible configuration.

In practice, incompatible configurations are specialized in specific concepts, one for each incompatibility type. This specialization allows to perform some reasoning and classification on partially annotated individuals.

Once these equivalent concepts are given by the expert, they are used in two ways: either to *build a decision tree* (which will be detailled in Sect. 3.3) or to *suggest a correction of the segmentation*. Indeed, during the segmentation/annotation process, incompatible configurations might appear due to either a segmentation error or a missing equivalent concept. In this case, the reasoner can be asked for the cause of the inconsistency which is then presented to the user for correction. The main advantages of this approach is that it allows us to ask the user to correct errors only in the regions that caused an incompatibility instead of having to explore the whole mesh/labelling.

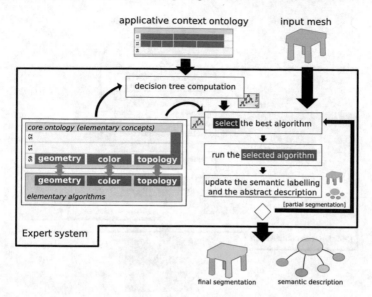

Fig. 6 Overview of the proposed method applied on the furniture domain

3.3 Expert System

In this section, we describe how the expert knowledge is used in our framework to efficiently segment and annotate an object. Figure 6 gives an overview on this framework.

3.3.1 Generation of a Decision Tree

One of the advantages of our approach is that it gives the possibility to easily compute a tree containing the order of the questions to ask to reach the solution in the most efficient way. To build this *decision tree*, we first use the equivalent concepts of each layer to build the set of possible configurations. For each layer and starting at the first layer S_0, the Cartesian product between all the properties of the layer is performed (Fig. 5, Shape × Orientation column). The reasoner is then used to classify the instances in equivalent concepts and incompatible candidates are removed (Fig. 5, Eq. Concept column). The remaining ones are then used in the above layer as semantic concepts in the Cartesian product to compute the new list of possible configurations.

Once the set of all possible configurations Ω is created, the idea is to split it according to the concept maximizing a criterion C. The choice of this concept gives us the question to ask and thus a node of the tree. For each possible answer, we then get the corresponding subset and look for the next concept maximizing C. This operation is iterated until only one possibility is left in each subset. In the resulting tree, the root thus corresponds to Ω and stores the first question to ask, each leaf is

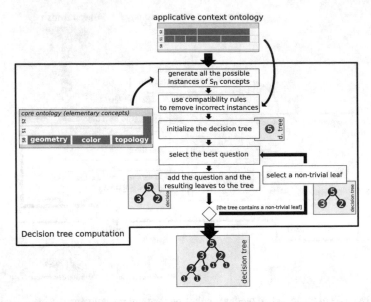

Fig. 7 Offline decision-tree generation from the ontology of the application domain

a possible configuration and intermediary nodes are subsets of Ω storing the next question to ask. Note that this step can be performed only once and offline in order to speed up the process. This procedure is illustrated in Fig. 7 and an example of complete decision tree is given in Fig. 8 where the criterion used to split Ω is the dichotomy. The reader will also find an illustration on the applicative context of furnitures in Sect. 5.1.1.

3.3.2 Optimized Segmentation Using the Decision Tree

The first inline step of our expert system is to browse the decision tree: starting from the root, the system gets the question to ask and runs the associated elementary algorithm on the input object (see Sect. 3.4 for the formulation and classification of these algorithms). The system then selects the child node corresponding to the algorithm's result and the process iterates until a leaf of the tree is reached, meaning that the semantics associated to the object is known as only one possible configuration remains. A complete run of the inline process is given in Sect. 5.1.2 on the applicative context of furnitures.

In some cases, the expert system might reach a leaf before every part of the mesh is segmented or annotated (because some concepts might be inferred from the presence/absence of others). Some supplementary algorithms can then be run on the mesh to confirm the global semantics and annotate the missing parts. This can again be done very efficiently by questioning the reasoner, which will give the expert system the missing concepts and thus the elementary algorithm to run.

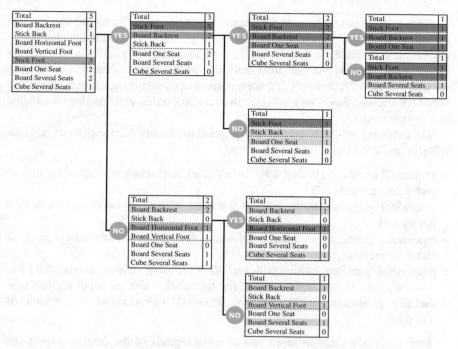

Fig. 8 Illustration of the generation of the decision tree. The *dashed lines* group the elementary semantic concepts into a common property. *Yellow* best question at a given node of the tree; *Orange* asserted concepts through previous question; *Blue* inferred concepts (because only one possibility *left* in the property)

3.4 Mesh Processing Formalisation

In this work, the core idea about the mesh processing part is to split the processing into *elementary algorithms*, each dedicated to the evaluation of one of the elementary semantic concepts. To produce a synthetic catalog of the algorithms, however, the signature of each algorithm must be defined both by the concept it handles and by the desired specific task.

3.4.1 Semantic Type Signatures of Algorithms

Each elementary algorithm is involved in the segmentation and labelling process and can realize the following tasks:

- looking for a given concept and segmenting the mesh in corresponding subparts,
- checking if an existing region satisfies or not a given concept,
- focusing on relative positions or links of several regions.

Since a region can be described by more than one concept (for example, a part can be both a "stick" and "vertical"), and since our goal is to have as many elementary algorithms as possible, we deduce that there also exist algorithms that do not split the given region, but only increase the knowledge on this region. Finally, and because we want to deal with concepts that are not necessarily involving a single region, we need to distinguish functions associated to unary properties, and functions associated to n-ary properties.

The following synthetic catalog is a proposal to identify each segmentation algorithm in the context of semantic labelling:

- semantical questions starting with *find all* (SF), with unary concepts (*e.g.* find all rectangles in region $\mathscr{A}$),
- semantical questions starting with *is it a* (SI), with unary concepts (*e.g.* is $\mathscr{B}$ a flat region),
- topological questions (T), identifying n-ary relations between regions (*e.g.* are $\mathscr{B}$ and $\mathscr{C}$ connected),
- topological questions starting with *find all* (TF), using an n-ary relation and 1 to $n - 1$ regions, to find regions satisfying the relation *wrt* the input regions (*e.g.* find all regions connected to the region $\mathscr{B}$; find all regions between the regions $\mathscr{B}$ and $\mathscr{C}$).

Each algorithm takes as input one or more regions of the original object and returns a set of regions, each enriched by a semantic description generated by the function, plus a score in [0; 1] to illustrate the matching between this region and the associated concept. This first synthetic catalog of possible semantic type signatures covers all the useful algorithms for mesh segmentation using semantic description, as illustrated in Sect. 3.3 and thereafter.

3.4.2 Minimal Set of Algorithms

Designing one algorithm of each type for each property can be very complex, in particular for concepts that are not very precisely defined. In particular, if a SI algorithm can be designed for each property, the SF algorithms are taking on them the complicated task of segmentation. Depending of the accuracy of the concept, it can be very complicated to design a corresponding SF algorithm.

Fortunately, the expert system described in the next paragraph is able to deal with a minimal set of algorithms that does not contain all the SF algorithms. Whenever a new *find all* algorithm is required, the system restricts itself to the existing ones. The only mandatory constraint is that all the regions of the objects must be segmentable. We describe in Sect. 4 an example of such a minimal set.

For n-ary properties we can proceed in a similar manner. T algorithms must be implemented for each n-ary properties. On the other hand, TF algorithms can be implemented, or only deduced using a combination of T and SF algorithms.

3.4.3 Region Description

The algorithms of our framework use regions of meshes as input, and can also produce regions in case of "find all" algorithm. A classical way to describe regions consists on using existing vertices, edges and triangles without refining the data structure. Moreover, a segmentation process consists in splitting a given object into subregions. Using triangles as elementary parts of a surface mesh is then a straightforward choice.

However, we know that the result of a segmentation step is sometimes uncertain, in particular on the segment borders. To handle this uncertainty, we choose to use in this work fuzzy maps to describe regions. A fuzzy map is defined as a function that associates to each triangle of the mesh a membership value in [0; 1]. This formulation gives a way to deal with contradictory segmentations such as partially overlapping regions.

4 Details of the Elementary Algorithms

We defined a minimal set of algorithms corresponding to the S_0 concepts described in Fig. 12 left.

4.1 Fuzzy Membership

Each of the defined algorithms is able to express its result using a score value in [0; 1] to illustrate the matching between the manipulated region and the corresponding concept. In this work, we choose to use trapezoidal membership functions (Klir and Yuan 1995) to compute this score. A trapezoidal membership function is defined by 4 real parameters $a_1 \leq a_2 \leq a_3 \leq a_4$ as described in Fig. 9.

The defined algorithms are described in the next section, with a common use of the trapezoidal membership functions as fuzzy rules: we first measure a geometric property on the manipulated region (such as an angle with the vertical axis, or a ratio between scales), then we use a specific fuzzy rule to obtain the final score.

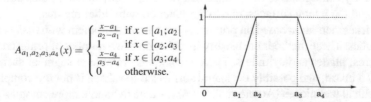

$$A_{a_1,a_2,a_3,a_4}(x) = \begin{cases} \frac{x-a_1}{a_2-a_1} & \text{if } x \in [a_1;a_2[\\ 1 & \text{if } x \in [a_2;a_3[\\ \frac{x-a_4}{a_3-a_4} & \text{if } x \in [a_3;a_4[\\ 0 & \text{otherwise.} \end{cases}$$

Fig. 9 Trapezoidal membership function A_{a_1,a_2,a_3,a_4}

We can notice here that more complex membership functions can be used for this description, but trapezoidal functions has been identified as rich enough to drive our algorithms in the applicative context described in Sect. 5.

4.2   Elementary Algorithms

For each of these concepts we designed an "is-a" algorithm. "Find-all" algorithms have been defined only for the shape properties, since this property domain is handling all the possible regions we want to address. This partial implementation of the "find-all" algorithms is motivated in Sect. 3.4.2, and we focus our work on the shape properties due to their efficiency to define strict boundaries.

4.2.1   "Is a" Algorithms

Each of the elementary properties has been translated into an "is a" algorithm. As described in Sect. 3.4, the input of these algorithms is defined by a fuzzy region F on a mesh $\mathcal{M}$.

Using elementary geometric algorithms, we designed "is a" algorithms to identify shape properties (cube, stick, board), vertical and horizontal orientations, compactness of a shape, and position properties *wrt* the vertical axis. The details of these algorithms presented in Sect. 6.2 are related with the extraction of an oriented bounding box, and some fuzzy rules to quantify the examined properties.

4.2.2   "Find all" Algorithms

The quality of the final segmentation and annotation is mainly related to the quality of the "find all" algorithms: in our framework, the location of the region and the accuracy of their boundaries depend only on these algorithms. We focus our implementation work on the algorithm able to detect boards, i.e. able to detect regions made up of two parallel planar regions and possibly lateral surfaces. Figure 10 give an example of board that illustrate possible specific configurations: the lateral surfaces may not be perfectly orthogonal to the planes, and the shape may contain holes both in the planes and the lateral surfaces, due to the junction with other regions.

Mesh segmentation based on primitives is a classical problem with many existing approaches. These methods are usually focusing on a few number of primitives, such as spheres, planes and cylinders. These techniques are able to segment each of the faces of a board, and possibly the lateral surfaces if the shape is not too complicated. Hierarchical techniques (Attene et al. 2006) are able to handle more complex shapes by providing a multi-level segmentation. Figure 11a illustrates the inability of these approaches to handle board shapes: a lateral surface of the seat has been selected as a part of the back rest.

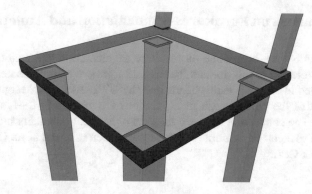

Fig. 10 A board is composed of two parallel planar regions (*pink*) with lateral surfaces (*blue*), possibly with holes due to connections with other regions (*leg*, *back*)

(a)   **(b)**

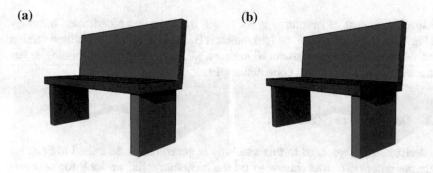

Fig. 11 **a** Hierarchical primitive fitting (Attene et al. 2006) on a bench. **b** What we expected

Alternative approaches are using first a primitive fitting procedure, then combine the detected primitives to generate thin-plates (Geng et al. 2010).

We have designed an alternative approach that performs the fitting of two parallel primitives at the same time, then defines the shape by adding the lateral surfaces. This approach avoid the over segmentation of planar regions without opposite parts, and could possibly be extended to non straight boards, by introducing a parametric model for parallel bended surfaces.

Section "Find all" Algorithms gives the details of our approach, which uses first an estimation of the local thickness of the object, then uses a growing process from the thin points to find the two sides of each region. Finally, a last step is applied to localize the side triangles.

5   Experiments on Furnitures Segmentation and Annotation

The experiments have been done on furniture segmentation and annotation, using basic shapes relevant to this domain (see Fig. 12 right). We implemented the expert system detailed in Sect. 3.3 using `Java` and the OWL API, and designed our prototype such that the purely mesh manipulations are written in `C++`, using `CGAL` (see Sect. 4). The connection between these two parts is done using a client/server paradigm via sockets. The complete software has been released as an Open-Source software under GPL.[2]

5.1   Applicative Context and Results of the Annotation

The specific context of furnitures is addressed by designing a dedicated ontology (see Fig. 12 left) using Protégé[3] and the reasoner HermiT on top of a simplified version of the elementary concepts introduced in Sects. 3.2.1 and 3.2.2. These ontologies has been released with our source code under GPL.[4]

5.1.1   Decision Tree Generation

The decision tree associated to this ontology is generated as described in Sect. 3.3 where the criterion C was chosen to be the dichotomy, i.e. we look for concepts allowing to split the set into two subset of same size.

5.1.2   Expert Segmentation Processing

Once the decision tree is generated, we can run our expert segmentation system on meshes. Fig. 13 gives the list of questions that are computed in order to segment and recognize the first bench in Fig. 12. The other images in Fig. 12 illustrate the segmentation and annotation process using the same expert ontology with various meshes.

Figure 14 illustrates the consequence of a rule (Sect. 3.2.4) that detects a configuration that is not part of the possible furnitures described in the expert system: a board is identified in the foot part, and the object is classified as incompatible (concept `FootIncompatibleFurnitures`).

---

[2] Software and ontologies of our work are available online: http://odds.jmfavreau.info/.

[3] Protégé, an ontology editor: http://protege.stanford.edu/.

[4] See Footnote 2.

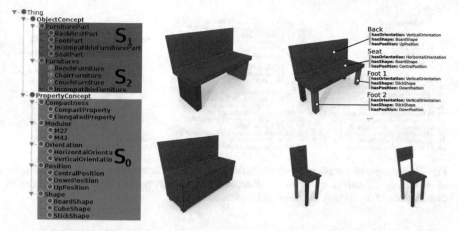

Fig. 12 Ontology and meshes used for our experiments on furniture segmentation and annotation. *Right* result of the segmentation and annotation on 5 objects from the furniture domain. Result of the annotation: *orange* ⇒ backrest, *blue* ⇒ seat, *other colors* ⇒ feet

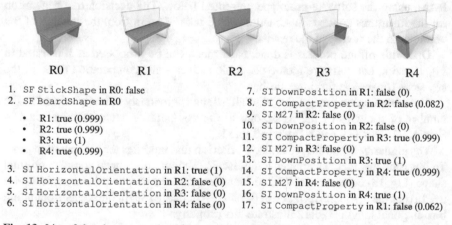

| R0 | R1 | R2 | R3 | R4 |

1. SF StickShape in R0: false
2. SF BoardShape in R0

   - R1: true (0.999)
   - R2: true (0.999)
   - R3: true (1)
   - R4: true (0.999)

3. SI HorizontalOrientation in R1: true (1)
4. SI HorizontalOrientation in R2: false (0)
5. SI HorizontalOrientation in R3: false (0)
6. SI HorizontalOrientation in R4: false (0)

7. SI DownPosition in R1: false (0).
8. SI CompactProperty in R2: false (0.082)
9. SI M27 in R2: false (0)
10. SI DownPosition in R2: false (0)
11. SI CompactProperty in R3: true (0.999)
12. SI M27 in R3: false (0)
13. SI DownPosition in R3: true (1)
14. SI CompactProperty in R4: true (0.999)
15. SI M27 in R4: false (0)
16. SI DownPosition in R4: true (1)
17. SI CompactProperty in R1: false (0.062)

Fig. 13 List of the elementary algorithms and corresponding answers generated by our expert system to segment and annotate a bench mesh. SF: semantic *find all*, SI: semantic *is it a*

5.2 Evaluation of the Approach

In this section, we first give time information on the experiments we performed, providing computation times for each of the main stages of our pipeline. In the next section, we present some comparison facts with existing work. Finally, we present some high level complexity elements considering the usage of the ontology.

1. SF StickShape in R0

- R1: true (1)
- R2: true (0.999)
- R3: true (1)
- R4: true (1)
- R5: true (1)
- R6: true (1)

2. SI VerticalOrientation in R1: true (1)
3. SI VerticalOrientation in R2: true (1)
4. SI VerticalOrientation in R3: true (1)
5. SI VerticalOrientation in R4: true (1)
6. SI VerticalOrientation in R5: true (1)
7. SI VerticalOrientation in R6: true (1)
8. SI DownPosition in R1: true (1)
9. SI DownPosition in R2: true (1)
10. SI DownPosition in R3: true (1)
11. SI DownPosition in R4: true (1)
12. SI DownPosition in R5: false (0)

13. SI DownPosition in R6: false (0)
14. SI CompactProperty in R5: false (0)
15. SI M27 in R5: false (0)
16. SI CompactProperty in R6: false (0)
17. SI M27 in R6: false (0)
18. SF BoardShape in R0- { R1, R2, R3, R4, R5, R6 }

- R7: true (0.999)
- R8: true (0.999)
- R9: true (1)

19. SI VerticalOrientation in R7: false (0)
20. SI VerticalOrientation in R8: false (0)
21. SI VerticalOrientation in R9: true (1)
22. SI UpPosition in R9: true (0.858)
23. SI CompactProperty in R7: true (0.999)
24. SI M43 in R7: false (0)
25. SI CompactProperty in R8: true (0.999)
26. SI M43 in R8: true (0.956)
27. SI CentralPosition in R8: true (0.699)

Fig. 14 Example of a mesh which is not valid *wrt* the expert knowledge, the corresponding list of elementary algorithms and answers produced by our expert system, and the final diagnosis: ChairFurniture, FootIncompatibleFurnitures

5.2.1 Computation Time

The presented computations has been done on an Intel i-7 2.00 GHz with 8 Go of RAM, using the software prototype presented below. The decision tree generation for the furnitures context illustrated in Fig. 8 takes 11.45 s from the loading of the ontology to the saving of the results.

Once this offline process is done, each mesh can be processed as illustrated in Fig. 13 using our software prototype. Table 1 gives some computation times of the key steps for several furniture meshes.

The computation time of the "find all" steps is obviously strongly related to the number of triangles of the mesh, due to the computation technics introduced in Sect. 4.2.2. This fact can be seen in Table 1.

The planarity of the shapes is also a criterion that modifies the computation time. To illustrate this fact, we applied a multiscale random noise on an complex original shape (Fig. 16) composed of 75 520 triangles gathered in 4 board-like regions. We used a root mean square deviation of 0.5 % (respectively 1 %) of the object's bounding box diagonal length. Table 2 illustrate this property.

Table 1 Computation time of the expert system for segmentation and annotation of several meshes

Mesh	#subparts	#triangles	Find all (s)	Is a (ms)	Total (s)
bench	4	36	0.188	0.259	2.270
bench2	6	52	0.216	0.319	2.712
chair	6	52	0.231	0.326	3.096
chair2	8	82	0.276	0.509	2.376
couch	6	52	0.291	0.264	2.295
curved-bench	29	2542	1.038	6.167	6.743
strange-object2	9	100	0.337	0.695	2.502

Table 2 Computation time of the segmentation process on a 75,520 triangles mesh with different levels of noise, with our method and with (Attene et al. 2006)

Noise	0 %	0.5 %	1 %
Our method	3.40 s	5.25 s	7.01 s
Fitting primitives	24 s	6 s	6 s

It gives the computation time of the "find all" segmentation process that segments the shape into two board regions, one stick region and one cube region.

We used the same computer to run the implementation of Attene et al. (2006) provided by the author. This existing work is not exactly analogous, since it provides a complete hierarchical representation of fitted primitives, and do not gives a way to obtain a solution to the board finding problem. However, it gives a good intuition about the state of the art in primitive detection.

The computation times given in Table 2 are illustrating the fact that the two methods share the same order of magnitude.

5.2.2 Comparison with Existing Work

Comparing our current work with existing methods can be separated into two sides: segmentation and annotation.

In our current implementation, the segmentation part is mainly driven by the "find all" board described in the previous section. As we underline at the beginning of this section, the closest approach is the thin-plates mesh model splitting approach (Geng et al. 2010). This reference work does not contain any complexity or computation time evaluation, but the fact that we fit simultaneously the two planes of our boards necessarily reduces the complexity of our approach, since our model avoids the segmentation of single planes that are not accompanied by another parallel surface. For example, a thin board with a profile composed of n segments will be composed of $n + 2$ planes. A classical approach will detect $n + 2$ planes, while the proposed method will detect only a pair of planes. The n other planes will be part of the detected border, without subdivision. Moreover, our method is able to handle thin-plates with complex boundary shapes (as illustrated in Fig. 16), that are not handled by classical fitting primitive approaches.

The other "is a" algorithms are not strongly novel. The novelty consists here on the categorisation and review described in section "Is a" Algorithms: to the best of our knowledge, no other work provides such a comprehensive analysis.

The annotation part of our framework consists both of the "is a" algorithms that we mentioned above and the expert system described in section "Find all" Algorithms. This work can be compared with existing works such as Feng and Pan (2013) and Attene et al. (2009), but in these two approaches the complexity of the annotation is handled by either the user or context-specific algorithms. As a result, we are not aware of any other significant existing work which we could compare to.

Table 3 Number of segmentation steps for a complete annotation and segmentation

Object	#subparts	S. then I.		Naive S&I		Our method	
		Preprocessing		-		-	
		# SF	# SI	# SF	# SI	# SF	# SI
Bench 1	4	0	28	3	20	2	15
Bench 2	6	0	42	3	30	3	14
Couch	6	0	42	3	30	2	13
Chair 1	6	0	42	3	30	2	14
Chair 2	8	0	56	3	40	2	22

5.2.3   On the Usage of the Ontology

In this section, we present a short comparison of our framework with two alternative approaches, in case the ontology is not exploited to drive the expert system. Experimental results of this comparison are summarized in Table 3. The first column, which is entitled *S. then I.* corresponds to an approach where a first segmentation preprocessing is done to split regions, then each region is labeled using the semantic concepts. The second column which is entitled *Naive S&I* corresponds to an approach where the initial split is produced using segmentation algorithms dedicated to shape detection, then by using the complementary algorithms to fully identify the regions. The last column corresponds to our approach. For each method, we detailed the number of semantic *"find all"* (SF) and semantic *"is it a"* (SI) algorithms required to segment and annotate the 5 objects shown in Fig. 12 right.

We choose to distinguish the 2 kinds of algorithms, because the complexity of each algorithm family is significantly different: a SF algorithm will require to browse all the given regions (possibly the whole mesh), and will have to extract the subparts from it. In comparison, a SI algorithm will only have to validate or not a feature on a given region. Minimizing the number of SF runs is thus the main goal of a segmentation and annotation process.

For each region of an object, the expert of the ontology uses 5 range concepts (shape, position, orientation, modulor, compactness) that implies 12 elementary semantic concepts. The number of SI algorithms in the *S. then I.* has been estimated counting for each subregion of the mesh one SI per elementary semantic concept (minus 1 per range concept that can be deduced). The *Naive S&I* consist in running the SF algorithms dedicated to shapes, then run all the other SI algorithms on each of the segmented regions (all minus one per concept range). The number of SF and SI of our method comes from the trace of the experimental runs (see an example in Fig. 13).

We first compare our work to a *S. then I.* approach, where a segmentation preprocessing is applied before the annotation. We cannot quantitatively compare this approach with ours, but since we use the expert knowledge to reduce the number of

SF algorithms in our approach, we can deduce that our SF computations are almost equivalently expensive as the preprocessing stage of the *S. then I.* approach.[5] The number of SI algorithms is also strongly reduced with our approach.

The second approach considered for comparison is a *Naive S&I* approach, where all the SF algorithms of the shape range are run, then all the complementary SI algorithms are run. The number of SF algorithms is at least preserved by our approach and sometimes reduced, and many SI runs are saved thanks to our framework.

These first results motivate the relevance of our method with respect to the existing approaches: mixing segmentation and annotation steps is a good approach to reduce the theoretical complexity of the global algorithm, by reducing the number of segmentation and identification algorithms to run. Moreover, it gives a global framework to extend the existing approaches to more complex contexts.

6 Conclusion and Future Work

In this paper, we presented a new framework for efficient segmentation and annotation of meshes. It is composed of two blocks: a multi-layer ontology gathering the semantics and a segmentation part allowing to detect elementary geometrical, chromatical and topological concepts. The main advantage of our method is that it separates the domain knowledge with the processing allowing an expert to segment and annotate an object without knowledge in image or mesh processing. Another advantage is that using the expert knowledge, we are able to build a decision tree to perform an efficient search amongst the set of possible objects while being able to suggest segmentation and annotation corrections to the user if an impossible configuration is reached.

6.1 Limitations

We can identify two kind of limitations in this work. First, the elementary concepts and the associated algorithms which we designed are mainly focused on manufactured shapes. In other words, the first layer of our ontology does not effectively conceptualize other objects such as organic shapes. Furthermore, we assume that the input geometry is clean enough: as it is, our current implementation is not appropriate to treat digitalized meshes which may include a significant amount of noise.

Moreover, we would like to underline here that the original goal of this project was to design a framework able to handle shapes in a well defined domain. In particular, our pipeline will consider any unexpected configuration of the shape as a default, and will highlight it as an incompatible configuration. This behavior is motivated by

---

[5]The best strategy for a preprocessing can be to choose the smallest range concept, then run for each elementary concept a SF algorithm.

closed world contexts, such as industrial inspection for example, where the context is fully controlled, but can be a limitation for applications in an less constrained context.

6.2 Future Work

The ontology designed for this experiment is very basic, and we forsee significant extensions in our future work. In particular, the n-ary properties will be integrated in order to express more realistic constraints between subparts of the objects.

The results which we presented in Sect. 5 are using basic elementary algorithms only. In a near future, we plan to introduce more flexible and robust approaches to detect features as described in Mortara et al. (2004) and Laga et al. (2013). The next algorithms which we plan to introduce will be methods to handle curved shapes and relative positions of the objects.

Besides the algorithmic aspects, the approach which we presented here gives some interesting tracks on the semantic side. One of the challenges will be to replace the current dichotomical algorithm selection with a more elaborated strategy that better adapts to the situation at hand. A first criterion to consider could be a weighting system that favours algorithms with a small computational time, or to include the accuracy of the algorithms. These weights will be introduced in the decision tree computation in order to design an expert system that handles the question of efficiency.

In addition, applying our approach on meshes acquired from low resolution devices will complicate the job of the segmentation algorithms. It will probably generate incoherent subregions, with overlappings or unlabelled parts. In fact the use of fuzzy maps solves a part of the problem, but we still have an open question: how to adjust an existing partial segmentation? Our framework is a good candidate to provide a specific answer to this problem, since the expert knowledge contains information about the expected configurations. One possible extension of this work could be to introduce adjustment algorithms for each elementary concept, that will be able to refine a first segmentation using the global knowledge of a specific domain.

Finally, a long term extension of this work will be to introduce it into a machine learning system, where an existing training set of shapes will be used to either deduce an ontology from scratch or find a suitable extended ontology based on a fundamental one. This extended framework will be a possible challenge of the 3D Shape Retrieval Contest (SHREC) organized each year in the mesh segmentation community.

Acknowledgments Marco Attene thanks the EU FP7 Project IQmulus for having supported his contributions in this research. Jean-Marie Favreau thanks BPI-France and FEDER Auvergne via the FUI AAP 14 Project 3DCI for having supported his contributions in this research.

Appendix: Details of the Segmentation Algorithms

In Sect. 4 we introduced a series of elementary algorithms to segment and identify regions of a requested object. We already introduced in Sect. 3.4.1 the two families of algorithms: "Is a" algorithm to label an already segmented region, and "Find all" algorithms to extract regions corresponding to a specific property. In this section, we present the implementation details of the core algorithms we've introduced to handle the experiments on Furnitures (see Sect. 5).

"Is a" Algorithms

Shape properties are associated with dedicated "is a" algorithms that follow a common pipeline. First we compute an approximation of the minimal volume oriented bounding box using ApproxMVBB,[6] a C++ extension of the work of (Barequet and Har-Peled 2001) with many efficient preprocessing steps.

Using the three lengths $l_0 \geq l_1 \geq l_2$ of this bounding box, we wrote fuzzy rules to express the following descriptions:

- in a cube, l_1 is almost equivalent to l_0 and l_2,
- in a board, l_0 and l_1 are obviously longer than l_2,
- in a stick, l_0 is obviously longer than l_2, and l_1 is almost equivalent to l_2.

Vertical and horizontal orientations are well defined for sticks and boards, using the main directions of the minimal volume oriented bounding box. Let v_0 be the axis associated to the largest side of the box (*w.r.t.* its area), and let v_1 be the axis associated to the smallest side of the box.

We wrote fuzzy rules to express the following descriptions if the region is a board:

- if v_0 is almost parallel to the up-down axis, the region is horizontal,
- if v_0 is almost orthogonal to the up-down axis, the region is vertical.

If the region as been identified as a stick, we introduce the following descriptions:

- if v_1 is almost parallel to the up-down axis, the stick is vertical,
- if v_1 is almost orthogonal to the up-down axis, the stick is horizontal.

The **compactness of a shape** is defined by comparing the two first lengths l_0 and l_1 of the bounding box. If l_0 is sufficiently longer than l_1 the shape is elongated. If these two lengths are almost equivalent, the region has a compact shape.

In this work, we assume that the input mesh has as correct scale (in our case, 1 unit corresponding to 1 meter) and is correctly oriented. Existing methods such as (Fu et al. 2008) are available to automatically estimate the orientation of a manufactured object.

---

[6]https://github.com/gabyx/ApproxMVBB.

This property has been used to define two kinds of properties relative to the height of the regions. First we used the relative vertical positions comparing the highest, the central and the lowest points of the region with respect to the equivalent points of the object, to define algorithms able to identify **position properties**: up position, down position and central position.

To define the global position of a region, we compared the lowest point of the object with the highest point of the region. In his work Le Corbusier defined standard sizes for furnitures and buildings, based on the golden ratio (Corbusier 2000). In this work, we defined two **Modulor properties**: M27 and M43, corresponding respectively to bench and chair classical height.

"Find all" Algorithms

We introduced in Sect. 4.2.2 the motivations to design a board segmentation algorithm. To achieve this goal, we designed an original approach that performs the fitting of two parallel primitives at the same time, then it defines the shape by adding the lateral surfaces.

Figure 15 illustrates our board segmentation algorithm, defined as follows:

- let t be a triangle of a board (Fig. 15a), and $\vec{n}$ its outward normal,
- find the opposite triangle t' using a ray in the opposite direction of $\vec{n}$ (Fig. 15b),
- fit two parallel planes to these triangles (Fig. 15c),
- starting from t and t', grow iteratively the two regions by selecting only triangles if they fit with one of the two planes (Fig. 15d), and readjust the model at each step,
- when the growing process is finished, we stop the process if the two sides are not similar enough (areas significantly different, or too small overlapping),
- otherwise, we consider the adjacent triangles as initial triangles of the lateral surface growing process (Fig. 15e),
- for each new triangle, find the closest edge e in the boundary of the parallel surfaces (Fig. 15f),
- consider the virtual facet orthogonal to the planes starting from e, and compare it with the new candidate, using the distance between barycenters and the angle between normals.

The initial triangles are selected by first computing a simplified version of the Shape Diameter Function (SDF) (Shapira et al. 2008) where only one ray is used, since we are manipulating CAD models and meshes that represent manufactured objects. Then we sort triangles by ascending SDF value. For each non-visited triangle, we run our board segmentation algorithm. Finally, we run the "is a" algorithms to decide if it is as board shape, a stick shape or a cube shape.

Figure 16 left illustrates the detection of board, stick and cube in a basic shape composed of 75 520 triangles, with a computation time of 3.4 s on a Intel i-7 2.00GHz.

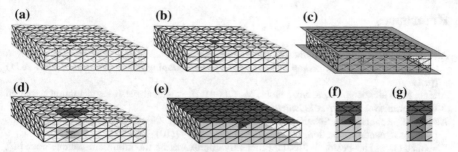

Fig. 15 Details of the "find all" board algorithm, starting from a single triangle (**a**), growing a region by fitting two parallel planes (**a–d**), then adding the lateral surface (**f–g**)

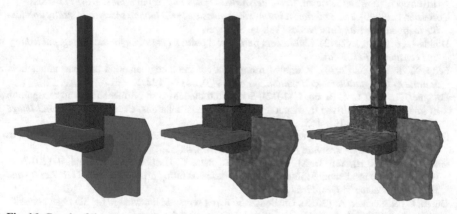

Fig. 16 Result of the segmentation: boards in *blue* and *green*, stick in *pink*, cube in *yellow*. *Light regions* corresponds to bad fitting scores. *Left* original shape. *Middle* and *right* two different level of noise

Table 4 Result of the segmentation: semantic labelling and score

Region	Label	Score (original)	Score (noise 0.5 %)	Score (noise 1 %)	#triangles
Cube	Cube	0.999966	0.983023	0.939295	45,568
Vertical board	Board	0.99997	0.944411	0.86583	14,592
Stick	Stick	0.999915	0.974192	0.926152	6,656
Horizontal board	Board	0.99996	0.95555	0.876332	8,704

We applied a multiscale random noise on the original shape with a root mean square deviation of 0.5 % (respectivly 1 %) of the object's bounding box diagonal length.

During the fitting process, a fitting score is computed for each triangle. The final score of the region (Table 4) is estimated using the mean of these scores, multiplied by the fuzzy result of the corresponding "is a" algorithm.

References

Albrecht, S., Wiemann, T., Günther, M., & Hertzberg, J. (2011). Matching CAD object models in semantic mapping. In *Workshop Semantic Perception, Mapping and Exploration (ICRA 2011)* (p. 1).

Attene, M., Falcidieno, B., & Spagnuolo, M. (2006). Hierarchical mesh segmentation based on fitting primitives. *The Visual Computer, 22*(3), 181–193.

Attene, M., Robbiano, F., Spagnuolo, M., & Falcidieno, B. (2009). Characterization of 3D shape parts for semantic annotation. *Computer-Aided Design, 41*(10), 756–763.

Barequet, G., & Har-Peled, S. (2001). Efficiently approximating the minimum-volume bounding box of a point set in three dimensions. *Journal of Algorithms, 38*, 91–109.

Camossi, E., Giannini, F., & Monti, M. (2007). Deriving functionality from 3D shapes: Ontology driven annotation and retrieval. *Computer-Aided Design and Applications, 4*(6), 773–782.

Corbusier, L. (2000). *The modulor: A harmonious measure to the human scale, universally applicable to architecture and mechanics* (Vol. 1). Springer.

Dodds, L., & Davis, I. (2012). *Linked data patterns: A pattern catalogue for modelling, publishing, and consuming linked data.*

Feng, X., & Pan, X. (2013). A unified framework for mesh segmentation and part annotation. *Journal of Computational Information Systems, 9*(8), 3117–3128.

Fouquier, G., Atif, J., & Bloch, I. (2012). Sequential model-based segmentation and recognition of image structures driven by visual features and spatial relations. *Computer Vision and Image Understanding, 116*, 146–165.

Fu, H., Cohen-Or, D., Dror, G., & Sheffer, A. (2008). Upright orientation of man-made objects. In *ACM transactions on graphics (TOG)* (Vol. 27, p. 42). ACM.

Geng, C., Suzuki, H., Yan, D.-M., Michikawa, T., Sato, Y., Hashima, M., & Ohta, E. (2010). A thin-plate CAD mesh model splitting approach based on fitting primitives. In *EG UK Theory and Practice of Computer Graphics* (pp. 45–50).

Gurau, C., & Nüchter, A. (2013). Challenges in using semantic knowledge for 3D object classification. In *KI 2013 Workshop on Visual and Spatial Cognition*, p. 29.

Hassan, S., Hétroy, F., & Palombi, O. (2010). Ontology-guided mesh segmentation. In *FOCUS K3D Conference on Semantic 3D Media and Content*.

Hudelot, C., Atif, J., & Bloch, I. (2008). Fuzzy spatial relation ontology for image interpretation. *Fuzzy Sets System, 159*, 1929–1951.

Klir, G., & Yuan, B. (1995). *Fuzzy sets and fuzzy logic* (Vol. 4). Upper Saddle River: Prentice Hall.

Laga, H., Mortara, M., & Spagnuolo, M. (2013). Geometry and context for semantic correspondence and functionality recognition in manmade 3D shapes. *ACM Transactions on Graphics (TOG), 32*(5), 150.

Maillot, N., & Thonnat, M. (2008). Ontology based complex object recognition. *Image and Vision Computing, 26*(1), 102–113.

Mortara, M., Patané, G., Spagnuolo, M., Falcidieno, B., & Rossignac, J. (2004). Plumber: A method for a multi-scale decomposition of 3D shapes into tubular primitives and bodies. In *ACM Symposium on Solid modeling and applications* (pp. 339–344).

Othmani, A., Meziat, C., & Lomenie, N. (2010). Ontology-driven image analysis for histopathological images. In *International Symposium on Visual Computing (ISVC 2010)*.

Seifert, S., Thoma, M., Stegmaier, F., Hammon, M., Kramer, M., Huber, M., Kriegel, H., Cavallaro, A., & Comaniciu, D. (2011). Combined semantic and similarity search in medical image databases. In *SPIE Medical Imaging*.

Shapira, L., Shamir, A., & Cohen-Or, D. (2008). Consistent mesh partitioning and skeletonisation using the shape diameter function. *The Visual Computer, 24*(4), 249–259.

Shi, M., Cai, H., & Jiang, L. (2012). An approach to semi-automatic semantic annotation on Web3D scenes based on an ontology framework. In *Intelligent Systems Design and Applications (ISDA 2012)* (pp. 574–579).

Symonova, O., Dao, M.-S., Ucelli, G., & De Amicis, R. (2006). Ontology based shape annotation and retrieval. In *European Conference on Artificial Intelligence (ECAI 2006)*.

Ontology Alignment Using Web Linked Ontologies as Background Knowledge

Thomas Hecht, Patrice Buche, Juliette Dibie, Liliana Ibanescu and Cassia Trojahn dos Santos

Abstract This paper proposes an ontology matching method for aligning a source ontology with target ontologies already published and linked on the Linked Open Data (LOD) cloud. This method relies on the refinement of a set of input alignments generated by existing ontology matching methods. Since the ontologies to be aligned can be expressed in several representation languages with different levels of expressiveness and the existing ontology matching methods can only be applied to some representation languages, the first step of our method consists in applying existing matching methods to as many ontology variants as possible. We then propose to apply two main strategies to refine the initial alignment set: the removal of different kinds of ambiguities between correspondences and the use of the links published on the LOD. We illustrate our proposal in the field of life sciences and environment.

T. Hecht (✉) · J. Dibie · L. Ibanescu
UMR518 MIA-Paris, INRA - AgroParisTech, 75231 Paris Cedex 05, France
e-mail: thomashecht95@gmail.com

J. Dibie
e-mail: Juliette.Dibie@agroparistech.fr

L. Ibanescu
e-mail: Liliana.Ibanescu@agroparistech.fr

P. Buche
INRA & LIRMM, 2 Place Pierre Viala, 34060 Montpellier Cedex 2, France
e-mail: Patrice.Buche@supagro.inra.fr

C.T. dos Santos
IRIT & UTM2, 5 Allées Antonio Machado, 31058 Toulouse Cedex 9, France
e-mail: Cassia.Trojahn@irit.fr

© Springer International Publishing Switzerland 2017
F. Guillet et al. (eds.), *Advances in Knowledge Discovery and Management*,
Studies in Computational Intelligence 665, DOI 10.1007/978-3-319-45763-5_11

1   Introduction

Ontologies are nowadays used as a common and standardized vocabulary for representing concepts and relations from a particular domain (e.g. life-science, geography). The Linked Open Data (LOD) cloud[1] contains more and more data sources published and linked together on the Web. Publishing and linking scientific data on the Web using ontologies for describing them should facilitate scientific data sharing, such as giving access to data from specific disciplines or data produced within specific geographic regions (Bizer 2013).

When a new ontology, the source ontology, is published on the LOD, first, the 'target' ontologies, i.e. ontologies from similar domains with similar concepts, has to be identified among the already published ontologies in order to access new entities (concepts, properties or instances) and data sources. The source ontology can then be linked with each target ontology by finding an alignment (i.e. a set of correspondences) between entities. Different approaches have been proposed for the Ontology Matching task (Shvaiko and Euzenat 2013; Bernstein et al. 2011; Rahm 2011; Euzenat and Shvaiko 2007) and a systematic evaluation on data sets from different domains has been carried out over the last ten years by the Ontology Alignment Evaluation Initiative (OAEI).[2]

In this paper, we propose an ontology matching method for aligning a source ontology with different target ontologies already linked and published on the LOD. An ontology can be either a thesaurus, an ontology or an ontological and terminological resource, expressed in different representation languages. Our method is based on the principle of *alignment refinement*: starting from a set of input alignments generated by several existing ontology matching methods, we propose to apply different strategies in order to refine this initial alignment set. One of our strategies is to exploit the links between the target ontologies, already published on the LOD.

We illustrate our method in the field of life sciences and environment. In this field, several thesauri have been created and published on the LOD. The two largest ones are AGROVOC[3] and NALT.[4] AGROVOC was created in the 1980s by FAO (Food and Agriculture Organization of the United Nations) as a structured multilingual thesaurus for agriculture, forestry, fishery, food and related fields (such as environment). It is available in 19 languages, with about 40,000 terms in each language (Caracciolo et al. 2012). NALT is a bilingual thesaurus comparable with AGROVOC in terms of covered domain and maintained by USDA (United States Department of Agriculture). It is currently composed of approximately 91,000 terms in English and Spanish. For instance, the vocabulary of AGROVOC is currently linked to 15

---

[1] http://linkeddata.org.
[2] http://oaei.ontologymatching.org.
[3] http://aims.fao.org/standards/agrovoc/about.
[4] http://agclass.nal.usda.gov/agt.shtml.

international resources like GeoNames,[5] DBpedia[6] and GEMET.[7] In addition, 13,390 terms of AGROVOC are currently aligned with NALT (Caracciolo et al. 2012). In this paper, we focus on the alignment of an ontological and terminological resource NARYQ (n-ary Relations between Quantitative experimental data) (Buche et al. 2013) with AGROVOC and NALT, in order to publish it on the LOD. NARYQ contains about 1,100 concepts structured into several sub-domains, such as food products, microorganisms and packaging.

This paper is organised as follows. Section 2 describes our method for aligning an ontology with linked ontologies on the LOD. Section 3 discusses the results of our experiments in the field of life sciences and environment. Section 4 presents related work and, finally, Sect. 5 concludes the paper and presents our perspectives.

2 Ontology Matching Method with Linked Ontologies

In this section, we present our matching method for aligning a source ontology with two target and linked ontologies. Our method is designed to align ontologies, thesauri or ontological and terminological resources, possibly described in different representation languages with different levels of expressiveness (e.g. OWL DL,[8] SKOS[9]). An Ontological and Terminological Resource (OTR) (Reymonet et al. 2007; Roche et al. 2009; McCrae et al. 2011) is a hybrid model that combines a conceptual component and a terminological component: a concept is associated with a set of terms, each term denoting the concept with different lexical functions (e.g. synonyms, abbreviations, etc.). In the following, for the sake of simplicity and the paper's readability, we abusively use ontology for either ontology, thesaurus or OTR.

The proposed method relies on the refinement of a set of input alignments generated by existing ontology matching methods. Our aim is therefore to be able to apply as much existing matching methods as possible in order to generate as much candidate correspondences as possible. Since the existing ontology matching methods can only be applied to some particular representation languages and the ontologies to be aligned can be expressed in several representation languages with different levels of expressiveness, we propose to apply matching methods on different variants of the ontologies to be aligned. A variant of an ontology corresponds to its expression in a given representation language. The first step of our matching method consists in aligning variants of the source ontology O_s with variants of the two target ontologies O_t^1 and O_t^2 using existing ontology matching methods. It allows the production of an initial set of alignments. In the second step, different refinement strategies are applied to this initial set of alignments, including the exploitation of the links

---

[5]http://www.geonames.org.

[6]http://dbpedia.org.

[7]http://www.eionet.europa.eu/gemet.

[8]http://www.w3.org/TR/owl-guide.

[9]http://www.w3.org/TR/2009/REC-skos-reference-20090818.

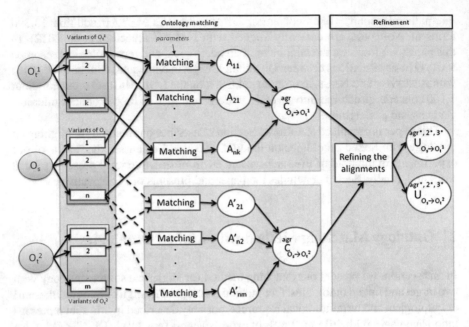

Fig. 1 Overview of our matching method

defined on the LOD between the target ontologies. Figure 1 gives the overview of our matching method, which is detailed in the next two subsections.

2.1 First Step: Ontology Matching

The first step of our method consists in aligning the source ontology O_s with each one of the two target ontologies O_t^1 and O_t^2.

2.1.1 Ontology Variants

The source and target ontologies to be aligned can be thesauri, ontologies or OTR and may be expressed in different representation languages. Since the existing matching methods are usually designed for one particular representation language, we propose to associate to each ontology a set of *variants*, defined in the following:

Definition 1 (*Set of variants of an ontology*) The *set V_O of variants of an ontology O* is composed of its transformations in different representation languages $L_1, L_2, \ldots$. It contains the original version O^{orig} of the ontology.

$V_O = \{O^{orig}, O^{L_1}, O_1^{L_2}, O_2^{L_2}, \ldots\}$, where $O_j^{L_i}$ is the jth transformation of the ontology O using the representation language L_i.

The aim of using variants is twofold. First, matching tools are designed to deal with specific input models (OWL ontologies in most cases). Diversifying the kinds of input, we are able to produce more candidate correspondences by using more tools. Second, representing resources using different constructors (OWL and SKOS) allows the encoded knowledge to be exploited in different ways. On the one hand, tools can take advantage of OWL models for better exploiting automated reasoning. On the other hand, the lexicalisation of concepts is better expressed in SKOS models than in OWL models. For instance, in classical SKOS to OWL transformations, both skos:prefLabel and skos:altLabel are mapped to rdfs:label, where skos:altLabel are often used to represent synonyms, but also to refer to related terms. As a consequence, without introducing variants to catch this semantic richness, we could lose this information, which can instead be useful for tools able to deal with the specificities of SKOS.

2.1.2 Matching the Ontology Variants

The ontology matching process takes as input two ontologies and produces as output a set of correspondences between the entities of these two ontologies. According to Euzenat and Shvaiko (2007), this process can be defined as follows:

Definition 2 (*Matching process* (Euzenat and Shvaiko 2007)) The *matching process* is a function f that, applied to two ontologies O_s and O_t and an (optional) initial alignment A^{orig}, produces a directed alignment $A^f_{O_s, O_t}$ between the two ontologies $(O_s \rightarrow O_t)$. This process can use matching parameters p (e.g. weights, thresholds) and external resources r (e.g. common knowledge and domain specific thesauri):

$$A^f_{O_s, O_t} = f(O_s, O_t, A^{orig}, p, r)$$

Definition 3 (*Correspondence* (Euzenat and Shvaiko 2007)) Let us consider two ontologies O_s and O_t, a *correspondence* c^f resulting from a matching process f is a relation r between the two entities e_s and e_t, denoted $c^f = \langle id, e_s, e_t, r, n \rangle$, such that: $c^f \in A^f_{O_s, O_t}$; $e_s \in O_s$ and $e_t \in O_t$; $r \in \{\equiv, \sqsubseteq, \sqsupseteq\}$; n is the confidence level (in general, $n \in [0, 1]$) indicating the degree of confidence that the relation r holds between e_s and e_t.

Since the structural and the lexical information of the ontologies are exploited in different ways by the matching processes, the use of the most expressive variant of an ontology does not guarantee the best results. Therefore, the first step of our matching method consists in launching several matching processes on several variants of the ontologies to be aligned. Let us consider the source ontology O_s, one of the target ontologies O_t and the set of matching processes $F = \{f_1, f_2, \ldots\}$, which are launched to align the ontologies O_s and O_t, each matching process f_i is launched on a pair of variants (O_s^j, O_t^k) where $O_s^j \in V_{O_s}$ and $O_t^k \in V_{O_t}$ and generates the following alignment (i.e. set of correspondences):

$$A^{f_i}_{O^j_s, O^k_t} = f_i(O^j_s, O^k_t, \emptyset, p, r) \tag{1}$$

The result of our matching method between the source ontology O_s and one of the two target ontologies is a set of sets of correspondences, denoted $\overset{agr}{C}_{O_s \to O_t}$ for *aggregated set*, generated by each matching process of F on each pair of ontology variants. This set, which comes from the concatenation of results from several matching processes on several ontology variants, is denoted:

$$\overset{agr}{C}_{O_s \to O_t} = \bigoplus_{i,j,k} A^{f_i}_{O^j_s, O^k_t} \tag{2}$$

The total number of matching processes launched in order to obtain an alignment between the source ontology O_s and each one of the two target ontologies O^1_t and O^2_t is:

$$|V_{O_s}| \times |V_{O^1_t}| \times |F| + |V_{O_s}| \times |V_{O^2_t}| \times |F| \tag{3}$$

2.2   Second Step: Refining the Alignments

The second step of our matching method consists in refining the sets of sets of correspondences $\overset{agr}{C}_{O_s \to O^1_t}$ and $\overset{agr}{C}_{O_s \to O^2_t}$. These two sets of sets contain many correspondences (suggesting a good coverage) but also a lot of noise (i.e. incorrect correspondences) that has to be reduced.

In order to improve the quality of the correspondences found in the first step, we propose two refinement methods: the first one allows the identification of the potentially correct correspondences (see Sect. 2.2.1), the second refinement method allows the deletion of the correspondences considered as ambiguous and therefore potentially incorrect (see Sect. 2.2.2). Finally, we present in Sect. 2.2.3 our refinement process.

2.2.1   Identification of Potentially Correct Correspondences

We distinguish two ways to identify potentially correct correspondences. When redundancies occur between correspondences that have been generated from at least two distinct matching methods, we assume that these correspondences can be considered as having more chances to be correct. We will retain them in a separate set, denoted $\overset{recT}{C}_{O_s \to O_t}$ for *recovering set*. These correspondences will be presented to the user as potentially correct correspondences.

Definition 4 (*Recovering set*) Let us consider two matching processes f_1 and f_2 applying two distinct matching methods for aligning two ontologies O_s and O_t, the recovering set $\overset{recT}{C}_{O_s \to O_t}$ is defined as follows:

If $c^{f_1} = \langle id_1, e_s^1, e_t^1, r_1, n_1 \rangle \wedge c^{f_2} = \langle id_2, e_s^2, e_t^2, r_2, n_2 \rangle \wedge e_s^1 = e_s^2 \wedge e_t^1 = e_t^2 \wedge r_1 = r_2$

then $c^{f_k} \in \overset{recT}{C}_{O_s \to O_t}$, where $c^{f_k} = \begin{cases} c^{f_1} & \text{if } n_1 \geq n_2 \\ c^{f_2} & \text{otherwise} \end{cases}$

Example 1 Let us also consider the correspondence c_1 generated by a matching process f_1 on the variants $\text{NARYQ}^{OWL-SKOS}$ of the source ontology NARYQ, and the variant AGROVOC^{SKOS} of the target ontology AGROVOC (see Sect. 3.1). Let us also consider the correspondence c_2 generated by a matching process f_2 which applies another matching method as the one used in the matching process f_1 on the variants $\text{NARYQ}^{OWL-SKOS}$ and AGROVOC_2^{OWL}. We have:

$$c_1 = \langle id_1, \text{sheep}, c\_8854, \equiv, 0.95 \rangle, c_1 \in A^{f_1}_{\text{NARYQ}^{OWL-SKOS}, \text{AGROVOC}^{SKOS}}$$

$$c_2 = \langle id_2, \text{sheep}, c\_8854, \equiv, 0.75 \rangle, c_2 \in A^{f_2}_{\text{NARYQ}^{OWL-SKOS}, \text{AGROVOC}_2^{OWL}}$$

The correspondences c_1 and c_2 generated by two distinct matching methods can be considered as redundant. The *recovering set* $\overset{recT}{C}_{\text{NARYQ} \to \text{AGROVOC}}$ contains the correspondence c_1 with the highest confidence level.

The second way of identifying potentially correct correspondences relies on the same assumption as above, i.e. 'comparable' correspondences can be considered as having more chances to be correct. Let us consider that there exists an alignment $A^{LOD}_{O_t^1 \to O_t^2}$ defined on the LOD between the target ontologies O_t^1 and O_t^2, correspondences are said 'comparable' if an entity of the source ontology O_s is aligned, by an equivalence relation, with two distinct but linked on the LOD entities of the target ontologies O_t^1 and O_t^2. These correspondences will be kept in two separate sets, denoted $\overset{LOD}{C}_{O_s \to O_t^1}$ and $\overset{LOD}{C}_{O_s \to O_t^2}$ as *LOD recovering sets*. These sets will be presented to the user as sets of potentially correct correspondences.

Definition 5 (*LOD recovering set*) Let us consider $A^{LOD}_{O_t^1 \to O_t^2}$ the result of a matching process between two ontologies O_t^1 and O_t^2 from the LOD, a set of matching processes $F^1 = \{f_1^1, f_2^1, \ldots\}$ applied to two ontologies O_s and O_t^1, and a set of matching processes $F^2 = \{f_1^2, f_2^2, \ldots\}$ applied to O_s and O_t^2, the *LOD recovering sets* $\overset{LOD}{C}_{O_s \to O_t^i}, i \in [1, 2]$, are defined as follows:

If $\exists c \in A^{LOD}_{O_t^1 \to O_t^2} \wedge c = \langle id, e_t^1, e_t^2, \equiv, n \rangle \wedge c^{f_i^1} \in A^{f_i^1}_{O_s, O_t^1} \wedge c^{f_j^2} \in A^{f_j^2}_{O_s, O_t^2} \wedge$
$c^{f_i^1} = \langle id_1, e_s^1, e_t^1, \equiv, n_1 \rangle \wedge c^{f_j^2} = \langle id_2, e_s^2, e_t^2, \equiv, n_2 \rangle \wedge e_s^1 = e_s^2 \wedge e_t^1 \neq e_t^2,$
then $c^{f_i^1} \in \overset{LOD}{C}_{O_s \to O_t^1}$ and $c^{f_j^2} \in \overset{LOD}{C}_{O_s \to O_t^2}$.

Hence, a correspondence $c^{f_i^1}$ belongs to the LOD recovering set $\overset{LOD}{C}_{O_s \to O_t^1}$, if (i) the entity source e_s ($e_s = e_s^1 = e_s^2$) is aligned with an entity target e_t^1, (ii) there exists a correspondence $c^{f_i^2}$ such that the entity source e_s is aligned with an entity target e_t^2, (iii) there exists on the LOD a correspondence c linking e_t^1 and e_t^2.

Example 2 Let us consider the correspondence c_3 generated by the matching process f_1 of Example 1 and the correspondence c_4 generated by the matching process f_3 on the variants $\text{NARYQ}^{OWL-SKOS}$ and NALT^{OWL}. We have:

$c_3 = \langle id_3, \text{surimi}, c\_33271, \equiv, 0.87 \rangle, c_3 \in A^{f_1}_{\text{NARYQ}^{OWL-SKOS},\text{AGROVOC}^{SKOS}}$

where c_33271 is a concept of AGROVOC;

$c_4 = \langle id_4, \text{surimi}, c\_40365, \equiv, 0.92 \rangle, c_4 \in A^{f_3}_{\text{NARYQ}^{OWL-SKOS},\text{NALT}^{OWL}}$

where c_40365 is a concept of NALT.

Let us also consider that: $\exists c \in A^f_{\text{AGROVOC},\text{NALT}}, c = \langle id_c, c\_33271, c\_40365, \equiv, 0.96 \rangle$.

Then, we have: $c_3 \in \overset{LOD}{C}_{\text{NARYQ} \to \text{AGROVOC}}$ and $c_4 \in \overset{LOD}{C}_{\text{NARYQ} \to \text{NALT}}$.

2.2.2 Deletion of Ambiguous Correspondences

We distinguish three types of ambiguity between correspondences. The first type covers the correspondences obtained from the same matching method launched on different variants of the source and target ontologies. The correspondences of this type have the same source entity, the same target entity and the same relation. We propose to remove ambiguities of type 1 by keeping the correspondence with the highest confidence level.

Definition 6 (*Ambiguous correspondences of type 1*) Let us consider two matching processes f_1 and f_2 applying the same matching method to align two ontologies O_s and O_t (with O_s^j and O_t^k its respective variants), two correspondences c^{f_1} and c^{f_2}, from the sets $A^{f_1}_{O_s^{j_1},O_t^{k_1}}$ and $A^{f_2}_{O_s^{j_2},O_t^{k_2}}$, are *ambiguous according to type 1* if:

$$c^{f_1} = \langle id_1, e_s^1, e_t^1, r_1, n_1 \rangle \wedge c^{f_2} = \langle id_2, e_s^2, e_t^2, r_2, n_2 \rangle \wedge e_s^1 = e_s^2 \wedge e_t^1 = e_t^2 \wedge r_1 = r_2.$$

The set of sets of non ambiguous correspondences according to type 1 is:

$$\overset{agr*}{C}_{O_s \to O_t} = \bigoplus_{i,j,k}(A^{f_i}_{O_s^j,O_t^k} \setminus \{c^{f_k}\}) \text{ where } c^{f_k} = \begin{cases} c^{f_1} & \text{if } n_1 \geq n_2 \\ c^{f_2} & \text{otherwise} \end{cases}$$

Remark 1 Let us remember that when redundancies occur between correspondences generated by two distinct matching methods, these correspondences are considered as potentially correct (see Definition 4).

Example 3 Let us consider the correspondence c_1 generated by the matching process f_1 of Example 1. Let us also consider the correspondence c_5 generated by a matching process f_4 using the same matching method as the one used in the matching process f_1 but on the variant NARYQ^{SKOS} and the variant AGROVOC^{SKOS}. We have:

$c_1 = \langle id_1, \text{sheep}, c\_8854, \equiv, 0.95 \rangle, c_1 \in A^{f_1}_{\text{NARYQ}^{OWL-SKOS}, \text{AGROVOC}^{SKOS}}$

where c_8854 corresponds to the concept 'caprins' in AGROVOC.

$c_5 = \langle id_5, \text{sheep}, c\_8854, \equiv, 0.88 \rangle, c_5 \in A^{f_4}_{\text{NARYQ}^{SKOS}, \text{AGROVOC}^{SKOS}}$

The set of sets $\overset{agr*}{C}_{\text{NARYQ}\rightarrow\text{AGROVOC}}$ of non ambiguous correspondences according to type 1 only contains the correspondence c_1 with the highest confidence level.

The second type of ambiguity covers the correspondences in which an entity of the source ontology O_s is aligned, by an equivalence relation, with two distinct entities of the target ontology O_t. We propose, in this case, to keep only the most relevant correspondence, i.e. the one that has *a priori* the highest confidence level. However, considering the fact that these correspondences were not generated by the same matching method, their confidence degrees are not comparable. Therefore, we propose to compute a similarity measure *sim* on the two correspondences to be compared, which is independent on the matching methods used to generate them. This similarity measure can rely, for instance, on syntactic similarity measures implemented in the Alignment API (David et al. 2011). Here, we use the following syntactic measures: Hamming distance, Levenshtein distance, n-grams and Jaro and Jaro-Winkler to compute the similarity between all the labels, in a given language, of the two entities. The *sim* measure is the average of the computed similarities.

Definition 7 (*Ambiguous correspondences of type 2*) Let us consider a set of matching processes $F = \{f_1, f_2, \ldots\}$ applied to two ontologies O_s and O_t, two correspondences c^{f_i} and c^{f_j} are *ambiguous according to the type 2* if:

$$c^{f_i} = \langle id_1, e_s^1, e_t^1, \equiv, n_1 \rangle \wedge c^{f_j} = \langle id_2, e_s^2, e_t^2, \equiv, n_2 \rangle \wedge e_s^1 = e_s^2 \wedge e_t^1 \neq e_t^2.$$

The set of sets of non ambiguous correspondences of type 2 is:

$$\overset{agr2*}{C}_{O_s \rightarrow O_t} = \bigoplus_{i,j,k} (A^{f_i}_{O_s^j, O_t^k} \backslash \{c^{f_k}\}) \text{ where } c^{f_k} = \begin{cases} c^{f_i} & \text{if } sim(e_s^1, e_t^1) \leq sim(e_s^2, e_t^2) \\ c^{f_j} & \text{otherwise} \end{cases}$$

Remark 2 We only consider the equivalence relation here, because with other relations, the correspondences are not necessarily ambiguous, i.e. both of the correspondences can, in some cases, be considered as correct.

Example 4 Let us consider the correspondence c_1 generated by the matching process f_1 of Example 1 and the correspondence c_6 generated by the matching process f_2 of Example 1. We have:

$c_1 = \langle id_1, \text{sheep}, c\_8854, \equiv, 0.95 \rangle, c_1 \in A^{f_1}_{\text{NARYQ}^{OWL-SKOS}, \text{AGROVOC}^{SKOS}}$

$sim(\text{sheep}, c\_8854) = 0.815$, where c_8854 corresponds to the concept 'caprins' in AGROVOC;

$c_6 = \langle id_6, \text{sheep}, c\_9214, \equiv, 0.65 \rangle, c_6 \in A^{f_2}_{\text{NARYQ}^{OWL-SKOS}, \text{AGROVOC}_2^{OWL}}$

$sim(\text{sheep}, c\_9214) = 0.621$, where c_9214 corresponds to the concept 'goat' in AGROVOC.

The set of sets $\overset{agr2*}{C}_{\text{NARYQ}\rightarrow\text{AGROVOC}}$ of non ambiguous correspondences according to type 2 only contains the correspondence c_1 with the highest similarity measure.

Finally, the third type of ambiguity covers the correspondences where two distinct entities from the source ontology O_s are aligned, by an equivalence relation, with the same entity of the target ontology O_t. We propose, in this case, to keep the most relevant correspondence, i.e. the one with the highest similarity measure sim.

Definition 8 (*Ambiguous correspondences of type 3*) Let us consider a set of matching processes $F = \{f_1, f_2, \ldots\}$ applied to two ontologies O_s and O_t, two correspondences c^{f_i} and c^{f_j} are *ambiguous according to type 3* if:

$$c^{f_i} = \langle id_1, e_s^1, e_t^1, r_1, n_1 \rangle \wedge c^{f_j} = \langle id_2, e_s^2, e_t^2, r_2, n_2 \rangle \wedge e_s^1 \neq e_s^2 \wedge e_t^1 = e_t^2 \wedge r_1 = r_2.$$

The set of sets of non ambiguous correspondences of type 3 is defined as:

$$\overset{agr3*}{C}_{O_s \rightarrow O_t} = \bigoplus_{i,j,k}(A^{f_i}_{O_s^j, O_t^k} \setminus \{c^{f_k}\}) \text{ where } c^{f_k} = \begin{cases} c^{f_i} \text{ if } sim(e_s^1, e_t^1) \leq sim(e_s^2, e_t^2) \\ c^{f_j} \text{ otherwise} \end{cases}$$

Example 5 Let us consider the correspondence c_1 generated by the matching process f_1 of Example 1 and the correspondence c_6 generated by the matching process f_1. We have:

$c_1 = \langle id_1, \text{sheep}, c\_8854, \equiv, 0.95 \rangle, c_1 \in A^{f_1}_{\text{NARYQ}^{OWL-SKOS}, \text{AGROVOC}^{SKOS}}$

$sim(\text{sheep}, c\_8854) = 0.815$, where c_8854 corresponds to the concept 'caprins' in AGROVOC;

$c_7 = \langle id_7, \text{ewe}, c\_8854, \equiv, 0.55 \rangle, c_7 \in A^{f_1}_{\text{NARYQ}^{OWL-SKOS}, \text{AGROVOC}^{SKOS}}$

$sim(\text{ewe}, c\_8854) = 0.722$.

The set of sets $\overset{agr3*}{C}_{\text{NARYQ}\rightarrow\text{AGROVOC}}$ of non ambiguous correspondences according to type 3 only contains the correspondence c_1 with the highest similarity measure.

2.2.3 The Refinement Process

Figure 2 gives the overview of our refinement process, detailed in Sects. 2.2.1 and 2.2.2.

The set obtained by the union of the two recovering sets defined in Sect. 2.2.1 is denoted by:

$$U_{O_s \rightarrow O_t} = \overset{recT}{C}_{O_s \rightarrow O_t} \cup \overset{LOD}{C}_{O_s \rightarrow O_t} \tag{4}$$

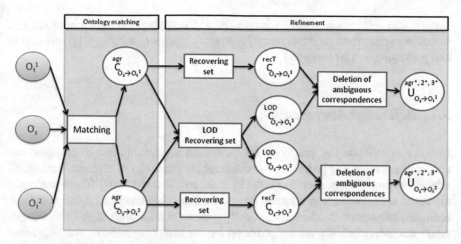

Fig. 2 Overview of our refinement process

We define the set of *potentially correct and non ambiguous correspondences* between a source ontology O_s and a target ontology O_t as follows.

Definition 9 The set of *potentially correct and non ambiguous correspondences* between a source ontology O_s and a target ontology O_t, denoted $U_{O_s \to O_t}^{agr*,2*,3*}$, is the set obtained after the removing of ambiguities of types 1, 2 and 3 as defined in Definitions 6, 7 and 8 from the set $U_{O_s \to O_t}$ given in Eq. 4.

3 Experiments

We illustrate in this section our matching method described above for aligning a source ontology NARYQ (presented in Sect. 3.1) with each of the two target ontologies AGROVOC and NALT. In the following, the alignment of NARYQ with AGROVOC will be denoted NARYQ → AGROVOC and the alignment of NARYQ with NALT: NARYQ → NALT.

3.1 The Source Ontology NARYQ

The ontology NARYQ (n-ary Relations between Quantitative experimental data) has been created for representing n-ary relations between quantitative experimental data (see Buche et al. (2013)). The characteristics of this ontology are the following: (i) it is an OTR; (ii) the labels are available in French and in English; (iii) it is represented in OWL DL and SKOS; and (iv) the conceptual component contains about 1,100 concepts structured into several sub-domains, the most important one in number being food products ($\approx$460 concepts), microorganisms ($\approx$180 concepts) and packaging ($\approx$150 concepts).

3.2 Reference Alignments

In order to evaluate the quality of the generated alignments and to compare the results of the matching processes, we consider the measures of precision and recall adapted to the ontology matching task (Euzenat and Shvaiko 2007). These measures are based on a comparison between an automatically generated alignment A and a reference alignment R. The automatically generated alignment A is in this paper either the individual alignments provided by the matching tools or the alignment resulting from our approach. The construction of a complete reference alignment R was not possible because it is a time-consuming task and it is difficult to find and to involve experts from the domain. Hence, we have built two partial manually validated reference alignments, denoted $\overline{R}^+_{\text{AGROVOC}}$ for the alignment NARYQ $\rightarrow$ AGROVOC, and $\overline{R}^+_{\text{NALT}}$ for the alignment NARYQ $\rightarrow$ NALT.

For each ontology and for every concept, we extracted their annotations (e.g. skos:prefLabel, skos:altLabel, rdfs:label, rdfs:comment) in English and in French as well as their structural elements (e.g. skos:broader, rdfs:subClassOf). A first alignment was created using SMOA (Stoilos et al. 2005) (*String Metric for Ontology Alignment*), a syntactic similarity metric for ontology matching. Using this metric and the equivalence relation $\equiv$, an alignment with 1,453 correspondences was generated, which was then manually validated by two experts in a double-blind process, and finally re-conciliated, i.e. the experts reached an *a posteriori* consensus. This first expert validation was performed in four hours using a visualisation tool developed for this specific task. It produced 318 validated correspondences in $\overline{R}_{\text{AGROVOC}}$ and 394 validated correspondences in $\overline{R}_{\text{NALT}}$, among which 233 concepts from NARYQ were aligned with both concepts from AGROVOC and concepts from NALT. In order to enrich these first generated reference alignments, an additional set of potentially correct correspondences was generated using our matching method and was validated by two experts. We therefore obtained two new and enriched reference alignments, denoted $\overline{R}^+_{\text{AGROVOC}}$ and $\overline{R}^+_{\text{NALT}}$.

- $\overline{R}^+_{\text{AGROVOC}}$ has 368 validated correspondences, with 361 concepts of NARYQ aligned with concepts of AGROVOC.
- $\overline{R}^+_{\text{NALT}}$ has 428 validated correspondences, with 424 concepts of NARYQ aligned with concepts of NALT.
- 303 concepts of NARYQ are aligned with both concepts of AGROVOC and concepts of NALT.

The alignments $\overline{R}^+_{\text{AGROVOC}}$ and $\overline{R}^+_{\text{NALT}}$, though partial, are used in the following as reference alignments.

3.3   Experimental Protocol

Several matching processes were launched on several variants of the ontologies to be aligned in order to generate as much candidate correspondences as possible. We first present the ontology variants and then the selected matching processes.

3.3.1   The Ontology Variants

The variants of NARYQ are:

$$V_{\text{NARYQ}} = \{\text{NARYQ}^{OWL-SKOS}, \text{NARYQ}^{OWL}, \text{NARYQ}^{SKOS}\}$$

where the original version of NARYQ, denoted $\text{NARYQ}^{OWL-SKOS}$, is defined using OWL2 DL and SKOS. We also use two variants of NARYQ. The variant NARYQ^{OWL} was generated from its conceptual component by transforming both skos:prefLabel and skos:altLabel into rdfs:label, while the variant NARYQ^{SKOS} was generated using the labels of its terminological component and transforming the conceptual hierarchy into a SKOS hierarchy. The variants of AGROVOC are:

$$V_{\text{AGROVOC}} = \{\text{AGROVOC}^{SKOS}, \text{AGROVOC}_1^{SKOS}, \text{AGROVOC}_2^{OWL}, \text{AGROVOC}_3^{OWL}\}$$

AGROVOC^{SKOS} includes AGROVOC in all languages and we used its version downloaded in April 2013 from the official Web site.[10] AGROVOC_1^{SKOS} is a much smaller version, in English Only, available on the same Web site. The variant AGROVOC_2^{OWL} was used into the 2007 OAEI campaign. The variant AGROVOC_3^{OWL} was generated from AGROVOC^{SKOS} using a SKOSParser.[11] The variants of NALT are:

$$V_{\text{NALT}} = \{\text{NALT}^{SKOS}, \text{NALT}^{OWL}\}$$

where the original version NALT^{SKOS} was downloaded in April 2013 from the official Web site[12] and the variant NALT^{OWL} was generated from NALT^{SKOS} using the same SKOSParser.

---

[10]http://aims.fao.org/access-agrovoc.

[11]http://oaei.ontologymatching.org/2007/SKOSParser.pdf.

[12]http://agclass.nal.usda.gov.

3.3.2 The Matching Processes

Two available ontology matching tools, implementing different matching approaches with good results in the 2011 and 2012 OAEI campaigns (Aguirre et al. 2012), were selected: LogMap[13] (Jiménez-Ruiz and Grau 2011) and Aroma[14] (David 2007).

Aroma makes use of the association rule paradigm and a statistical measure assessing the implication intensity of the rules. The matching approach is divided into three steps: (1) pre-processing: each ontology entity, i.e. classes and properties, is represented by a set of terms—bag of words; (2) discovery of association rules between entities, and (3) post-processing: cleaning and enhancing the resulting alignment (i.e. deduction of equivalence relations, suppression of cycles in the alignment graph, suppression of redundant correspondences, and enhancement of the alignment by using equality and string similarity-based methods). Aroma is able to deal with both SKOS and OWL variants. This is not the case for LogMap, which encounters problems to parse SKOS variants.

LogMap adopts an approach based on logical reasoning and inconsistency repair techniques. The matching method follows five main steps: (1) lexical indexation of labels of entities and their lexical variations; (2) structural indexation based on interval labeling schema for representing extended class hierarchies; (3) computation of initial anchor correspondences by intersecting the lexical indexes of entities; (4) iterative mapping repair and discovery, by filtering out logical inconsistencies in the mappings computed so far and by computing new mappings using string-based similarity method; and (5) ontology overlapping estimation, where ontology fragments overlap in both input ontologies.

Among 24 matching processes launched for aligning NARYQ and AGROVOC (see Eq. 3), only 9 produced non-empty alignments. From over 12 matching processes launched for aligning NARYQ and NALT, only 4 produced non-empty alignments. In the following, we assume that a correspondence is 'acceptable' if it has a confidence level greater than or equal to 0.5, a threshold empirically defined. The initial sets of alignments generated using the two matching tools (see Eq. 2) are: $\overset{agr}{C}_{\text{NARYQ} \rightarrow \text{AGROVOC}}$, denoted $\overset{agr}{C}_{\text{AGROVOC}}$, with 3,196 correspondences, and $\overset{agr}{C}_{\text{NARYQ} \rightarrow \text{NALT}}$, denoted $\overset{agr}{C}_{\text{NALT}}$, with 1,676 correspondences.

3.4 Experimental Results

3.4.1 Individual Results

Table 1 presents the results obtained from each matching tool, with respect to our two partial reference alignments $\overline{R}^{+}_{\text{AGROVOC}}$ and $\overline{R}^{+}_{\text{NALT}}$ (see Sect. 3.2). We can observe

---

[13] http://www.cs.ox.ac.uk/isg/projects/LogMap/.

[14] http://aroma.gforge.inria.fr/.

that thanks to variants we are able to overcome the limitations of tools in deadling with specific input models and hence we are able to produce more candidate correspondences. While LogMap is not able to generate alignments for the pairs involving SKOS variants (e.g. NARYQ^{SKOS}, AGROVOC^{SKOS} and NALT^{SKOS}), Aroma produces a set of correspondences with intermediary scores.

Table 2 presents the best scores obtained by the two matching tools and extracted from Table 1. These results are, in fact, an approximation, as they were computed using the partial reference alignments, which may affect the accuracy of the results. The values of each line of Table 2 represent the best score obtained for each indicator (number of correct correspondences, precision, recall or F-measure) by the matching tools. #* corresponds to the *highest number of good correspondences*; P^* corresponds to the *best precision*; R^* corresponds to the *best recall*; and $F\text{-}m^*$ corresponds to the *best F-measure*.

Table 1 Individual results for the matching tools

Alignment		Matching tools									
		LogMap					Aroma				
		# tot	#	P	R	F-m	# tot	#	P	R	F-m
$\text{NARYQ}^{OWL-SKOS}$	AGROVOC^{SKOS}	–	–	–	–	–	459	288	0.627	0.783	0.696
	AGROVOC_1^{SKOS}	–	–	–	–	–	386	288	0.746	0.783	0.764
	AGROVOC_2^{OWL}	203	180	0.887	0.489	0.630	–	–	–	–	–
	AGROVOC_3^{OWL}	185	167	**0.903**	0.454	0.604	–	–	–	–	–
	NALT^{SKOS}	–	–	–	–	–	476	**359**	0.754	0.839	0.794
	NALT^{OWL}	417	334	**0.801**	0.780	0.791	–	–	–	–	–
NARYQ^{OWL}	AGROVOC^{SKOS}	–	–	–	–	–	–	–	–	–	–
	AGROVOC_1^{SKOS}	–	–	–	–	–	–	–	–	–	–
	AGROVOC_2^{OWL}	312	269	0.862	0.731	0.791	1311	228	0.174	0.620	0.272
	AGROVOC_3^{OWL}	341	**300**	0.880	**0.815**	**0.846**	+	+	+	+	+
	NALT^{SKOS}	–	–	–	–	–	–	–	–	–	–
	NALT^{OWL}	456	356	0.781	**0.832**	**0.805**	–	–	–	–	–
NARYQ^{SKOS}	AGROVOC^{SKOS}	–	–	–	–	–	915	268	0.293	0.728	0.418
	AGROVOC_1^{SKOS}	–	–	–	–	–	1131	212	0.187	0.576	0.283
	AGROVOC_2^{OWL}	–	–	–	–	–	–	–	–	–	–
	AGROVOC_3^{OWL}	–	–	–	–	–	–	–	–	–	–
	NALT^{SKOS}	–	–	–	–	–	1011	327	0.323	0.764	0.454
	NALT^{OWL}	–	–	–	–	–	–	–	–	–	–

\# tot indicates the total number of correspondences
\# indicates the number of correct correspondences
+ indicates that the tool generated empty alignments
– indicates that the tool was not able to deal with the input

Table 2 Best scores of alignments obtained by the two matching tools

Alignment	#*	P^*	R^*	$F\text{-}m^*$
$\text{NARYQ} \rightarrow \text{AGROVOC}$	300	0.90	0.82	0.85
$\text{NARYQ} \rightarrow \text{NALT}$	359	0.80	0.83	0.81

Table 3 Evaluation of NARYQ → AGROVOC with respect to $\overline{R}^+_{\text{AGROVOC}}$

Set	# total	# good	P	R	F-m
$C^{agr*}_{\text{AGROVOC}}$	1583	366	0.23	0.99	0.37
$C^{recT}_{\text{AGROVOC}}$	582	354	0.61	0.96	0.74
C^{LOD}_{AGROVOC}	336	254	0.76	0.69	0.72
U_{AGROVOC}	620	363	0.58	0.99	0.73
$U^{agr*,2*,3*}_{\text{AGROVOC}}$	447	344*	0.77	0.93*	0.84

Table 4 Evaluation of NARYQ → NALT with respect to $\overline{R}^+_{\text{NALT}}$

Set	# total	# good	P	R	F-m
C^{agr*}_{NALT}	850	415	0.49	0.97	0.65
C^{recT}_{NALT}	480	368	0.77	0.86	0.81
C^{LOD}_{NALT}	337	255	0.76	0.59	0.67
U_{NALT}	551	404	0.73	0.94	0.82
$U^{agr*,2*,3*}_{\text{NALT}}$	400	348	0.87*	0.81	0.84*

3.4.2 Results of Our Approach

Table 3 presents the evaluation of NARYQ → AGROVOC generated by different refinement methods, with respect to the partial reference alignment $\overline{R}^+_{\text{AGROVOC}}$. On the last row of Table 3, the symbol ⋆ indicates, for the indicator of the column, a better result than the best result of the matching tools for the same indicator presented in Table 2.

Table 4 presents the evaluation of NARYQ → NALT generated by different refinement methods, with respect to the partial reference alignment $\overline{R}^+_{\text{NALT}}$. On the last row of Table 4, the symbol ⋆ indicates, for the indicator of the column, a better result than the best result of the matching tools for the same indicator presented in Table 2.

3.4.3 Discussion

As we can notice in Tables 3 and 4 and as we might expect, (1) increasing the set of alignments allows the recall to be improved for most of the produced alignment sets,[15] and (2) combining the different methods of refinement gives the best results in terms of precision (set $U^{agr*,2*,3*}$). Comparing these results with the best scores obtained by the two matching tools (Table 2), we obtained very promising results. Our approach obtains similar results in terms of F-measure for NARYQ → AGROVOC,

---

[15] Since ambiguous correspondences according to type 1 produce only noises, the evaluation of the best recall is done considering C^{agr*} and not C^{agr}.

while it increases F-measure for NARYQ → NALT. Our approach outperforms the best result in terms of recall for NARYQ → AGROVOC and in terms of precision for NARYQ → NALT. This performance produces very encouraging results.

Most matching tools apply strategies for combining different basic methods (i.e. lexical, structural, etc.) within a matching process and for filtering their results (threshold, weighted aggregation, etc.) (see Euzenat and Shvaiko (2007)). Our encouraging results can be explained by the fact that we propose in this paper to refine the sets of alignments produced by different matching methods in two different ways. First, we have identified three types of ambiguity to be resolved in order to refine the set of correspondences by deleting some of them. Second, we propose two new methods for discriminating and improving the sets of correspondences. In the first method, the redundant correspondences generated by at least two matching processes applying distinct matching methods are considered as potentially correct. The second method exploits the alignments defined on the LOD in order to reinforce the validity of some correspondences (i.e. correspondences allowing a same entity to be aligned with two distinct but linked entities on the LOD are considered as potentially correct). Another original aspect of our approach consists in exploiting ontology variants, taking advantage of the characteristics of the ontologies, which can be ontologies, thesauri, OTR and can be expressed in different representation languages with different levels of expressiveness. This gives us the ability to cover a wide and diverse range of resources.

4 Related Work

A key aspect of our proposal is the use of published links on the LOD as background knowledge for refining the results of a matching process. Similar works in this direction have been proposed in recent years in the literature, encouraged by the increasing number of available data sets on the LOD cloud. In Nikolov et al. (2009), a schema matching approach which uses existing instance-level coreference links, defined in third-party repositories, as background knowledge is proposed. It aims at generating schema-level correspondences to assist the instance coreference resolution process. Rather than producing strict equivalence or subsumption relations, the algorithm produces fuzzy correspondences representing degrees of overlap between different ontologies. In Pernelle and Sais (2011), an approach that addresses both link discovery and ontology alignment is proposed, where the results of the link discovery step are exploited to improve the results of the ontology alignment step and *vice versa*. In Parundekar et al. (2012), the proposal consists of (a) generating more expressive concepts from those already present in the ontologies (i.e. exploiting the space of concepts defined by value restrictions), and (b) aligning these extended concepts by exploiting the links between instances on the LOD. Contrary to our approach, these proposals consider a single representation of ontologies (i.e. OWL) and focus on links between instances.

For dealing with the specificity of community-created LOD data sets, a system for finding schema-level links is proposed in Jain et al. (2010). It computes alignments (not limited to equivalence relations) with the help of noisy community-generated data available on the Web, i.e. Wikipedia and Wikipedia category hierarchy. The idea of using Wikipedia category hierarchy, together with a rule-based verification approach, has also been exploited in Grütze et al. (2012), where a holistic matching approach aims at aligning simultaneously multiple schemes on the LOD. In Cruz et al. (2011), an extended version of the AgreementMaker system is proposed, aiming at handling subsumption relations and improving its performance when dealing with LOD ontologies. For each source and target concepts, the algorithm searches across several LOD ontologies for all concepts that are defined as subclasses, before applying matching strategies. Contrary to our approach, these works exploit other relations than equivalence and focus on the schema-level of LOD ontologies instead of exploiting the links between them. Contrary to the proposals described above, these latter works do not exploit the instance level within the schema-matching process.

Mochol and Jentzsch (2008) and Steyskal and Polleres (2013) propose, like us, to reuse existing tools and to combine their results to align two ontologies. In Mochol and Jentzsch (2008) a set of rules to select appropriate methods for a given pair of ontologies to be aligned is proposed. This selecting process is based on the background information describing the available approaches and the input properties of the ontologies. In Steyskal and Polleres (2013), an iterative method based on voting is proposed, where at every round, the correspondences accepted by the majority of tools are considered as valid. However, these works do not exploit the alignments on the LOD.

With regards to combining multiple alignments, different approaches have been proposed. In Ghoula et al. (2014), an approach for normalising, combining and integrating alignments from multiple sources is proposed, where a correspondence can be associated to a set of relations and confidence levels. The algebra defined in Euzenat (2008) was applied in order to implement operators like union, composition and intersection. The approach can be applied regardless of the formalism used to represent the ontologies to be aligned. As we do, the normalisation step allows the removing of 'concurrent' (i.e. ambiguous) correspondences. However, we do not apply any combining operator, while they do not exploit LOD alignments. In Lee et al. (2007), a library of matching components is made available and the user can select which components are to be used within a matching process, how they have to be combined together (average, minimum, maximum, weighed sum, decision trees, etc.), and how the correspondences are finally extracted (from a selection based on thresholding to formulate the selection as an optimisation problem over a weighted bipartite graph). The approach involves well automatic tuning of matching systems in order to find a tuning that optimises the performance of them. Here, we propose a different way for combining multiple alignments. In Eckert et al. (2009), alignments generated from different matching systems are used as training data for a classifier that learns which combination of results provides the best indication of a correct correspondence. The multiple matchers are treated as a black-box. The assumption on which the approach relies is similar to ours: by using multiple matchers one can benefit from the high degree of precision of some matchers and at the same time the broader coverage of other matchers. In Spiliopoulos and Vouros (2012), combining multiple matchers is seen as a problem of maximising the social welfare within a group of interacting agents. Different agents computing alignments using specific methods and considering a specific kind of ontology entity, interact with each other and share constraints on the validity of the correspondences in order to reach an agreement. Although we do not aim at reaching a consensus between matchers, a correspondence is more likely to be correct if it is accepted by more than one matcher, which are not dedicated to find correspondences between specific ontology entities.

Finally, with respect to matching of terminologies in several languages, Mougin and Grabar (2013) adopts a notion of refining that is close to ours. The authors present a cross-language approach for matching two biomedical terminologies (MedDRA and SNMI). From a set of correspondences computed using lexical methods, the incorrect correspondences are filtered out using the notion of semantic groups, which correspond to the partition of UMLS concepts. If the semantic groups which belong to UMLS concepts of MedDRA and SNMI terms are not the same, the correspondence is considered as incorrect. Then, they compute the number of correspondences which are common to different languages (correspondences which are more likely to be correct) and suppress the ambiguities by eliminating correspondences found in only one language.

5 Conclusion and Perspectives

In this paper, we have proposed a new ontology matching method which can raise one of the challenges of ontology matching stated in Shvaiko and Euzenat (2013): matching with background knowledge. In a first step, our matching method allows to generate many correspondences using and combining existing methods for aligning ontologies, thesauri and OTR expressed in different representation languages. Then, it allows a discrimination of the correspondences by removing some ambiguities and by exploiting the redundancy and existing alignments on the LOD, in order to identify a subset of potentially correct correspondences which will be submitted to the user for validation.

This proposal is a preliminary work for publishing ontologies on the LOD. Ontology matching allows a source ontology not only to be enriched with new concepts and/or terms, but also to be linked with existing and in use ontologies on the LOD in order to contribute to the data sharing in the target domain.

In order to improve our process of refinement, we plan in the short term, (i) to evaluate our approach using another data set, such as the OAEI Library task, which offers variants of their thesauri and for which we can find LOD alignments between them, (ii) to take into account the expressiveness of ontology variants to suppress the ambiguities of type 1; (iii) to exploit other relations than the equivalence in the treatment of ambiguities of type 2—instead of removing ambiguous correspondence of type 2, we plan to propose a methodology based on reasoning for choosing the best correspondence between the ambiguous ones: by removing, for instance, correspondences which introduce a logical inconsistency; (iv) to remove the ambiguities between correspondences by defining new relations—we can, for instance, use an algebra to define a new correspondence with a new relation which combines the relations involved in the ambiguous correspondences; (v) to study how to use the subsumption relation in order to facilitate the identification of potentially correct correspondences; (vi) to study how to use relations between concepts (e.g. their domains and ranges) and their matching results in order to suppress the ambiguities and/or to identify the potentially correct correspondences. In the long term, we plan to exploit indirect alignments between different sources on the LOD to improve the discrimination on the set of correspondences. We also plan to extend our approach to take into account more complex entities such as units of measurement and n-ary relations.

References

Aguirre, J., et al. (2012). Results of the ontology alignment evaluation initiative 2012. In *Proceedings of 7th ISWC Workshop on Ontology Matching (OM)* (p. 73115).

Bernstein, P. A., Madhavan, J., & Rahm, E. (2011). Generic schema matching, ten years later. *PVLDB, 4*(11), 695–701.

Bizer, C. (2013). Interlinking scientific data on a global scale. *Data Science Journal, 12*, GRDI6–GRDI12.

Buche, P., et al. (2013). Intégration de données hétérogènes et imprecise guide par une resource termino-ontologique. application au domaine des sciences du vivant. *RSTI série Revue dIntelligence Artificielle, 27*(4–5), 539–568.

Caracciolo, C., Stellato, A., Rajbhandari, S., Morshed, A., Johannsen, G., Keizer, J., et al. (2012). Thesaurus maintenance, alignment and publication as linked data: The AGROVOC use case. *IJMSO, 7*(1), 65–75.

Cruz, I. F., Palmonari, M., Caimi, F., & Stroe, C. (2011). Towards "on the go" matching of linked open data ontologies. In *Workshop on Discovering Meaning On the Go in Large Heterogeneous Data 2011 (LHD-11), Barcelona, Spain, July 16, 2011*.

David, J. (2007). *AROMA: une méthode pour la découverte d'alignements orientés entre ontologies partir de règles d'association*. Ph.D. thesis, Université de Nantes.

David, J., Euzenat, J., Scharffe, F., & Trojahn dos Santos, C. (2011). The alignment api 4.0. *Semantic Web, 2*(1):310.

Eckert, K., Meilicke, C., & Stuckenschmidt, H. (2009). Improving ontology matching using meta-level learning. In *The semantic web: research and applications* (Vol. 5554, pp. 158–172). Lecture notes in computer science. Berlin, Heidelberg: Springer.

Euzenat, J. (2008). Algebras of ontology alignment relations. In *International Semantic Web Conference* (Vol. 5318). Lecture notes in computer science. Heidelberg: Springer.

Euzenat, J., & Shvaiko, P. (2007). *Ontology matching* (Vol. 18). Heidelberg: Springer.

Ghoula, N., Nindanga, H., & Falquet, G. (2014). Opérateurs de gestion des alignements de ressources de connaissances hétérogènes (to be completed).

Grütze, T., Böhm, C., & Naumann, F. (2012). Holistic and scalable ontology alignment for linked open data. In C. Bizer, T. Heath, T. Berners-Lee, & M. Hausenblas (Eds.), *WWW2012 Workshop on Linked Data on the Web, Lyon, France, 16 April, 2012* (Vol. 937). CEUR workshop proceedings. CEUR-WS.org.

Jain, P., Hitzler, P., Sheth, A. P., Verma, K., & Yeh, P. Z. (2010). Ontology alignment for linked open data. In *Proceedings of the 9th International Semantic Web Conference on The Semantic Web - Volume Part I* (pp. 402–417). Berlin, Heidelberg: Springer.

Jiménez-Ruiz, E., & Grau, B. C. (2011). LogMap: Logic-based and scalable ontology matching. In *The Semantic WebISWC 2011* (pp. 273–288). Springer.

Lee, Y., Sayyadian, M., Doan, A., & Rosenthal, A. S. (2007). eTuner: Tuning schema matching software using synthetic scenarios. *The VLDB Journal, 16*(1), 97–122.

McCrae, J., Spohr, D., & Cimiano, P. (2011). Linking Lexical resources and ontologies on the semantic web with lemon. In G. Antoniou, M. Grobelnik, E. P. B. Simperl, B. Parsia, D. Plexousakis, P. D. Leenheer, & J. Z. Pan (Eds.), *ESWC (1)* (Vol. 6643, pp. 245–259). Lecture notes in computer science. Springer.

Mochol, M., & Jentzsch, A. (2008). Towards a rule-based matcher selection. In A. Gangemi & J. Euzenat (Eds.), *Knowledge engineering: Practice and patterns* (Vol. 5268, pp. 109–119). Lecture notes in computer science. Berlin, Heidelberg: Springer.

Mougin, F., & Grabar, N. (2013). Using a cross-language approach to acquire new mappings between two biomedical terminologies. In *Artificial intelligence in medicine* (Vol. 7885, pp. 221–226). Lecture notes in computer science. Berlin, Heidelberg: Springer.

Nikolov, A., Uren, V., Motta, E., & Roeck, A. (2009). Overcoming schema heterogeneity between linked semantic repositories to improve coreference resolution. In *Proceedings of the 4th Asian Conference on The Semantic Web, ASWC 2009* (pp. 332–346). Berlin, Heidelberg: Springer.

Parundekar, R., Knoblock, C. A., & Ambite, J. L. (2012). Discovering concept coverings in ontologies of linked data sources. In P. Cudré-Mauroux, et al. (Eds.), *International Semantic Web Conference (1)* (Vol. 7649, pp. 427–443). Lecture notes in computer science. Springer.

Pernelle, N. & Sais, F. (2011). LDM: Link discovery method for new resource integration. In M.-E. V. Zoé Lacroix & Edna Ruckhaus (Eds.), *Fourth International Workshop on Resource Discovery, Heraklion, Grèce* (Vol. 737, pp. 94–108).

Rahm, E. (2011). Towards large-scale schema and ontology matching. In Z. Bellahsene, A. Bonifati & E. Rahm (Eds.), *Schema Matching and Mapping* (pp. 3–27). Springer.

Reymonet, A., Thomas, J., & Aussenac-Gilles, N. (2007). Modelling ontological and terminological resources in OWL DL. In *OntoLex 2007—Workshop at ISWC07, Busan, South-Korea*

Roche, C., Calberg-Challot, M., Damas, L., & Rouard, P. (2009). Ontoterminology—a new paradigm for terminology. In J. L. G. Dietz (Ed.), *KEOD* (pp. 321–326). INSTICC Press.

Shvaiko, P., & Euzenat, J. (2013). Ontology matching: state of the art and future challenges. *IEEE Transactions on Knowledge and Data Engineering, 25*(1), 158–176.

Spiliopoulos, V., & Vouros, G. (2012). Synthesizing ontology alignment methods using the max-sum algorithm. *IEEE Transactions on Knowledge and Data Engineering, 24*(5), 940–951.

Steyskal, S., & Polleres, A. (2013). Mix 'n' match: An alternative approach for combining ontology matchers. In R Meersman, H. Panetto, T. S. Dillon, J. Eder, Z. Bellahsene, N. Ritter, P. D. Leenheer & D. Dou (Eds.), *On the Move to Meaningful Internet Systems: OTM 2013 Conferences - Confederated International Conferences: CoopIS, DOA-Trusted Cloud, and ODBASE 2013, Graz, Austria, September 9–13, 2013. Proceedings* (Vol. 8185, pp. 555–563). Lecture notes in computer science. Springer.

Stoilos, G., Stamou, G., & Kollias, S. (2005). A string metric for ontology alignment. In *The Semantic Web—ISWC 2005* (pp. 624–637). Springer.

LIAISON: reconciLIAtion of Individuals Profiles Across SOcial Networks

Gianluca Quercini, Nacéra Bennacer, Mohammad Ghufran
and Coriane Nana Jipmo

Abstract Social Networking Sites, such as Twitter and LinkedIn, are clear examples
of the impact that the Web 2.0 has on people around the world, because they target
an aspect of life that is extremely important to anyone: social relationships. The key
to building a social network is the ability of finding people that we know in real life,
which, in turn, requires those people to make publicly available some personal information, such as their names, family names, locations and birth dates, just to name a
few. However, it is not uncommon that individuals create multiple profiles in several
social networks, each containing partially overlapping sets of personal information.
As a result, the search for an individual might require numerous queries to match the
information that is spread across many profiles, unless an efficient way is provided to
automatically integrate those profiles to have an holistic view of the information on
the individual. This calls for efficient algorithms for the determination (or *reconciliation*) of the profiles created by an individual across social networks. In this paper,
we build on a previous research of ours and we describe LIAISON (reconciLIAtion
of Individuals profiles across SOcial Networks), an algorithm that uses the network
topology and the publicly available personal information to iteratively reconcile profiles across n social networks, based on the existence of individuals who disclose
the links to their multiple profiles. We evaluate LIAISON on real large datasets and
we compare it against existing approaches; the results of the evaluation show that
LIAISON achieves a high accuracy.

G. Quercini (✉) · N. Bennacer · M. Ghufran · C. Nana Jipmo
LRI, CentraleSupélec, University of Paris-Saclay, 91192 Gif-sur-Yvette, France
e-mail: gianluca.quercini@lri.fr

N. Bennacer
e-mail: nacera.bennacer@lri.fr

M. Ghufran
e-mail: mohammad.ghufran@lri.fr

C. Nana Jipmo
e-mail: coriane.nanajipmo@lri.fr

© Springer International Publishing Switzerland 2017
F. Guillet et al. (eds.), *Advances in Knowledge Discovery and Management*,
Studies in Computational Intelligence 665, DOI 10.1007/978-3-319-45763-5_12

229

1   Introduction

A social network is a set of individuals and their relationships. In a broader sense, the term social network also refers to a website, such as Twitter and LinkedIn, which enables individuals to create a personal page, or *profile*, and establish links to the profiles of their acquaintances and friends. The key to building a social network is the ability of finding people that we know in real life, which in turn requires those people to make publicly available on their profiles some personal information, such as their names, family names, locations and birth dates, just to name a few. Several surveys showed that Social Networking Services (SNSs) users tend to share many of their personal data, including sensitive information, such as home addresses and phone numbers (Gross and Acquisti 2005; Little et al. 2011; Stutzman 2006).

However, it is not uncommon that an individual creates multiple profiles in different SNSs, each disclosing sets of personal information that are unlikely to be identical, though they might overlap. Indeed, profile information might not be updated regularly and is not necessarily created at the same time. Moreover, the differences between two profiles of an individual might reflect the fact that they are created in SNSs that target different aspects of her life. For instance, information on her career is more likely to be found on her LinkedIn profile than her Twitter profile, as LinkedIn is mainly used for professional networking. As a result, finding a person based on a limited knowledge of her personal information might require several manual searches across social networks, which is obviously annoying and time-consuming. It would be useful to create a global profile that provides a holistic view of the personal information of an individual by automatically integrating all her profiles. This calls for efficient methods for automatically determining (or *reconciling*) the profiles that are created by an individual across different SNSs, which is the focus of our paper.

Building on our previous research (Bennacer et al. 2014a, b), in this paper we describe LIAISON (reconciLIAtion of Individuals profiles across SOcial Networks), an algorithm for the reconciliation of profiles across n distinct social networks, based on the personal information that is publicly available in the profiles and their links to other profiles. Basically, LIAISON determines whether two profiles refer to the same individual, and creates a special link between the two that we term a *cross-link*, by using a set of rules that compare the values of their *attributes* (e.g., names, nicknames, locations). Instead of applying the rules to all profile pairs, which would be unfeasible for large networks, LIAISON obtains a subset of *candidates* from all pairs of profiles that are already connected by a cross-link. More specifically, given two profiles v and w connected by a cross-link, candidates are selected from the set of the profile pairs (a, b) such that a is linked to v and b is linked to w. Next, LIAISON applies the rules to all candidate pairs and determines the ones that are to be connected by a cross-link because they refer to the same individual. The discovered cross-links are used to obtain new candidates and iterate the algorithm until no more candidates are found.

Compared to the previous version of our algorithm, LIAISON:

- Defines a rule to compare the values of the attribute LOCATIONS, which provides an important clue as to whether two profiles refer to the same individual.
- Reconciles the profiles in n networks without the need of comparing social networks pairwise. This results in a more efficient reconciliation through transitive closure on the discovered cross-links.
- Uses a better strategy for the selection of candidates, which results in a considerable reduction of the number of candidates and a dramatic improvement of the performances.
- Relaxes the constraint that two profiles a and b must belong to different social networks to be considered as candidates. As a result, LIAISON can identify individuals that have multiple profiles within the same social network.

In summary, the following are key contributions of our paper:

- We define rules that compare the values of a set of attributes to determine whether two profiles refer to the same individual. Unlike the existing rule-based approaches (i) we consider that all attributes are equally important, which relieves us from assigning each attribute an empirical and, inevitably, arbitrary weight, (ii) we study the combined contribution of the different attributes, when used in the same rule and (iii) each rule is assigned a *confidence* which represents the number of attributes whose values are considered to be equal or similar by the rule; the higher this number is, the higher the confidence is that the two profiles refer to the same individual.
- LIAISON reconciles new profiles in an iterative way, and more profiles are propagated by transitive closure across all social networks. To the best of our knowledge, no existing method is iterative in this sense.
- We evaluate LIAISON on two real datasets. The first consists of four social networks (Flickr, LiveJournal, Twitter and YouTube) and includes around 2 million profiles and 21 million links. The second consists of around 60,000 profiles and 29,000 links obtained from LinkedIn and Twitter. The evaluation and the comparison against two existing approaches show that LIAISON achieves a high precision with good performances.

The remainder of the paper is organized as follows. We survey the research work that is related to ours in Sect. 2 and we introduce basic concepts and notation in Sect. 3. In Sect. 4 we describe the attributes and the similarity measures used to compare their values and we detail the algorithm in Sect. 5. The evaluation results are then presented in Sect. 6, followed by concluding remarks in Sect. 7.

2 Related Work

Numerous solutions have been proposed to the problem that we study in this paper. Interestingly, two of them focus only on the nickname of an individual as a way to reconcile different profiles, based on the observation that individuals tend to use the same or a similar nickname across distinct social networks (Perito et al. 2011; Zafarani and Liu 2009). Although in our evaluation we confirm this observation, we also consider other attributes, in order to reconcile profiles of individuals who choose to use unrelated nicknames.

The use of the attributes to reconcile profiles across distinct social networks has been largely investigated (Raad et al. 2010; Rowe 2009; Cortis et al. 2012; Malhotra et al. 2012; Carmagnola and Cena 2009; Goga et al. 2013; Golbeck and Rothstein 2008; Motoyama and Varghese 2009). Two approaches describe each pair of profiles as a vector of scores, which represent the similarity between the values of the attributes, and use machine learning techniques to determine whether they can be reconciled (Malhotra et al. 2012; Motoyama and Varghese 2009). While the results are promising, both approaches need a training set, which is not easy to determine. In fact, a careful analysis of the available data is necessary to create a training set that is representative of all possible situations where profile pairs can be reconciled or not. Moreover, a model trained on a given pair of social networks might not be generalizable to other networks, which implies that a training set should be created for each network pair.

Some social networks allow the exportation of profiles that are described with the Friend of a Friend ontology (FOAF); the advantage is that standard Semantic Web techniques, such as OWL reasoning, can be used to reconcile profiles (Golbeck and Rothstein 2008; Rowe 2009). However, these techniques are applied to a limited set of attributes, and in particular to those, such as the email, that are likely to identify an individual uniquely.

Similarly to us, Carmagnola et al. determine the profile attributes that are more likely to identify an individual uniquely, by assigning them an *importance factor* (Carmagnola and Cena 2009). The importance factor is used to weigh the similarity score that is computed between two profiles that have similar attributes. Our approach goes a step further and uses the pairs of profiles that are reconciled to iteratively reconcile new profiles. Moreover, our evaluation is based on a real large social internetwork, while theirs uses different closed user-adaptive systems. The key difference is that in Web social networks often individuals are reluctant to disclose their real identities, while in closed user-adaptive systems they feel that their privacy is less threatened; as a result, data in social networks are likely to be erroneous and messy, which constitutes a real challenge. Some researchers also propose the computation of semantic similarity between profile attributes (Cortis et al. 2012; Raad et al. 2010). Although these approaches are original, they provide little (50 user profiles (Raad et al. 2010)) or no evaluation.

Some authors proposed to go beyond the profile attributes and investigated the possibility of using the network properties (Bartunov et al. 2012; Buccafurri et al. 2012; Jain et al. 2013; Narayanan and Shmatikov 2009). The approach proposed by Buccafurri et al. considers that two profiles are similar, and therefore likely to refer to the same individual, if they have similar nicknames and the profiles to which they are linked are recursively similar (Buccafurri et al. 2012). This approach presents two major drawbacks. First, profiles associated with dissimilar nicknames are ignored and discarded with no further analysis, although they might very well refer to the same person; second, the discovered associations between profiles are not used to re-iterate the algorithm and discover new associations. Our approach overcomes these two limitations. Besides considering the network structure, Jain et al. also propose to use the content that an individual publishes in the form of short texts (Jain et al. 2013). This approach has the merit of exploring the use of the content and the shared connections to reconcile profiles. However, the experiments reveal that this information is not very effective alone, as only 4 out of 543 profiles are reconciled correctly. An elegant approach that combines profile attributes and network by using conditional random fields is proposed in (Bartunov et al. 2012). The key advantage is that it is robust to the absence of profile and/or network information and therefore can also be applied to cases where no profile information is available except the network, although with a significant drop in recall. The disadvantage is that the proposed model needs training data, which, as recalled before, might not be easy to find. Finally, Narayanan et al. consider the case of anonymized networks where little or no profile attributes are available and only the network structure can be exploited (Narayanan and Shmatikov 2009). They propose a method that first selects a small set of seed profiles in both networks that are highly likely to belong to the same individual. Then, new reconciled profiles are propagated iteratively by using the seed. This is similar in spirit to our approach. However, since they only use the network structure the accuracy of their approach is quite low compared to ours.

Finally, social network aggregation systems, such as *FriendFeed* (2007) or *Plaxo* (2002), provide a platform for people to manage their own profiles but they make no attempt at automatically discovering profiles linked to an individual across social networks. *Spokeo* (2006) seems to be quite accurate in finding personal information from different sources (not necessarily social networks), but it shows its limits when it comes to aggregating them. To the best of our knowledge, there is no existing tool that is able to automatically reconcile profiles across social networks.

3 Background

We define a *social internetwork* as a collection of n distinct social networks and we model it as a directed graph. Its nodes correspond to the profiles of the individuals or, with an abuse of language, to the individuals themselves. A *profile* consists of a set of attributes (e.g., nickname, name, email address...), which are usually described in a Web page created by an individual, and a uri, identifying that page on the Web. A link in a social internetwork connects either two nodes referring to two distinct individuals, in which case we call it a *friendship link*, or two nodes that refer to the same individual, and we call it a *cross-link*.

Formally, a social internetwork with n social networks is a directed labelled graph defined as follows:

$$\mathcal{G} = \langle \bigcup_{i=1}^{n} V_i, \bigcup_{i=1}^{n} E_i, \bigcup_{i,j=1}^{n} E_{i,j} \rangle$$

where:

- V_i is the node set of the social network i. Since the social networks are distinct, $V_i \cap V_j = \emptyset, \forall i \neq j$. Each node $v_i \in V_i$ is the profile of an individual in the social network i. A is the set of the attributes disclosed in a profile, with the exception of REALNAME, all attributes could be multivalued, and each attribute $a \in A$ in the profile v_i could have no or many value(s) in $P_a(v_i)$.
- E_i is the set of friendship links, which are identified by the label $friend$. Each link $(v_i, friend, u_i) \in E_i$ represents a friendship link from the individual v_i to the individual u_i within the social network i. We denote by $friends(v_i) = \{u_i | (v_i, friend, u_i) \in E_i \vee (u_i, friend, v_i) \in E_i\}$ the set of all friend's profiles of v_i in the social network i.
- $E_{i,j}$ is the set of cross-links, which are identified by the label me. A cross-link (v_i, me, v_j) joins two nodes v_i and v_j that refer to the same individual, either in the same network (*intra-network* cross-link) or across two different networks (*inter-network* cross-link). By definition, this type of link is symmetrical and transitive. For instance, Bob might indicate in his Flickr (fk) profile, represented by the node v_{fk}, the *uri* of his LiveJournal (lj) profile, represented by the node v_{lj}, and in this page he declares the *uri* of his Twitter (tw) profile, represented by the node v_{tw}. In this case,

  - $E_{fk,lj} = \{(v_{fk}, me, v_{lj}), (v_{lj}, me, v_{fk})\}$,
  - $E_{tw,lj} = \{(v_{tw}, me, v_{lj}), (v_{lj}, me, v_{tw})\}$ and
  - $E_{tw,fk} = \{(v_{tw}, me, v_{fk}), (v_{fk}, me, v_{tw})\}$.

The problem of reconciling the profiles referring to the same individual across social networks is the problem of discovering the missing cross-links in a social internetwork and is formalized as follows:

Input: $\mathscr{G} = \langle \bigcup\limits_{i=1}^{n} V_i, \bigcup\limits_{i=1}^{n} E_i, \bigcup\limits_{i,j=1}^{n} E_{i,j} \rangle$

Output: $\mathscr{G} = \langle \bigcup\limits_{i=1}^{n} V_i, \bigcup\limits_{i=1}^{n} E_i, \bigcup\limits_{i,j=1}^{n} E_{i,j} \bigcup\limits_{i,j=1}^{n} D_{i,j} \rangle$ where

$D_{i,j} = \{(v_i, me, v_j) | v_i \in V_i, v_j \in V_j \wedge (v_i, me, v_j) \notin E_{i,j}\}$ is the set of the discovered cross-links.

For the sake of simplicity, in the remainder of the paper we will denote $\bigcup\limits_{i,j=1}^{n} E_{i,j}$ as E_{me} and $\bigcup\limits_{i,j=1}^{n} D_{i,j}$ as D_{me}.

In order to determine whether a cross-link exists between two nodes v_i and v_j, LIAISON compares the values of their attributes, based on the observation that if v_i and v_j refer to the same person, the values of their attributes are likely to be equal or similar. A major challenge here is the choice of the pairs of nodes to compare. Evidently, a comparison between all possible pairs is both computationally unfeasible and unnecessary. It is unfeasible because real social networks normally consist of millions of nodes. For example, the sample of four social networks on which we evaluate LIAISON contains around 2 million nodes, that is around 4×10^{12} pairs; if we assume that the comparison of each pair takes 0.1 ms, LIAISON would take 12 years to complete! It is unnecessary because a previous research pointed out that the individuals that have multiple profiles tend to be connected with friends who also have multiple profiles; moreover, when two friends both have multiple profiles, they are frequently friends in multiple networks (Golbeck and Rothstein 2008).

Based on this observation, LIAISON obtains a subset of node pairs to compare, which we term the *candidate set*, from the set $friends(v_i) \times friends(v_j)$ for each (v_i, v_j) such that a cross-link between v_i and v_j already exists in E_{me}. Next, LIAISON uses a set of rules to compare the attribute values of each candidate pair and discover new cross-links which are added to the set D_{me}. Finally, LIAISON iterates the candidate selection for each *me* pair in D_{me} to discover further cross-links until no more candidates can be determined.

In the next sections we describe in greater details the attributes that LIAISON considers to compare two nodes, the similarity measures used to compare the values of each attribute as well as the candidate selection strategy, which differs from the one adopted in the previous version of LIAISON (Bennacer et al. 2014a).

4 Attribute Comparison

In all major social networks the values of some attributes are publicly accessible as per default privacy policy and/or left accessible by the individuals. It is therefore natural to analyze these data to establish new cross-links.

Building on a previous research, which identified the attributes that are generally publicly available in 12 of the most important social networks (Krishnamurthy and Wills 2009), we focus our attention on the following set A of attributes: NICK-NAMES, REALNAME (which includes first name and last name), LOCATIONS, EMAILS, PROFILES (links to social network profile pages) and WEBSITES (links to other web pages). Any two values $p_a(v_i) \in P_a(v_i)$ and $p_a(v_j) \in P_a(v_j)$ of an attribute $a \in A$ are compared with a similarity measure, which assigns a score between 0, when the values are dissimilar, and 1, when the values are identical. Two attribute values *match* if their score is greater than θ_a, where θ_a is a threshold value. We now describe each attribute in more detail.

4.1 NICKNAMES

Denoted as u, the *nickname* (or *username*) is always publicly accessible, as it is the only way to uniquely identify an individual within a social network, and is generally a part of the URI of the web page that hosts the profile. Studies have shown that individuals tend to use the same nickname, or a similar one, when registering different profiles (Perito et al. 2011; Zafarani and Liu 2009); as a result, the similarity of two nicknames is best represented by their *Levenshtein* distance, which for two strings is defined as the minimum number of single character edits (insertion, deletion and substitution) needed to change one word into the other (Buccafurri et al. 2012; Perito et al. 2011; Zafarani and Liu 2009). Therefore, the similarity of two nickname values $p_u(v_i)$ and $p_u(v_j)$ is computed as:

$$1 - \frac{Levenshtein Distance(p_u(v_i), p_u(v_j))}{length_{max}(p_u(v_i), p_u(v_j))}$$

where $length_{max}$ is the number of characters of the longest string. As an example, the Levenshtein distance between the nickname *cospics* of the Flickr profile at www. flickr.com/photos/cospics and the nickname *cos* of the LiveJournal profile at www. livejournal.com/users/cos/profile is 4, because we need to suppress the phrase "pics", composed of four characters, to obtain the second nickname from the first. As a result, their similarity score is 0.43.

4.2 REALNAME

Denoted as n, the first and family names are also present in most of the networks we came across, but their values cannot be trusted as much as the nicknames. Indeed, in some social networks, such as LiveJournal, the profile of a person is almost entirely public and consequently individuals do not feel confident in revealing their real names. Moreover, names are often ambiguous, and do not generally identify an individual uniquely. As a result, we do not expect the name of an individual to reveal many cross-links, unless it is in combination with other attributes. The similarity of two names $p_n(v_i)$ and $p_n(v_j)$ is computed with the *Jaccard* similarity measure as $\frac{|N_i \cap N_j|}{|N_i \cup N_j|}$, where N_i and N_j are the sets of the words that compose $p_n(v_i)$ and $p_n(v_j)$ respectively. For example, if $p_n(v_i)$ is "Barack Obama" and $p_n(v_j)$ is "Barack Hussein Obama", then $N_i = \{Barack, Obama\}$, $N_j = \{Barack, Hussein, Obama\}$ and their similarity is $\frac{2}{3}$. The reason why we select the Jaccard measure instead of the Levenshtein distance is that generally social networks do not force their users to specify their first names before the last names. Moreover, some individuals might specify their middle names in a profile, while omitting them in another. Therefore, a comparison between "Barack Obama" and "Obama Barack" would give a Levenshtein distance of 10, although the two strings are equivalent, while Jaccard gives a score of 1.

4.3 LOCATIONS

Denoted as l, the information about the current location and/or birthplace of an individual can often be found in social network profiles. While the location poses more challenges compared to other attributes, it provides a useful indicator to strengthen or discard the hypothesis that two profiles refer to the same individual. The main problem is that in a profile the location is specified with a *toponym* (e.g., "Paris") which is often ambiguous, as there are multiple locations, or *interpretations*, for a given toponym (e.g., "Paris, France", "Paris, Texas, USA", "Paris, Ontario, Canada").

Intuitively, two toponyms have a strong similarity if they have a high degree of overlap within a low number of interpretations. For example, the overlap between the interpretations of the pair of toponyms ("Paris", "Paris, France") is the same as the overlap of the pair of toponyms ("Paris, Ile-de-France", "Paris, France"),[1] the overlap containing the interpretation corresponding to the capital of France. However, in the first case the toponym "Paris" contains many possible interpretations (corresponding to all locations named "Paris" around the world), while in the second case the two toponyms have only one possible interpretation; as a result, the similarity score of the second pair of toponyms should be higher than the similarity score of the first pair. Hence, two toponyms which represent accurate geographical locations and have a strong overlap within their administrative tiers like country, state, and city are deemed highly similar. On the contrary, two locations which have a diverse set of interpretations and/or have a small degree of match within the administrative divisions are deemed weakly similar. The challenge is to be able to measure the similarity of two toponyms by taking into account the overlap of their interpretations and their ambiguity.

OpenStreetMap[2] service (OSM) exposes a web-service to query for toponyms. For a given query, the service returns the most relevant possibilities and also a hierarchical break down of the administrative divisions (like country, state, city, post code etc.). Furthermore, the results are ordered by *importance*, a numerical value ranking the pertinence of the results with respect to the search query. We utilize this service to collect information about possible interpretations for the toponyms to quantify similarity between them.

4.3.1 Representation of OSM Query Answers

The query's results set R_l obtained for a toponym l from OSM is represented as a weighted-tree, which we term the *interpretation tree*, with a maximum depth of 3 (*country*, *state* and *city*). Since the OSM service has a finer-grained division of toponyms, some of them are merged together (e.g., *city* and *town* are both considered under the category *city*). Each branch represents a unique geographic location and has an *importance* associated with it (shown in brackets). Each administrative level is assigned an empirical weight w, to reflect its strength in determining whether two profiles refer to the same individual. In other words, the fact that the location attribute of two profiles mention the same country does not provide as strong an evidence as

---

[1] Ile-de-France denotes the region of Paris.
[2] http://www.openstreetmap.org/.

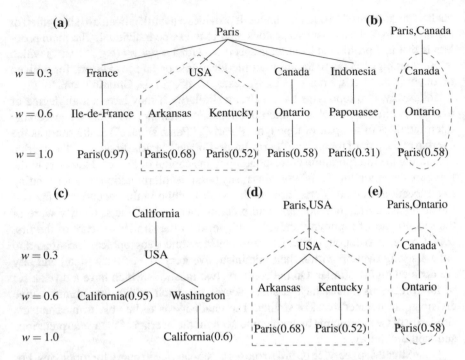

Fig. 1 Representation of results examples for toponyms: **a** "Paris"; **b** "Paris, Canada"; **c** "California"; **d** "Paris, USA"; and, **e** "Paris, Ontario"

to whether they refer to the same individual as the fact that they mention the same city. As a result, the country is given a lower score than the city.

This tree is used to compute the *dis-ambiguity score* combining the *importance* (i_r) and the granularity (w_r) of a result $r \in R_l$. Formally, the *dis-ambiguity score* of r can be expressed by $w_r \times i_r$. In Fig. 1a, the interpretation "Paris, France" of the toponym "Paris" gets a higher *dis-ambiguity score* than "Paris, Canada", as the importance given by OSM to "Paris, France" is higher. The first answer of the query "California" will get a lower *dis-ambiguity score* (0.57) than the second answer (0.6) as the granularity of "California" in Washington is more precise. In this case the granularity influences the score more.

4.3.2 Location Similarity Measures

In order to compare two toponyms l_1 and l_2, their interpretation trees obtained from R_{l_1} and R_{l_2} are constructed and compared using their overlapping $R^\cap_{l_1,l_2}$. We denote as $n_{l_1}, n_{l_2}, n_{l_1,l_2}$ and n the *dis-ambiguity score* of $R_{l_1}, R_{l_2}, R^\cap_{l_1,l_2}$ and $R^\cup_{l_1,l_2}$, respectively; they are defined as follows:

Table 1 Similarity measures applied to OSM query answers for toponym pairs

	Location 1	Location 2	W.Support	Ochiai
1.	San Diego, USA	San Diego	0.4	0.87
2.	Houston, Texas, USA	Houston	0.56	0.77
3.	Canada	Toronto, Canada	0.11	0.28
4.	Orlando, Florida, USA	Florida	0.23	0.63
5.	Wausau, Wisconsin, USA	Wausau, WI	1	1
6.	Los Angeles, USA	Los Angeles	0.4	0.85
7.	Argentina	Argentina, Buenos Aires, Junn	0.08	0.19
8.	Montreal, Canada	Montreal, Quebec	0.76	0.69
9.	United States, USA	Puerto Rico	0.13	0.3
10.	Apeldoorn, Netherlands	Deventer	0.13	0.3
11.	Bengaluru, India	Bangalore, India	1	1
12.	Utrecht, Netherlands	Amersfoort, The Netherlands	0.26	0.6
13.	New York City, USA	Brooklyn, NY, USA	1	1

$$n_{l_1} = \sum_{r \in R_{l_1}} \frac{w_r \times i_r}{I} \qquad n_{l_2} = \sum_{r \in R_{l_2}} \frac{w_r \times i_r}{I}$$

$$n_{l_1,l_2} = \sum_{r \in R_{l_1,l_2}^{\cap}} \frac{w_r \times i_r}{I} \qquad n = \sum_{r \in R_{l_1,l_2}^{\cup}} \frac{w_r \times i_r}{I}$$

$I = \sum_{r \in R_{l_1,l_2}^{\cup}} i_r$ is the normalization coefficient computed by considering the union of R_{l_1} and R_{l_2}. For example, the overlap between the trees of the locations "Paris" and "Paris, USA" in Fig. 1a, d is represented by the red dotted line. The *dis-ambiguity score* value of $R^{\cap}_{"Paris-Paris,USA"}$ and $R^{\cap}_{"Paris-Paris,Canada"}$ are 0.39 and 0.18, respectively. Measuring the similarity between two toponyms l_1 and l_2 depends on the *dis-ambiguity score* of their overlap. To this extent, we investigated two similarity measures, namely the weighted *Support* measure and the *Ochiai* measure (Ochiai 1957), a variation of the cosine similarity.

$$S_{W-Support} = \frac{1}{\sqrt{n}} \times \frac{n_{l_1,l_2}}{n}$$

$$S_{Ochiai} = \frac{n_{l_1,l_2}}{\sqrt{n_{l_1} \times n_{l_2}}}$$

The value of these similarity measures are between 0, if the overlapping between R_{l_1} and R_{l_2} is empty, and 1, if $R_{l1} = R_{l2}$ for Ochiai, whereas weighted Support is more strict as the necesary condition for this measure to yield 1 is that both the inputs have exacly one interpretation that match exactly. In fact, this measure penalizes the similarity if the toponyms are ambiguous.

Table 1 shows the similarity measures applied to OSM query answers for some examples of toponym pairs collected from a real dataset (Malhotra et al. 2012). Several observations can be made from the examples shown in this table. Firstly, the proposed method is able to handle orthographic differences in inputs and predicts an exact match when the inputs are orthographically different but are in-fact non-ambiguous and refer to the same geographic location (examples 5, 11, 13). Secondly, it is able to assign low similarity in cases when the granularity of the two inputs is different (and thus ambiguous) as can be observed in examples 3, 7 and 9. It is also able to assign a low similarity score when the granularity is the same but the geographic locations are different as in examples 10 and 12. Lastly, the weighted support measure gives a lower similarity value when the inputs have many interpretations. This can be attributed to the term $(1/\sqrt{n})$ which penalizes the score as the number of interpretations increases.

4.4 EMAILS

Denoted as e, EMAILS is a multi-valued attribute whose values correspond to the different email addresses disclosed by an individual. The email address is a very sensitive attribute, because it could identify a person uniquely. If two profiles are associated with the same email address, there are high chances that the two profiles refer to the same individual. It is certainly possible that two individuals share the same email address, as in the case of people that work within the same organization. But these are particular cases, and in general email addresses can be trusted to uniquely identify a person. The only problem is that only a small percentage of people grant public access to their email addresses. In order to compare the values of the attribute e of two profiles, we need to determine whether one of the email addresses of a profile is identical to one of the email address of the other profile. In other words, two email addresses match if they are identical.

4.5 WEBSITES and PROFILES

WEBSITES (w) and PROFILES (p) are two multi-valued attributes whose values are URLs to respectively general web pages and profile pages in social network sites. We aim at investigating the contribution of the two attributes separately, because profile pages are usually more "personal" than links to generic web pages; as a result, the fact that two profiles share the same link(s) to profile pages is likely to be a stronger evidence as to whether they refer to the same individual than the fact that they share the same links to generic web pages. We say that two values of the attribute WEBSITES or PROFILES match if they are identical.

5 LIAISON

LIAISON consists of two steps, the *candidate selection* and the *cross-link determination*; the first obtains a subset of profile pairs to compare, the second determines the pairs that have to be connected through a new cross-link. The two steps are iterated until no more candidates are found. In the remainder of the section, we describe the two steps in more detail and we comment on the pseudocode of the algorithm.

5.1 Candidate Selection

The previous version of LIAISON (Bennacer et al. 2014a, b) selects a subset of node pairs to compare, which we term the *candidate set*, by considering all pairs in $friends(v) \times friends(w)$ for each (v, w) such that $(v, me, w) \in E_{me}$ (the cross-link between v and w exists) or $(v, me, w) \in D_{me}$ (the cross-link between v and w is discovered by LIAISON). Although this procedure considerably reduces the number of candidate pairs, compared to considering all possible pairs, it still selects around 1 billion candidates in one of the datasets that we use in our evaluation, which yields to more than 27 h of computation.

For this reason, in the current version of LIAISON we opt for a more efficient solution. We observe that in the vast majority of the cases, most of the pairs in $friends(v) \times friends(w)$ have dissimilar attribute values and no cross-link will ever be established by LIAISON between them. Therefore these pairs are not good candidates and should not be selected as such; in other words, only the pairs whose attribute values are identical or similar should be selected as candidates. But how do we determine these pairs without comparing all possible pairs in $friends(v) \times friends(w)$? We do so by selecting appropriate data structures as we explain later in Sect. 5.3.

5.2 Cross-Links Determination

In order to determine whether two profiles v and w refer to the same individual, we defined a set of rules based on the attributes introduced in Sect. 4. Each rule considers the contribution of one or several attributes. We assume that the higher the number of attributes that match, based on the defined similarity measures, the higher the probability for two profiles to refer to the same individual. We therefore define the *order k* of a rule as the number of attributes that the rule uses. The rule with the highest confidence is the one that uses all the attributes ($k = |A|$). The rules with the lowest confidence are those that use just one attribute ($k = 1$).

Let $match(P_a(v), P_a(w))$ be the predicate which is true when two values $p_a(v)$ and $p_a(w)$ match, based on the similarity measure defined for the attribute a. A rule with the order k, or $k-$rule, $\mathscr{R}^k$ is defined as follows:

$$\mathscr{R}^k(v, w) = \begin{cases} \bigwedge_{a \in A} match(P_a(v), P_a(w)) & \text{if } k = |A| \\ \bigvee_{B \in [A]^k} \bigwedge_{a \in B} match(P_a(v), P_a(w)) & \text{if } 1 \le k < |A| \end{cases}$$

where $[A]^k$ is the set of all subsets of A with k elements.

Two profiles v and w are considered to refer to the same individual if and only if $\bigvee_{1 \le k \le |A|} \mathscr{R}^k(v, w)$ is true; in this case, the link (v, w) is added to the set of discovered cross-links D_{me}. Note that the rules are applied by decreasing values of k, therefore if $\mathscr{R}^K(v, w)$ is *true*, for some value K, no rule with order lower than K is applied. Each discovered cross-link (v, w) is assigned a *confidence score* $conf_{v,w}$ that is equal to the order of the first rule that is found to be true. More formally:

$$conf_{v,w} = \max_k \{\mathscr{R}^k(v, w) = true\}$$

Beside applying the rules, LIAISON computes the *transitive closure* of the links in $E_{me} \cup D_{me}$ to discover more cross-links; the idea is that a cross-link (v, w) is added to D_{me} by transitivity if there is at least one path of cross-links (or *cl-path*) between node v and node w. The challenge here is the determination of the confidence score of a cross-link (v, w) discovered by transitive closure, because the cross-links of one particular *cl-path* do not necessarily have the same confidence score and there might be more than one *cl-path* between v and w. Intuitively, for a given *cl-path* from v to w, the confidence score of the cross-link (v, w) cannot be higher than the minimum of the confidence scores of the cross-links that are part of p; in fact, this would be tantamount to stating that (v, w) has a confidence score higher than the confidence score of one of the cross-links that contributed to its discovery. Formally, we define the confidence score $conf_{v \to w}$ of a *cl-path* from v to w as:

$$conf_{v \to w}(v_1 = v \to v_2 \to \dots \to v_n = w) = \min_{i=1,..,n-1} conf_{v_i, v_{i+1}}$$

If there is more than one *cl-path* linking v to w, LIAISON sets the confidence score of the cross-link (v, w) as the maximum of the confidence scores of all such paths. In fact, the existence of more than one *cl-path* between v and w gives more credit to the existence of the cross-link (v, w) and therefore the confidence score of (v, w) should be the highest possible. The same reasoning is applied when a cross-link (v, w) is discovered both by a rule and by transitivity.

5.3   The Algorithm

In order to better explain how LIAISON works, we use as a running example a small social internetwork $\mathscr{G}$ consisting of four social networks, namely Twitter, Flickr, LiveJournal, and YouTube, as shown in Fig. 2. Each social network includes a set of profiles identified by names and connected by friendship links that are depicted as solid arrows. In order to detect cross-links in $\mathcal{G}$, LIAISON assumes that some cross-links, connecting profiles that refer to the same individual, already exist; in the figure, these cross-links are represented as dashed lines, while the dotted and dash-dotted lines refer to cross-links that are missing and are discovered by LIAISON, by using the rules and the transitive closure respectively. For the purpose of the example, we assume that two profiles that are identified by the same name refer to the same individual.

As a first step, LIAISON computes the transitive closure of the existing cross-links, which leads to the discovery of the cross-link number 1 between the profiles of Alice in LiveJournal and YouTube. The existing cross-links and the newly discovered one are added to a queue $\mathcal{Q}$. LIAISON repeats the following procedure until $\mathcal{Q}$ is empty: it removes the first cross-link $e = (v, w)$ in $\mathcal{Q}$, it uses e to create a *candidate set* C_e, consisting of the cartesian product of the sets $friends(v)$ and $friends(w)$, and applies the rules to determine the pairs of nodes in C_e to be connected by a new cross-link; the newly discovered cross-links are added to the queue $\mathcal{Q}$.

In our example, LIAISON considers the cross-link between the profiles of Lisa in Flickr and LiveJournal and creates the candidate set {(Bob, Alice), (Bob, Ben), (Mark, Alice), (Mark, Ben), (Alice, Alice), (Alice, Ben)}; after applying the rules to each pair in the candidate set, LIAISON unveils, and adds to $\mathcal{Q}$, the cross-link number 2 between the profiles of Alice in Flickr and LiveJournal. Once the queue $\mathcal{Q}$ is empty, meaning that there are no more cross-links to obtain further candidates, the transitive closure of the cross-links is computed again, which results in the discovery of the two cross-links 3 and 4. These two cross-links are added to $\mathcal{Q}$ and a new

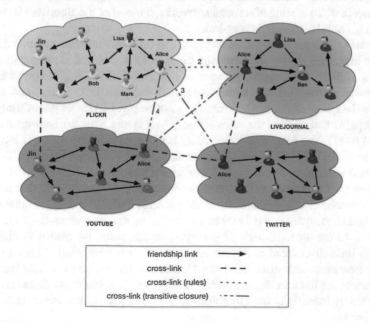

friendship link	⟶
cross-link	– –
cross-link (rules)	· · · · ·
cross-link (transitive closure)	· – · –

Fig. 2 Description of the algorithm on a small social internetwork

---

Algorithm 1: LIAISON algorithm

Data: $\mathscr{G} = < \cup_i V_i , \cup_i E_i , E_{me} >$
Result: $\mathscr{G}' = < \cup_i V_i , \cup_i E_i , E_{me} \cup D_{me} >$
1   $D_{me} \leftarrow transitiveClosure(E_{me}); it \leftarrow 1; D_{it} \leftarrow \emptyset; enqueue(E_{me} \cup D_{me}, \mathscr{Q});$
2 **while** $\mathscr{Q}$ *is not empty* **do**
3   **do**
4       $e \leftarrow dequeue(\mathscr{Q});$
5       $C_e \leftarrow candidateSelection(e);$
6   **foreach** $c_e \in C_e$ **do**
7   **if** $c_e \notin E_{me} \cup D_{me} \cup D_{it}$ **then**
8          $k \leftarrow applyRules(c_e);$
9   **if** $k \geq 1$ **then**
10           $D_{it} \leftarrow D_{it} \cup \{c_e\};$
11           $enqueue(\{c_e\}, \mathscr{Q});$
12   **while** $\mathscr{Q}$ *is not empty*;
13     $D_c \leftarrow \emptyset;$
14   **foreach** $k \in [|A|, 1]$ **do**
15       $D_c \leftarrow transitiveClosure_k(E_{me} \cup D_{me} \cup D_{it});$
16       $D_{it} \leftarrow D_{it} \cup D_c;$
17       $enqueue(D_c, \mathscr{Q});$
18     $D_{me} \leftarrow D_{me} \cup D_{it}; it \leftarrow it + 1; D_{it} \leftarrow \emptyset$

---

iteration is started over. LIAISON stops when the transitive closure at end of a given iteration does not discover any other cross-link.

We now describe the algorithm in greater detail (cf. Algorithm 1). Given a social internetwork $\mathscr{G}$, consisting of n social networks, the goal of the algorithm is to create the set D_{me} of all discovered cross-links.

In order to determine the candidate pairs, LIAISON maintains a queue $\mathscr{Q}$ of cross-links, as explained above, and saves the cross-links discovered at the iteration it in a set D_{it}. Initially the transitive closure of the cross-links in E_{me} is computed, and the discovered cross-links are added to D_{me}; all the cross-links in E_{me} and D_{me} are added to the queue $\mathscr{Q}$ (Line 1). While $\mathscr{Q}$ is not empty, LIAISON goes through two main steps; it first applies the rules (Line 3 to 12) and then the transitive closure (Line 14 to 17). The first cross-link e in $\mathscr{Q}$ is removed from the queue (Line 4) and is used to obtain a candidate set C_e, consisting of pairs of nodes (Line 5). The rules are applied to each candidate $c_e \in C_e$ for which a cross-link does not exist (Line 6 to 8); here the function *applyRules* returns the order $k > 0$ of the rule that is true, if any, 0 otherwise. The candidates for which at least one rule is true are added to the set D_{it} and to the queue $\mathscr{Q}$ (Line 9 to 11), meaning that new cross-links are found. Lines 3 to 12 are repeated until $\mathscr{Q}$ is empty. At this point, the transitive closure of the cross-links discovered at the current iteration it is computed (Lines 14 to 17) and any new cross-link discovered is added to D_{it} and the queue $\mathscr{Q}$. At the end of the iteration, all links in D_{it} are added to D_{me} and if some cross-links have been discovered by transitivity (and therefore $\mathscr{Q}$ is not empty), a new iteration is started over (Line 18).

We note that the function $transitiveClosure_k$ at Line 15 computes the transitive closure of the cross-links that have confidence score at least k; this function is invoked by decreasing values of k so that the confidence value of any discovered cross-link is computed as described in Sect. 5.2.

In this algorithm the selection of the candidates is a sensitive point, as the number of candidates obtained from a given cross-link (v, me, w) can be very large $(O(|friends(v)| \times |friends(w)|))$. In order to reduce the number of candidates, the idea is to store the values of each attribute a of all nodes $x \in friends(v)$ in a data structure I_a that allows for fast retrieval and obtain the nodes $y \in friends(w)$ that have identical or similar attribute values as the ones stored in I_a. I_a can be either a hash table, in case the values of the attribute a are compared through exact matching (e.g., the attribute WEBSITES), or a BK-Tree (Burkhard and Keller 1973), in case the values of a are compared through approximate matching (e.g., the attribute NICK-NAMES). Both data structures are known to have good performances when adding new values and retrieving existing ones. As a result, this procedure is much more efficient than considering all possible pairs in $friends(v) \times friends(w)$. Indeed, as our experiments reveal, we reduce the number of candidates from 1 billion to 76 million, without missing any cross-link.

The transitive closure cost is $O(|A| \times |\bigcup_i V_i| \times |\bigcup_i E_i \bigcup D_{me} \bigcup E_{me}|)$, knowing that $|\bigcup_i E_i \bigcup D_{me} \bigcup E_{me}| << |\bigcup_i V_i|$.

6 Evaluation Results

In order to evaluate our approach, we considered the dataset used by Buccafurri et al. in their experiments (Buccafurri et al. 2012). The original dataset includes a social internetwork with four social networks, namely LiveJournal, Flickr, Twitter and YouTube.[3] The graph is composed of 93,169 nodes, 145,580 friendship links and 503 cross-links, of which 474 inter-network and 29 intra-network. We note that the number of cross-links declared by Buccafurri et al. (2012) is 745, but this also includes duplicate links, which we removed.

After a careful analysis of the data, we found that many $friend$ links were missing between a large number of nodes, probably because they were added after the internetwork was crawled. Moreover, the only available profile attribute is the nickname. For this reason, we updated the internetwork by obtaining the missing information using the API of the four SNSs under evaluation. While we were at that, we also enriched the graphs by adding new nodes that are linked via a $friend$ link to the existing nodes. As a result, we obtained a much larger internetwork, whose properties are shown in Table 2. In total, we have more than 2 million nodes, more than 21 million links and 29 intra-network cross-links. In addition to that, we have 474 inter-network cross-links, whose distribution across the social networks is shown in Table 3.

---

[3] http://www.ursino.unirc.it/pkdd-12.html.

Table 2 Statistics on the social internetwork used in our evaluation

Network	Nodes	Links		
		friend	*intra − me*	Total
Flickr	1,814,405	15,415,083	0	15,415,083
LiveJournal	211,044	5,628,509	1	5,628,510
Twitter	8,842	19,008	13	19,021
YouTube	1,210	1,367	15	1,382
Total	**2,035,501**	**21,063,967**	**29**	**21,063,996**

Table 3 Cross-links between all pairs of social networks

Network	Flickr	LiveJournal	Twitter	YouTube
Flickr	0	148	29	12
LiveJournal	148	1	11	2
Twitter	29	11	13	272
YouTube	12	2	272	15

In the implementation of our approach, the social internetwork is stored in a Neo4j database,[4] which is particularly indicated to handle large graphs.

6.1 Evaluation of LIAISON

In our previous work, we described a preliminary evaluation aimed at identifying the attributes that are most useful to reconcile profiles, as well as tuning the thresholds θ_u and θ_n of the similarity measures used to compare respectively two nicknames and two names. That evaluation showed that:

- Any k-rule, with $k \geq 2$, discovers cross-links with high precision. In other words, if two profiles have at least two attributes with matching values, they are extremely likely to refer to the same person.
- The 1-rule using the attribute *nicknames* discovers cross-links with high precision if the threshold θ_u is set to 0.9
- The 1-rule using the attribute *names* leads to a high error rate, no matter how the threshold θ_u is set. Therefore, two profiles where the names match should not be considered as referring to the same person, unless other attribute values match.

We repeat a similar evaluation on a sample of our dataset to tune the threshold θ_l of the similarity measure used to compare two values of the attribute *locations*. Like the attribute *names*, the mere fact that two profiles disclose the same or similar

---

[4]www.neo4j.org/.

Table 4 Cross-links discovered by LIAISON by iteration and k. The total number of discovered cross-links (6,572) does not include the 6 links discovered by transitive closure from the existing links

Iteration	Method	k = 1	k = 2	k = 3	k = 4	k = 5	Total	Grand total
1	R	3,792	853	84	4	0	4,733	4,907
	Tc	161	13	0	0	0	174	
2	R	1,104	69	47	20	4	1,244	1,620
	Tc	373	2	1	0	0	376	
3	R	19	0	1	0	0	20	45
	Tc	25	0	0	0	0	25	
Total		5,474	937	133	24	4	6,572	6,572

locations is not conclusive as to whether they refer to the same person. Therefore, the attribute *locations* need to be used in combination with other attributes; moreover, we observed that good results are obtained by setting θ_l at 0.7.

Based on these observations, we run LIAISON by setting $\theta_u = \theta_n = 0.9$ and $\theta_l = 0.7$ on the social internetwork described above and including all rules except the 1-rule using the attribute *name* and the 1-rule using the attribute *location*. We conducted all the experiments on a computer equipped with a 64 bit Intel Xeon processor at 2.3 GHz, 16 GB RAM and running Debian Linux 6.0 (kernel version 2.6.32-5). The Java Heap size reserved for LIAISON is set to 9 GB. LIAISON discovered 6,578 links in 2 h, 11 min and 58 s after comparing 76,368,416 candidate pairs through 4 iterations; the average time taken to retrieve a set of candidates given a cross-link was 0.48 s, while only 0.5 ms were necessary on average to compare each pair of candidates. Considering that the candidate selection is repeated for each cross-link and there are 7,081 cross-links after running LIAISON (503 already existing plus the 6,578 discovered by LIAISON), LIAISON spent approximately 1 hour retrieving the candidates and 1 h to compare them. In total, the transitive closure needed only 2 s.

The number of cross-links discovered at each iteration is shown in Table 4. More precisely, the table shows the number of cross-links discovered by using the rules (R) and the transitive closure (Tc) at each iteration for each value of k. As expected, the number of discovered cross-links decreases while LIAISON progresses through the iterations. At the iteration 3, LIAISON discovers 25 cross-links by transitive closure, from which new candidates are found that are compared in the fourth iteration; since none of these candidates are found to be profiles referring to the same individual, LIAISON stops.

Most of the cross-links are discovered at the first iteration and by using the 1-rules, which clearly indicates that two profiles created by the same individual usually have little overlapping information.

Despite that, by using the value of just one attribute, LIAISON discovers 5,474 cross-links, most of which are correct, as discussed below. This result is particularly remarkable if we consider that LIAISON starts from a seed set of 503 cross-links, of which only 239 connect two nodes that have friendship links to other nodes and therefore can be used to obtain new candidates. We note also that the total number of discovered cross-links shown in the table (6,572) does not include 6 cross-links that are discovered by transitive closure from the existing links before the first iteration.

The cross-links discovered through transitive closure are considerably less than those discovered through the rules. One possible explanation lies in the nature of the internetwork itself, which, although large, is still a limited sample of 4 real social networks that combined have more than 500 million profiles. As a result, our internetwork might not contain all the profiles of an individual.

In order to evaluate the accuracy of the rules, we determined a ground truth by tagging each cross-link $(v, me, w) \in D_{me}$ as either *correct*, if v and w actually refer to the same individual, or *incorrect*, if they do not, or *undetermined*, if no decision can be taken. To this extent, we split D_{me} into four equal-size independent subsets, one for each author of this paper, who had to assign the proper tag to each cross-link, based on a visual inspection of the profile web pages of the individuals concerned. The visual inspection consisted in looking at every possible aspect of the profiles except the values of the attributes used by LIAISON, in particular: photos (especially in Flickr), textual content (especially in LiveJournal), information on web pages linked by the profile and retrieved from other social networks. Most of the time the information were enough to determine whether two profiles referred to the same individual; however, in some cases the available information is so scarce that no conclusive evidence as to whether the two profiles match can be found. In order to avoid errors in the ground truth, which would inevitably invalidate the results of our evaluation, we introduced the tag *undetermined*, which we assigned to all cross-links that we could not determine with certainty as either *correct* or *incorrect*. As a result, we determined three subsets of D_{me}: (i) C, the set of the cross-links tagged as *correct*; (ii) W, the set of the cross-links tagged as *incorrect* and (iii) U, the set of the cross-links tagged as *undetermined*. It took approximately 10 days to tag all cross-links in D_{me}. Based on this ground truth, we can compute the *precision* of LIAISON as $P = \frac{|C|}{|C|+|W|}$. We note that to obtain the recall, that is the ratio between the correct cross-links and the total number of profile pairs that actually refer to the same individual, we would need to tag all possible profile pairs in our dataset, which is clearly not feasible. In the next section, we will discuss the recall on another much smaller dataset, where the ground truth for every pair of profiles is already known.

Precision of LIAISON. The overall precision of LIAISON on the dataset across all iterations is 94 %, which is a good result, considering that most of the cross-links, either discovered through a rule or by tansitive closure, have confidence $k = 1$. The graph in Fig. 3 shows the precision obtained by LIAISON with respect to the value of the confidence k. As expected, the precision increases with the confidence and for both the cross-links discovered through the rules and the transitive closure is 100 % for $k \geq 3$. As for $k = 1$, the rules achieve a precision of 94 %, while the transitive closure is sensibly lower (73 %); thisis due to the fact, as opposed to the cross-links

Fig. 3 Precision with respect to the values of k

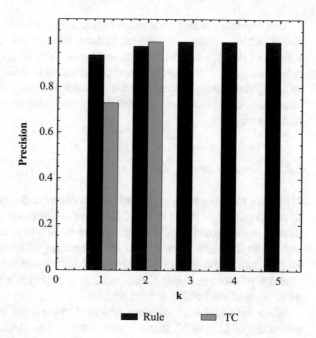

with $k \geq 2$, some cross-links with $k = 1$ are wrong (6 % of them) and they are propagated by the transitive closure.

6.2 Comparison with Existing Work

In this section we compare LIAISON against the approach proposed by Buccafurri et al. (2012) (BUCC), as we built our dataset on top of theirs, and the one described by Malhotra et al. (2012) (MAL), which is evaluated on a small dataset (60,000 nodes) consisting of two social networks.

6.2.1 Comparison Against BUCC

BUCC is evaluated by its authors by randomly selecting 160 existing cross-links, which are used to discover new cross-links. The final number of cross-links discovered by the algorithm is 22, of which 16 are correct, 2 are wrong and 2 undetermined (Buccafurri et al. 2012): the precision of their approach is therefore 80 %. We note that their algorithm also returns a set of 133 node pairs, which are classified as profiles not referring to the same person ("non-me" links); as a result, they also have the number of true and false negatives, which allows them to compute the overall accuracy, which is 85 %.

Our approach discovers a much higher number of cross-links with a better precision, which depends on several factors. First of all, our dataset is an enriched version of theirs, with more nodes and links. Secondly, LIAISON relies on a set of rules which considers the combined contribution of different attributes, while BUCC only exploits the nicknames and the network topology. Finally, LIAISON uses the cross-links that it discovers to obtain new candidates and thus more cross-links in an iterative way.

6.2.2 Comparison Against MAL

MAL uses machine learning techniques to compare the values of multiple attributes of two profiles (Malhotra et al. 2012). For the comparison, we use exactly the same dataset, which consists of a small sample of two popular social networks, Twitter and LinkedIn. Each network has 29,129 nodes with values for several attributes, no friendship links and 29,129 inter-network cross-links; we note that although the friendship links are missing, each node has an attribute whose value is the number of its connections, which is used by MAL.

Since no friendship links are provided, we cannot use the candidate selection procedure of LIAISON; Instead, we considered as candidates all the pairs of nodes (t, l), such that t belongs to Twitter and l belongs to LinkedIn and the values of at least one attribute are similar or identical. To avoid a comparison between all the possible pairs of profiles, we index the values of the attributes of the Twitter profiles by using hash tables and BK-trees. The attributes used by MAL are the nickname, the realname, the short description (a.k.a. "about me") which is often found on social network profiles, the location, the profile image and the number of friends. The 29,219 cross-links are not fed to LIAISON and are considered as ground truth. Moreover, since there are only two networks, we do not apply the transitive closure; we only run one iteration of the algorithm. The values of θ_n, θ_u and θ_l are set as before.

LIAISON discovers 9,210 cross-links, of which 9,134 are correct, in 3 min and 24 s; the overall precision is 99 % and the recall is 31 %. The graph in Fig. 4 shows the variation of the values of the precision with the values of k; consistently with the observations above, the precision increases with the confidence.

The reported precision of MAL on the same dataset is 64 % (Malhotra et al. 2012), which is considerably less than the one that LIAISON achieves, while no recall is given. As for the recall, we note that the low recall achieved by LIAISON is due to the fact that we tuned our rules to ensure that the discovered cross-links are correct with a very high precision, which is extremely important given that we propagate the discovered cross-links to discover new cross-links. We observe that different values of precision/recall can be obtained by tuning the thresholds for the attributes nicknames, realnames and locations differently. As for the realnames and locations, it is not clear how the values can be changed, because the values of both attributes are usually ambiguous and therefore not suitable for the task of reconciling profiles. The case for the nicknames is different, because the same individual tends to use similar nicknames across different profiles. Therefore, we played only with the value of θ_u

Fig. 4 Precision with
respect to the values of k

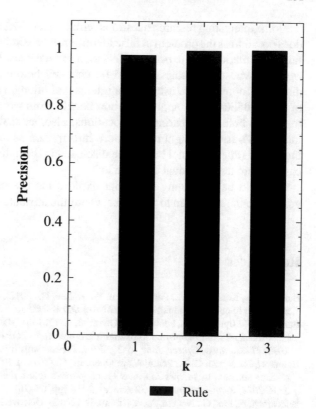

and we observed that the best value for it is 0.7, where the precision is 86 %, recall
is 49 % and the f-measure is 62 %.

7 Concluding Remarks

In this paper, we presented the LIAISON algorithm, built on previous research of ours,
to match profiles of individuals across several social networks by using the network
topology and the personal information that are publicly available in the profiles. We
thoroughly evaluated the algorithm on a large dataset of four real social networks,
which constitutes a real challenge, because data are likely to be erroneous and messy.
The evaluation and the comparison against two existing approaches showed the
robustness of our algorithm, as it achieves a high precision (94 %), and proved its
effectiveness in discovering a lot of cross-links in a large social internetwork with a
highly satisfactory time performance.

We note that our algorithm relies on attributes whose values are publicly available
on the profiles of the individuals. Other available attributes like photos could be easily
taken into account without any changes of the algorithm.

Some interesting research questions remain open. More specifically, our approach does not address the problem of false identities, where individuals disclose voluntarily false information so as to better protect their private lives. Also, the location is another attribute whose correctness should be checked before using it. In particular, the disclosed toponyms are usually ambiguous and often do not reflect the real location of the individuals. Our approach might benefit from techniques to disambiguate the toponyms before comparing two locations; also, an analysis of the locations of an individual's friends might reveal more information as to the current location of an individual (that might not be the one disclosed in the individual profile, especially in cases where the individual travels a lot).

It would be interesting to further explore the use of the network topology to generalize the algorithm to networks where the attribute values are anonymized.

References

Bartunov, S., Korshunov, A., Park, S., Ryu, W., & Lee, H. (2012). Joint link-attribute user identity resolution in online social networks. In *SNA-KDD Workshop*.

Bennacer, N., Jipmo, C. N., Penta, A., & Quercini, G. (2014a). Matching user profiles across social networks. In *Advanced Information Systems Engineering—26th International Conference, CAiSE 2014, Thessaloniki, Greece, June 16-20, 2014. Proceedings* (pp. 424–438).

Bennacer, N., Jipmo, C. N., Penta, A., & Quercini, G. (2014b). Réconciliation des profils dans les réseaux sociaux. In *14èmes Journées Francophones Extraction et Gestion des Connaissances, EGC 2014, Rennes, France, 28-32 Janvier, 2014* (pp. 65–76).

Buccafurri, F., Lax, G., Nocera, A., & Ursino, D. (2012). Discovering links among social networks. In *Machine learning and knowledge discovery in databases*, vol. 7524 (pp. 467–482). Lecture Notes in Computer Science. Berlin: Springer.

Burkhard, W. A., & Keller, R. M. (1973). Some approaches to best-match file searching. *Communications of the ACM, 16*(4), 230–236.

Carmagnola, F., & Cena, F. (2009). User identification for cross-system personalisation. *Information Science, 179*(1–2), 16–32.

Cortis, K., Scerri, S., Rivera, I., & Handschuh, S. (2012). Discovering semantic equivalence of people behind online profiles. In *Proceedings of the Resource Discovery (RED) Workshop, ser. ESWC*.

FriendFeed (2007). friendfeed.com.

Goga, O., Lei, H., Parthasarathi, S. H. K., Friedland, G., Sommer, R., & Teixeira, R. (2013). Exploiting innocuous activity for correlating users across sites. In *Proceedings of the 22nd International Conference on World Wide Web* (pp. 447–458). International World Wide Web Conferences Steering Committee.

Golbeck, J., & Rothstein, M. (2008). Linking social networks on the web with FOAF: A semantic web case study. *In AAAI* (Vol. 8, pp. 1138–1143).

Gross, R., & Acquisti, A. (2005). Information revelation and privacy in online social networks. In *Proceedings of the 2005 ACM Workshop on Privacy in the Electronic Society, WPES 2005* (pp. 71–80). New York, NY, USA: ACM.

Jain, P., Kumaraguru, P., & Joshi, A. (2013). @i Seek 'fb.me': Identifying users across multiple online social networks. In *WWW (Companion Volume)* (pp. 1259–1268).

Krishnamurthy, B., & Wills, C. E. (2009). On the leakage of personally identifiable information via online social networks. In *Proceedings of the 2nd ACM Workshop on Online Social Networks* (pp. 7–12). ACM.

Little, L., Briggs, P., & Coventry, L. (2011). Who knows about me?: An analysis of age-related disclosure preferences. In *Proceedings of the 25th BCS Conference on Human-Computer Interaction, BCS-HCI 2011* (pp. 84–87). Swinton, UK: British Computer Society.

Malhotra, A., Totti, L., Meira, W., Kumaraguru, P., & Almeida, V. (2012). Studying user footprints in different online social networks. In *International Workshop on Cybersecurity of Online Social Network (ACM ASONAM 2012)*.

Motoyama, M., & Varghese, G. (2009). I seek you: Searching and matching individuals in social networks. In *Proceedings of the Eleventh International Workshop on Web Information and Data Management* (pp. 67–75). ACM.

Narayanan, A., & Shmatikov, V. (2009). De-anonymizing social networks. In *30th IEEE Symposium on Security and Privacy* (pp. 173–187). IEEE.

Ochiai, A. (1957). Zoogeographic studies on the soleoid fishes found in japan and its neighbouring regions. *Bulletin of the Japanese Society of Scientific Fisheries, 22*(9), 526–530.

Perito, D., Castelluccia, C., Kaafar, M. A., & Manils, P. (2011). How unique and traceable are usernames? In *Privacy Enhancing Technologies* (pp. 1–17). Springer.

Plaxo (2002). plaxo.com.

Raad, E., Chbeir, R., & Dipanda, A. (2010). User profile matching in social networks. In *2010 13th International Conference on Network-Based Information Systems (NBiS)* (pp. 297–304). IEEE.

Rowe, M. (2009). Interlinking distributed social graphs. In *Linked Data on the Web Workshop, WWW2009*.

Spokeo (2006). spokeo.com.

Stutzman, F. (2006). An evaluation of identity-sharing behavior in social network communities. *iDMAa Journal, 3*(1).

Zafarani, R., & Liu, H. (2009). Connecting corresponding identities across communities. In *Third International AAAI Conference on Weblogs and Social Media*.

Clustering of Links and Clustering of Nodes: Fusion of Knowledge in Social Networks

Erick Stattner and Martine Collard

Abstract The extraction of knowledge from social networks is an area that has experienced significant growth in recent years. Indeed, thanks to the improvement of storage and calculation capacities, and the heterogeneity of data that can currently be extracted, much effort has been made to go beyond traditional knowledge, by proposing new kinds of patterns that take into account the context. However, while many works were interested in designing new patterns of knowledge or in optimizing existing approaches, few studies have been focused in merging patterns and on the useful knowledge emerging from such fusions. In this work, we focus on two network clustering approaches, able to extract two distinct kinds of patterns, and we seek to understand both the intersections that can exist between them and the knowledge that emerges from their fusion. The first is the classical nodes clustering approach that consists in searching for communities into a network. The second is the search for frequent conceptual links, a new link clustering approach that aims identifying frequent links between groups of nodes sharing common attributes. We propose a set of original measures that aim to evaluate the amount of shared information between these patterns when they are extracted from a same network. These measures are applied to three datasets and demonstrate the interest in simultaneously considering several sources of knowledge.

1 Introduction

The domain of knowledge extraction from social networks, also called *social network mining* (Getoor and Diehl 2005; Scott 2011), has experienced strong growth in recent years. While pioneering works have proposed various methods to address classical data mining tasks such as classification of nodes, prediction of links or clustering of nodes, recent approaches have attempted to go beyond traditional knowledge

E. Stattner (✉) · M. Collard
LAMIA Lab., University of the French West Indies, Pointe-à-Pitre, France
e-mail: erick.stattner@univ-ag.fr

M. Collard
e-mail: martine.collard@univ-ag.fr

© Springer International Publishing Switzerland 2017
F. Guillet et al. (eds.), *Advances in Knowledge Discovery and Management*,
Studies in Computational Intelligence 665, DOI 10.1007/978-3-319-45763-5_13

patterns by defining new kinds of knowledge suitable to the context (Manyika et al. 2011). Indeed, thanks to the improvement of storage and computation capacities, and the heterogeneity of data that can currently be extracted from online systems, more and more works have focused on approaches combining several sources of data, redefining traditional patterns of knowledge.

Clustering from social networks has been an active research area that has received a lot of contributions. Indeed, in natural or social systems, entities often tend to organize themselves in groups (Croft et al. 2008). For example, we observe that sharing common interests leads to the emergence of online communities through discussion forums or the exchange of messages or files. The detection of such groups is a good way to identify substructures that possibly have major roles in the targeted systems. Thus, the identification of these clusters and the comprehension of mechanisms underlying their formation are relevant challenges in many disciplines for uncovering relationships between the structure and the function into complex systems.

First clustering approaches exploited only the structure of the network in order to identify some particular patterns called *communities* (Radicchi et al. 2004; Fortunato 2009), namely groups of nodes densely connected. More recently, new approaches have attempted to combine both the network structure and the properties of nodes (Zhou et al. 2009; Stattner and Collard 2012b).

Nevertheless, the great majority of these works is conducted without taking into account the complementarity of the knowledge that can be acquired. Indeed, while many works were interested in designing new kinds of knowledge or in optimizing existing approaches, few studies have been focused in the fusion of patterns and the knowledge that could emerge from such fusions.

In this paper, we focus on two network clustering approaches and we seek to understand both the intersections that can exist between them and the useful knowledge emerging from their fusion. First, the classical *node clustering* approach that consists in searching for communities into a network. Second, the search for frequent conceptual links (FCL), a *link clustering* approach that exploits both the network structure and the properties of nodes to identify frequent links between groups of nodes sharing common attributes.

Our objective is to evaluate the potential relationships existing between FCL and communities for understanding how the patterns obtained with both approaches may overlap. For this purpose, we propose a set of original measures that aim to evaluate the amount of shared information between these patterns when they are extracted from the same network. These measures are then applied to three datasets (a proximity-based network, a product co-purchasing network and a phone call network) for demonstrating the interest to consider simultaneously several sources of knowledge. For each network, we provide several examples of the knowledge resulting from the fusion.

The paper is organized as follows. Section 2 presents the related works conducted on the identification of clusters. Section 3 describes the notions of communities and frequent conceptual links and discusses the questions raised when they are combined. Section 4 is devoted to the measures proposed to evaluate the quality of the fusion between communities and FCL. Section 5 presents the experimental results we have obtained by applying the measures to three datasets. Finally, Sect. 6 concludes and presents future directions.

2 Related Works

Numerous methods for identifying clusters from networks can be found in the literature (Riadh et al. 2009; Steinhaeuser and Chawla 2010; Yang et al. 2013). While these methods are all able to highlight groups from data arising from networks, we observe that some factors such as the extracted knowledge or the data used vary from one method to another. Several criteria can be used to classify these approaches. In this section, we present the two clustering methods addressed in this paper: the identification of communities and the search for frequent conceptual links, according to three main criteria.

(i) **Extracted knowledge**. The identification of communities and the search for frequent conceptual links provide two distinct kinds of patterns. On the one hand, the concept of community is currently the most common approach for *clustering nodes* in networks. It provides an information on groups of nodes most densely connected in the network (Newman 2006). The associated algorithms aim to partition the network in several connected components, called "*communities*", so that the nodes in each component have a high density of connection while nodes in different components have a lower link density (Fortunato 2009).

On the other hand, frequent conceptual links provide an information on groups of nodes most frequently connected in the network, in which each group is defined as a set of nodes sharing common attributes. Here, "*conceptual*" means that such a link is not a real social link, but represents a "*meta-link*" that is a set of social links between two groups of nodes considered as a concept according to the formal concept analysis area (Ganter et al. 2005). The set of frequent conceptual links extracted from a network provides a "*conceptual view*", namely a new network structure in which a node represents a group of nodes sharing common attributes and a link represents a frequent connection between two groups in the original network.

(ii) **Clustering criterion**. In the domain of group identification, the building of clusters may rely on various clustering criteria (Mangiameli et al. 1996; Lancichinetti et al. 2008). In traditional network clustering, approaches attempt to identify a network partition in which the number of inter-clusters links is maximized while the number of intra-clusters links is minimized. For this purpose, they use the criterion of *modularity* introduced by Newman (2006) to evaluate the quality of the partition. The modularity measures the density of links into a group and is commonly used as an optimisation function in some network clustering algorithms (Lehmann and Hansen 2007; Blondel et al. 2008). Some approaches perform clustering on networks by using different measures, such as those using Potts models (Kumpula et al. 2007).

The search for frequent conceptual links relies on the notion of *support*, well known about frequent itemsets (Agrawal and Srikant 1994). It allows to evaluate the percentage of links in the network connecting a group of nodes satisfying a given property A to another group of nodes satisfying a given property B. Thus, the higher the value of support is, the higher the amount of links connecting nodes satisfying A to nodes satisfying B is Stattner and Collard (2012b).

(iii) **Source of data**. In several applications, networks are modeled by links and nodes may have various kinds of associated attributes. Such networks are called "*information networks*" or "*networks with content*" (El Gamal and Kim 2011). For instance, in a telecommunication network, consumers (nodes) may be identified by attributes such as age, type of package, job status, etc. If the wide majority of network clustering approaches does not take into account the attributes of nodes, some recent works have proposed new definitions of community for including node properties in the clustering task (Yoon et al. 2011). These approaches aim to provide a semantic decomposition of the network by focusing on the "*densely connected groups of nodes with homogeneous attributes values*" as explained in Zhou et al. (2009).

The search for frequent conceptual links exploits both network structure and node attributes (Stattner and Collard 2012a). The extracting process involves two key steps: a clustering phase, that builds the concepts by grouping nodes with common attributes and an evaluation phase that exploits the network links to assess the frequency of links between concepts.

3 Towards a Fusion of Knowledge

This section describes formally the concepts of *communities* and *frequent conceptual links*. We first present each kind of pattern, then we discuss the useful knowledge resulting from their fusion.

First of all, let $G = (V, E)$ be a social network, where V is the set of nodes (vertices) and $E \subseteq V \times V$ the set of social links (edges).

3.1 Communities in Social Networks

We define C as the set of communities extracted from the network G. We assume that there are no overlapping communities, thus a node belongs to one and only one community. We denote $F : V \to C$, the function that returns, for a given node v, the community to which it belongs.

The communities are extracted in order to maximize the modularity Q defined as follows:

$$Q = \frac{1}{2|E|} \sum_{ij} [W_{ij} - \frac{k_{v_i} k_{v_j}}{2|E|}] \, \delta(F(v_i), F(v_j))$$

where W_{ij} represents the weight of the edge between nodes v_i and v_j, k_{v_i} corresponds to the degree of node v_i and the δ-function is equal to 1 if $F(v_i) = F(v_j)$ and 0 otherwise.

The method we use in our experiments is the algorithm proposed by Blondel et al. (2008), based on modularity optimization.

3.2 Frequent Conceptual Links in Social Networks

V is defined as a relation $R(A_1, \ldots, A_p)$ where each A_i is an attribute. Thus, each vertex $v \in V$ is defined by a tuple $(a_1, \ldots, a_p)$ where $\forall q \in [1...p]$, $v[A_q] = a_q$, the value of the attribute A_q in v and $|R| = p$.

An item is a logical expression $A = x$ where A is an attribute and x a value. The empty item is denoted $\emptyset$. An itemset is a conjunction of items for instance $A_1 = x$ and $A_2 = y$ and $A_3 = z$. An itemset which is a conjunction of k non empty items is called a k-itemsets.

Let m and sm be two itemsets. If $sm \subseteq m$, we say that sm is a sub-itemset of m and m is a super-itemset of sm. For instance $sm = xy$ is a sub-itemset of $m = xyz$.

Any itemset is a sub-itemset of itself.

We denote I_V the set of all itemsets built from V.

Let us consider G as a *unipartite directed graph*. Thus, for any itemset m in I_V, we denote V_m the set of nodes in V that satisfy m and we define:

- the *m-left-hand linkset* LE_m as the set of links in E that start from nodes satisfying m i.e.
  $LE_m = \{e \in E \; ; \; e = (a, b) \quad a \in V_m\}$
- the *m-right-hand linkset* RE_m as the set of links in E that arrive to nodes in V_m i.e
  $RE_m = \{e \in E \; ; \; e = (a, b) \quad b \in V_m\}$

Definition 1 (*Conceptual link*) For any two elements m_1 and m_2 in I_V, the *conceptual link* (m_1, m_2) of G is the set of links connecting nodes in V_{m_1} to nodes in V_{m_2}.

For instance, if m_1 is the itemset cd and m_2 is the itemset efj, the *conceptual link* $(m_1, m_2) = (cd, efj)$ includes all links in E between nodes in V that satisfy the property cd with nodes in V that satisfy the property efj.

Let L_V be the set of conceptual links of $G = (V, E)$ and (m_1, m_2) be any element in L_V.

$(m_1, m_2) = LE_{m_1} \cap RE_{m_2}$
$\qquad = \{e \in E \; ; \; e = (a, b) \quad a \in V_{m_1} \text{ and } b \in V_{m_2}\}$

Definition 2 (*Support of conceptual link*) We call *support* of any element $l = (m_1, m_2)$ in L_V, the proportion of links in E that belong to l.

$$supp(l) = \frac{|(m_1, m_2)|}{|E|}$$

For an itemset m and a conceptual link l, if $l = (\emptyset, m)$ or $l = (m, \emptyset)$ then $supp(l) = 0$.

Definition 3 (*Frequent Conceptual Link*) Given a real number $\beta \in [0 \ldots 1]$, a conceptual link l in L_V is *frequent* if its support is greater than a minimum link support threshold β,

$$supp(l) > \beta$$

Let FL_V be the set of frequent conceptual links (FCL) in $G = (V, E)$ according to a given link support threshold β.

$$FL_V = \bigcup_{m_1 \in I_V, m_2 \in I_V} \{(m_1, m_2) \in L_V \; ; \; \frac{|(m_1, m_2)|}{|E|} > \beta\}$$

Definition 4 (*Conceptual sub-link*) Let two any itemsets sm_1 and sm_2 be respectively sub-itemsets of m_1 and m_2 in I_V. The conceptual link (sm_1, sm_2) is called conceptual *sub-link* of (m_1, m_2).

Similarly, (m_1, m_2) is called conceptual *super-link* of (sm_1, sm_2).

We write $(sm_1, sm_2) \subseteq (m_1, m_2)$

Definition 5 (*Maximal frequent conceptual link*) Let β be a given link support threshold, we call *maximal frequent conceptual link* (MFCL), any frequent conceptual link l such as, there exists no super-link l' of l that is also frequent.

More formally, $\nexists l' \in FL_V$ such as $l \subset l'$.

MFCLs provide a conceptual view of the social network about groups of nodes that share common *internal* properties (or concepts according to the area of formal concept analysis (Ganter et al. 2005)) and that are the most connected. More precisely, the conceptual view is a graph structure in which each node is related to an itemset (i.e. group of nodes that satisfy this itemset), and each link corresponds to a MFCL. By this way, the conceptual view provides a semantic and reduced representation of the initial network. More precisely, the set of the maximal frequent conceptual links provides a conceptual and synthetic view of the social network in which only relevant links between groups of nodes are represented.

Definition 6 (*Conceptual view of the social network*) Let $G = (V, E)$ be a social network and β the minimum support threshold. We define G_β^*, the graph (M, L), as the conceptual view of the network G obtained with the link support threshold β.

- M is the set of itemsets, called "*meta-nodes*"
- L is the set of maximal frequent conceptual links.

3.3 Merging Communities and Frequent Conceptual Links

Figure 1 shows resulting patterns extracted by community extraction and search for frequent conceptual links methods from a reference network. We can observe that patterns extracted by both methods provide two very different kinds of knowledge. The identification of communities extracts **cluster of nodes** based on the density of internal links, while the search for frequent conceptual links extracts **clusters of links** based on their frequency in the network. Obviously considering simultaneously these two kinds of pattern can improve the knowledge of these structures. It also raises a variety of interesting questions on the organisation of the involved structures such as:

1. Are communities composed by a single meta-node, i.e. a unique property?
2. Do the meta-nodes contain nodes that belong to a same community?
3. Do the frequent conceptual links connect nodes belonging to a same community, or nodes belonging to different communities?

To answer these questions related to the fusion between both kinds of pattern, we present in the next section a set of interestingness measures designed to evaluate the quality of the merging. More precisely, the proposed measures evaluate the degree of inclusion of communities in meta-nodes, and inversely, the degree of inclusion of meta-nodes in communities.

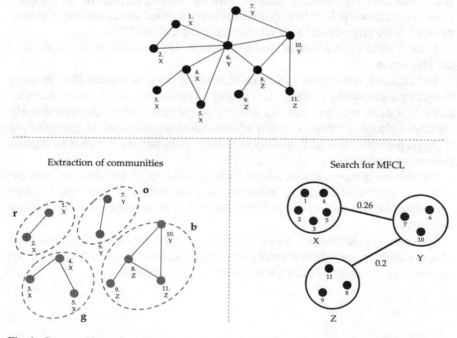

Fig. 1 Communities and maximal frequent conceptual links extracted from a reference network

4 Interestingness Measures

This section is devoted to the measures we propose for evaluating the intersections of both patterns: communities and frequent conceptual links.

4.1 Preliminaries

We remind that $G = (V, E)$ is a social network in which, V is the set of nodes and E the set of links. Cardinality of the sets V and E, respectively denoted $|V|$ and $|E|$ provides the number of nodes and the number of links.

C is the set of communities identified on the network by using a classical link-based clustering techniques (Blondel et al. 2008). Cardinality $|C|$ provides the total number of communities identified on the network.

We note V_c the set of nodes in V that belong to the community c, i.e. $V_c = \{v \in V \; ; \; F(v) = c\}$.

Finally, let $G_\beta^* = (M, L)$ be the conceptual view obtained by extracting maximal frequent conceptual links from the network G. The set M is the set of meta-node and $L \subseteq M \times M$ is the set of maximal frequent conceptual links. The extraction of MFCL from G can be performed by algorithms proposed in (Stattner and Collard 2012b). Let us specify that the computation time related to the extraction of frequent conceptual links exponentially increase with the number of links in the network. However, some works have been carried out in order to reduce the computation time by using some properties of node sets (Stattner and Collard 2013).

Let $m \in M$ be a given itemset. We remind that V_m is the set of nodes in V satisfying the property m.

In this paper, our objective is to understand the possible relationships between the patterns extracted with methods focusing on both communities and conceptual links. In a more semantic way, we investigate the relationships between densely connected groups of nodes (i.e. communities or clusters) and groups of nodes sharing common properties that are frequently connected in the whole network (i.e. frequent conceptual links).

For this purpose, three objects have to be considered: (i) *communities*, that are related to link-based clustering techniques and (ii) *meta-nodes* and (iii) *frequent conceptual links*, that refer to the patterns extracted by the conceptual links extraction techniques.

In this section, we present various measures, related to the homogeneity into each kind of objects to understand how communities are included in conceptual links, and inversely, how conceptual links are involved into communities.

4.2 Homogeneity Rate into a Community

The *homogeneity rate into a community*, noted H_c, is a measure that indicates, for a given community $c \in C$, its ability to aggregate nodes that belong to the same meta-node, i.e. a set of nodes sharing common properties. This measure corresponds to the fraction of meta-nodes that do not occur in the community c.

$$H_c = 1 - \frac{|\{m \in M ; \exists v \in V \ with \ F(v) = c \ and \ v \in V_m\}|}{|M|} \tag{1}$$

If $H_c = 0$, all meta-nodes are present in community c. More semantically, nodes in the community c satisfy all properties involved in conceptual links. Inversely, a high H_c value indicates that nodes in community c only belong to a small fraction of meta-nodes, i.e. nodes in community c tend to have similar properties.

For instance, the homogeneity rate in community r is $H_r = 0.6$, while the homogeneity rate in community b is $H_b = 0.3$ (see Fig. 2).

For considering weighting of a property into a community, we introduce $H_{c/m}$, the *homogeneity rate of a given meta-node m into a community c*. It corresponds to the fraction of nodes satisfying property m in community c.

$$H_{c/m} = \frac{|\{v \in V ; F(v) = c \ and \ v \in V_m\}|}{|\{v \in V ; F(v) = c\}|} \tag{2}$$

Thus if $H_{c/m} = 0$, nodes in meta-node m are not present in c. In a more semantic view, the nodes satisfying property m do not belong to community c. Inversely, when $H_{c/m}$ tends to 1, property m is satisfied by a high percentage of nodes in community c.

For instance, the homogeneity rate of meta-node X in community r is $H_{r/X} = 1$ (see Fig. 2). In the same way, the homogeneity rate of meta-node Z in community b is $H_{b/Z} = 0.75$.

As previously, the set of all $H_{c/m}$ values obtained for each pair (c, m) provides a $|C| \times |M|$ matrix.

Fig. 2 Meta-nodes (x, y and z) included into communities (r, g, o and b) from the example of Fig. 1

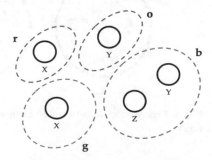

4.3   Homogeneity Rate into a Meta-node

The *homogeneity rate into a meta-node*, H_m, is a measure that indicates, for a given meta-node $m \in M$, its ability to aggregate nodes of the same community. It corresponds to the fraction of communities that do not occur in the meta-node m.

$$H_m = 1 - \frac{|\{c \in C \; ; \; \exists v \in V_m \; with \; F(v) = c\}|}{|C|} \quad (3)$$

Thus, if $H_m = 0$, all communities are represented in the meta-node m. In other words, all communities contain nodes satisfying property m. Inversely, when H_m tends to 1, only a small percentage of communities is present in m, i.e. the meta-node contains nodes of the same community.

For instance, regarding the example of Fig. 1 containing 4 communities, the homogeneity rate into meta-node X (see Fig. 3) is $H_X = 0.5$, while homogeneity rate into meta-node Z is $H_Z = 0.75$.

To take into account the weighting, we introduce $H_{m/c}$, the *homogeneity rate of a given community c into a meta-node m*. This measure indicates the fraction of nodes of community c, in the meta-node m.

$$H_{m/c} = \frac{|\{v \in V_m \; ; \; F(v) = c\}|}{|V_m|} \quad (4)$$

Thus, if $H_{m/c} = 0$, nodes of community c are not present in m. More semantically, nodes in cluster c does not satisfy the property m. Inversely, when $H_{m/c}$ tends to 1, m is mostly represented in community c.

For example, starting from the example of Fig. 1, the homogeneity rate of community r in meta-node X is $H_{X/r} = 0.4$ (see Fig. 3). Similarly, homogeneity rate of community b in meta-node Z is $H_{Z/b} = 1$.

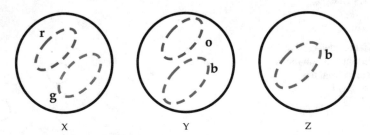

Fig. 3 Communities (r, g, o and b) included into meta-nodes (x, y and z) from the example of Fig. 1

4.4 Homogeneity Rate into a Conceptual Link

The *homogeneity rate H_l, into a conceptual link*, measures for a given frequent conceptual links $l = (m_1, m_2)$, its ability to connect nodes belonging to the same community. In other words, it indicates if nodes of the same community maintain a frequent conceptual link. It corresponds to the fraction of similar communities represented in both sides of the frequent conceptual links.

$T_1 = \{c \in C \; ; \; \exists v \in V \text{ with } F(v) = c \text{ and } v \in V_{m_1}\}$
$T_2 = \{c \in C \; ; \; \exists v \in V \text{ with } F(v) = c \text{ and } v \in V_{m_2}\}$

$$HL_l = \frac{|(T_1 \cap T_2)|}{|(T_1 \cup T_2)|} \tag{5}$$

Thus, for a given frequent conceptual link $l = (m_1, m_2)$, a low HL_l value indicates that nodes involved in both sides of the frequent conceptual link belong to different communities, while a high HL value indicates that a large amount of communities represented in meta-node m_1 are also represented in meta-node m_2.

For example, the homogeneity rate into the conceptual link (Z, Y) is $H_{(Z,Y)} = 0.5$ (see Fig. 4). In the same way, the homogeneity rate into the conceptual link (X, Y) is $H_{(X,Y)} = 0$.

As previously, we introduce $H_{l/c}$, *the homogeneity rate of a given community c into the frequent conceptual link* $l = (m_1, m_2)$. More precisely, $H_{l/c}$ measures the difference in representation of a community c in meta-nodes m_1 and m_2.

$$H_{l/c} = 1 - \frac{|H_{m_1/c} - H_{m_2/c}|}{max(H_{m_1/c}, H_{m_2/c})} \tag{6}$$

Thus, for a given frequent conceptual link $l = (m_1, m_2)$, the homogeneity rate $H_{l/c} = 1$ indicates that the fraction of nodes of community c in meta-nodes m_1 and m_2 of l

Fig. 4 Communities (r, g, o and b) included into frequent conceptual links from the example of Fig. 1

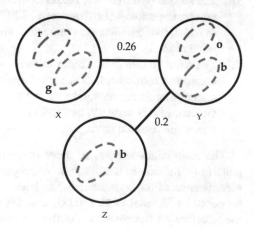

is similar. Inversely, $H_{l/c} = 0$ indicates that at least one of the meta-nodes does not contain nodes belonging to the community c.

For example, the homogeneity rate of community b into the frequent conceptual link (Z, Y) is $H_{(Z,Y)/b} = 1 - \frac{0.7}{1} = 0.3$.

5 Experimental Results

We have conducted various set of experiments to evaluate the quality of the fusion. The results obtained show that very homogeneous structures can be found, and demonstrate the interest to consider simultaneously several sources of knowledge and more particularly the communities and the frequent conceptual links.

Section 5.1 describes the datasets used and their main characteristics regarding the network structural properties, the communities and their size and the frequent conceptual links and their properties. Section 5.2 presents and discusses the results we have obtained by applying the interestingness measures proposed on the three datasets.

5.1 Testbed

Three datasets have been used in our experiments.

(i) The first (referred as EpiSims in the remaining of the paper) is a geographical proximity-based social network obtained with *EpiSims* (Barrett et al. 2008), a simulation tool that statistically reproduces the daily movements of individuals in the city of Portland. In this network, two individuals are connected when they were co-located in the same place during the simulation.

(ii) The second (referred as Amazon in the remaining of the paper) is a product co-purchasing network (Leskovec et al. 2007), extracted from the *Amazon* database, in which two products are connected when they were purchased together by a same user.

(iii) The third (referred as Communications in the remaining of the paper) is a connected subnetwork of a very large communication network provided by a local mobile telephony operator (Stattner 2014) in French West Indies and Guiana. In this network, two individuals are connected when a telephone call was made between them.

The main characteristics of these datasets and the properties of the extracted patterns (communities and frequent conceptual links) are described in Table 1. The identification of the communities has been performed with the *Louvain Algorithm* proposed by Blondel et al. (2008), a nodes clustering method that relies only on the structure on the network. As the Louvain Algorithm is non-deterministic we

Table 1 Main properties of the dataset used

		EpiSims	Amazon	Communications
General information	#Nodes	1043	5001	1705
	#Attributes	6	7	7
	#Links	2382	14981	1807
	Density	0,0044	0,0012	0,0012
	Coeff. Clust	0,7091	0,4874	0,0439
	#Composante	1	1	1
	Max degree	15	64	12
	Avg degree	4,5675	5,9912	2,1196
	Degree Distribution			
Communities	Modularity	0,864	0,883	0,945
	#Clusters	29	45	40
	Size of communities			
	Community size distribution			
Conceptual Links	#Meta-nodes	35	21	44
	#FCL	116	43	105
	Distribution size of meta-nodes			

focused on an extraction that represented a meaningful snapshot of communities. The extraction of maximal frequent conceptual links has been performed with the *MFCL-Min Algorithm* proposed in (Stattner and Collard 2012b). The minimum link support threshold β was set at 0.1, namely we keep only groups that contain at least 10 % of the network links.

(i) **The EpiSims network** is composed of 1043 nodes and 2382 links. Each node is identified by 6 attributes: (1) age class, i.e. $\lfloor \frac{age}{10} \rfloor$ (2) gender (1-male, 2-female), (3) worker (1-has a job, 2-has no job), (4) relationship to the head of household (1-spouse, partner, or head of household, 2-child, 3-adult relative, 4-other), (5) contact class (i.e. $\lfloor \frac{degree}{2} \rfloor$) and (6) sociability (i.e. 1-*coeff. clust.* > 0.5, 2-*else*). The network contains 29 communities, and 35 Meta-nodes and 116 frequent conceptual links have been identified.

Figure 5 shows the knowledge extracted from the Episims network: (a) Communities and (b) Conceptual view by keeping only FCL with $\beta \geq 0.2$ for more readability. In this figure, nodes that belong to a same community have an identical color. Moreover for simplicity, meta-nodes (properties) are denoted as follows:

$$(\text{<att 1>}, \text{<att 2>}, \ldots, \text{<att } n\text{>})$$

where <att i> corresponds to the value of the attribute i on the node. The character '*' means that the attribute may have any value.

For example, we can observe that the FCL $((*; 2; 2; *; *; *), (*; *; 2; *; *; *))$ has been identified with a support equals to 0.2. It indicates that 20 % of the links of the network connect women who has no job to individuals who has no job.

(ii) **The Amazon network** is composed of 5001 nodes and 14981 links. Each node is identified by 7 attributes: (1) product group (eg. Book, DVD, Video or Music), (2) number of similar co-purchased products (integer), (3) category (integer), (4) main category (eg. Literature and Fiction, Arts and Photography, Sport, ...), (5) sub cat-

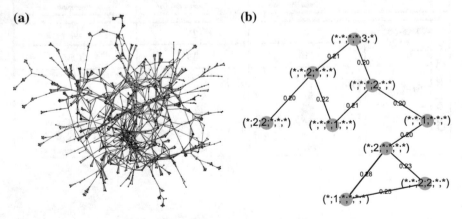

Fig. 5 Knolwedge from Episims network: **a** communities and **b** conceptual view with $\beta \geq 0.2$

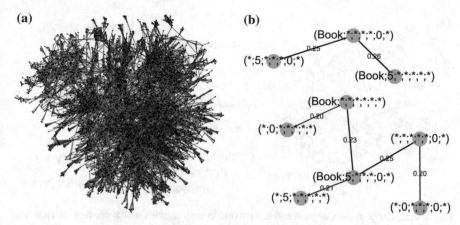

Fig. 6 Knolwedge from Amazon network: **a** communities and **b** conceptual view with $\beta = 0.1$

egory (like (5)), (6) number of reviews (integer) and (7) rating (integer between 1 and 5). The network contains 45 communities and 21 Meta-nodes, and 43 frequent conceptual links have been identified.

Figure 6 shows the knowledge extracted from the Amazon network: (a) Communities and (b) Conceptual view by keeping FCL links with $\beta \geq 0.2$ for more readability. We can observe that the FCL $((Book; *; *; *; *; 0; *), (Book; 5; *; *; *; *; *))$ is identified with a support of 0.26. It indicates that 26 % of the links of the Amazon network connect books that have no-review to books that are co-purchased with five similar products.

(iii) The Communication network is composed of 1705 nodes and 1807 links. The data have been processed to keep only calls between users, namely removing calls to voice mail, customer service, etc. Each node of this network is characterized by 7 attributes.

1. localisation ("Martinique", "Guadeloupe", "Guyane" or "Other"),
2. class of received calls number, i.e. $\lfloor \frac{\#received\ calls}{10} \rfloor$,
3. class of received average calls duration, i.e. $\lfloor \frac{rec.\ avg\ call\ duration}{10} \rfloor$,
4. class of outgoing calls number, i.e. $\lfloor \frac{\#outgoing\ calls}{10} \rfloor$,
5. class of outgoing calls average duration, i.e. $\lfloor \frac{out.\ avg\ calls\ duration}{10} \rfloor$,
6. class of number of SMS sent, i.e. $\lfloor \frac{\#SMS\ sent}{10} \rfloor$
7. class of number of SMS received, i.e. $\lfloor \frac{\#SMS\ received}{10} \rfloor$.

The network contains 40 communities and 44 Meta-nodes, and 105 frequent conceptual links have been identified.

Figure 7 shows the knowledge extracted from the Communication network: (a) Communities and (b) Conceptual view by keeping FCL links with $\beta \geq 0.2$ for more readability. We can observe that the FCL

$$(GUAD.; *; *; *; *; *; *), (GUAD.; *; 1; *; *; *; *)$$

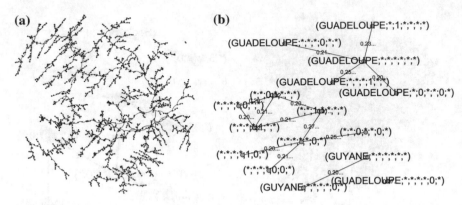

Fig. 7 Knowledge from Communication network: **a** communities and **b** conceptual view with
$\beta = 0.1$

is identified with a support equals to 0.23. It indicates that 23 % of the links of the network connect consumers located in Guadeloupe to consumers located in Guadeloupe
and having an average call duration comprised between 10 and 19 min.

Note that the datasets used are relatively small because of the difficulty for extracting FCL on large datasets. More particularly, in Stattner and Collard (2012b) it
has been shown that the computation time exponentially increases with the number of attributes. However, some recent works have focused on the optimisation
of the extraction process and have proposed various solutions to reduce the search
space (Stattner and Collard 2013).

5.2 Results

In our experiments, we apply the proposed measures to the three datasets with the goal
to identify homogeneous structures regarding the fusion of communities and frequent
conceptual links. For this purpose we focus, for each measure, to the distribution of
the values in order to highlight the amount of situations in which the measures are
maximized. Moreover, for each measure we give some examples of interesting fusion.

5.2.1 Meta-Nodes Inside Communities

As a first step, Fig. 8 shows, for each dataset, the distribution of the homogeneity rate
$H_{c/m}$ of a meta-node into a community. We remind that the homogeneity rate into a
community allows evaluating if a community consists only of nodes belonging to the
same meta-nodes, i.e. nodes satisfying common properties and involved in frequent
conceptual links as described previously in Fig. 2.

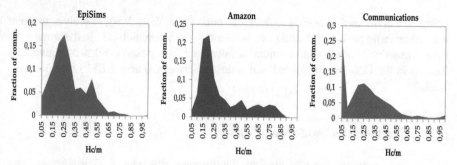

Fig. 8 Distribution of homogeneity rate $H_{c/m}$ of a meta-node into a community

We can observe that trends are very similar for the three networks. Indeed, for each dataset the vast majority of the homogeneity rates is rather low. For instance in the Episim network 91.13 % of the $H_{c/m}$ values are less than 0.5. In the Amazon network 81.16 % of the $H_{c/m}$ values are less than 0.5 and in the Communications network this proportion is 89.71 %. This result suggests that a strong proportion of communities are very heterogeneous in their structure, since they are not composed of nodes that belong to a same meta-node. Consequently, several attributes can be found in such communities.

However, our approach also allows highlighting that it exists a small percentage of communities which have a high homogeneity rate. For instance, in the EpiSims network 1.08 % of the $H_{c/m}$ values are higher than 0.75. These proportions are 7.93 % for the Amazon network and 4.09 % for the Communications network. This result indicates that it exists some communities very homogeneous since they are mainly composed of nodes belonging to a same meta-node, i.e. a group of nodes that share common attributes and that is involved in a frequent conceptual link.

Table 2 shows some interesting patterns regarding the $H_{c/m}$ measure. For example, 80 % of the nodes in the community 24 of the EpiSims network (see line 1 Table 2) is

Table 2 Examples of interesting patterns regarding $H_{c/m}$

Network	Community	Meta-Node	$H_{c/m}$
EpiSims	24 (10 nodes)	(*;*;2;*;3;*)	0.80
	13 (19 nodes)	(*;*;1;*;*;*)	0.73
	4 (25 nodes)	(*;1;*;*;*;*)	0.70
Amazon	39 (64 nodes)	(*;*;*;*;*;0;*)	0.87
	25 (50 nodes)	(Book;*;*;*;*;*)	0.84
	20 (46 nodes)	(*;5;*;*;*;*)	0.72
Communication	6 (31 nodes)	(GUADELOUPE;*;*;*;*;*;*)	1.00
	36 (59 nodes)	(GUYANE;*;*;*;*;*;*)	1.00
	29 (24 nodes)	(GUADELOUPE;*;*;*;*;0;*)	0.89

composed of nodes that belong to the meta-node (∗; ∗; 2; ∗; 3; ∗), namely a group of individuals who have no job and have between 5 and 6 connections. In the same way, the community 29 of the Communication network is composed to 89 % of individuals located in the Guadeloupe island and sending between 0 and 9 SMS (see last line Table 2).

5.2.2   Communities Inside Meta-Nodes

In a second study, we have focused on the homogeneity rate $H_{m/c}$ of a community into a meta-node. As previously, we show on Fig. 9 the distribution of this measure for each dataset. We remind that the homogeneity rate into a meta-node allows evaluating if a meta-node (i.e. a group of nodes that share common properties and that is involved in a frequent conceptual link) is solely composed of nodes that belong to a same community. In other words, this measure assesses whether the nodes that share common attributes are densely interconnected.

The values obtained here for the homogeneity rate $H_{m/c}$ are very low whatever is the dataset. For instance, for the EpiSims network $max(H_{m/c})$ is 0.11, while it is 0.08 for the Amazon network and 0.11 for the Communication network. This result suggests that situations in which meta-nodes are fully homogeneous are very rare. In other words, it seems to be unlikely that the nodes into a meta-node are densely interconnected. Obviously, we can assume that these results vary according to the nature of the network and the semantics of the links.

Table 3 shows some examples of patterns regarding the $H_{m/c}$ measure. For instance, the first line of the table indicates that in EpiSims network 11 % of nodes satisfying property (1; 1; 2; 2; ∗; ∗) belong to community 20. In other words little boys who are between 0 and 9 years old are involved in frequent conceptual links and they are densely connected. In the same way, 11 % of the set of subscribers located in French Guiana whose received calls have a duration between 0 and 9 min and who have sent between 0 and 9 SMS (i.e. $(GUYANE; ∗; 0; ∗; ∗; 0; ∗)$) belong to community 28 (see line 7 Table 3).

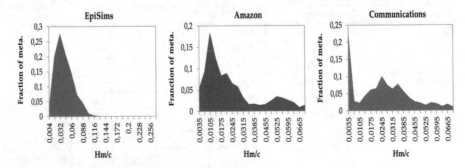

Fig. 9 Distribution of homogeneity rate $H_{m/c}$ of a community into a meta-node

Table 3 Examples of interesting patterns regarding $H_{m/c}$

Network	Meta-node	Community	$H_{m/c}$
EpiSims	(1;1;2;2;*;*) (119 nodes)	20	0.11
	(1;*;2;2;*;*) (194 nodes)	19	0.10
	(2;*;2;2;*;*) (196 nodes)	9	0.10
Amazon	(Book;0;*;*;*;0;*) (996 nodes)	35	0.08
	(*;*;*;*;General;*;*) (1071 nodes)	35	0.07
	(Music;*;*;*;*;*;*) (997 nodes)	35	0.07
Communication	(GUYANE;*;0;*;*;0;*) (212 nodes)	28	0.11
	(GUYANE;*;*;*;1;*;*) (300 nodes)	33	0.10
	(GUADELOUPE;*;1;*;1;*;*) (192 nodes)	10	0.09

5.2.3 Communities Inside Frequent Conceptual Links

In the last study, we have focused on the homogeneity rate $H_{l/c}$ of a community into a frequent conceptual link. Figure 10 shows the distribution of this measure for each dataset. We remind that the homogeneity rate into a frequent conceptual link measures the ability for a FCL to connect nodes that belong to same communities. It allows evaluating if a frequent conceptual link is composed, for right and left sides, to nodes belonging to a same community as described in Fig. 4.

We can observe that the trends are very similar for the three networks. Indeed, for each dataset the vast majority of the homogeneity rates obtained is rather high. For instance, in the EpiSims network 87.97 % of the $H_{l/c}$ values are greater than 0.75. In the Amazon and in the Communication networks, this proportion is respectively 76.24 and 71,5 %. This result suggests that frequent conceptual links tend to be very homogeneous, since equivalent percentages of nodes belonging to a same community are found at the both sides of the pattern. In the EpiSims network, only 15.15 % of the values obtained are less than 0.5. In the Amazon and the Communication networks, this percentage is respectively 6,56 and 10,62 %. This suggests that frequent

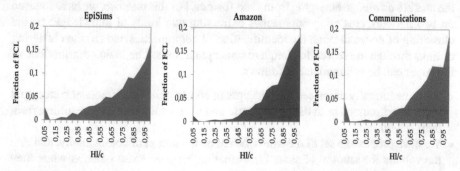

Fig. 10 Distribution of homogeneity rate $H_{l/c}$ of a community into a frequent conceptual link

Table 4 Examples of interesting patterns regarding $H_{l/c}$ (remind support of FCL: $\beta = 0.1$)

Network	Frequent conceptual link	Community	$H_{l/c}$
EpiSims	$((*;1;*;*;*;*),(*;2;*;*;3;*))$	13	1.0000
	$((*;1;*;*;*;*),(*;2;1;*;*;*))$	23	0.9995
	$((*;2;2;2;*;*),(*;1;*;*;*;*))$	7	0.9962
Amazon	$((*;*;*;*;General;*;*),(Book;*;*;*;*;0;*))$	22	0.9999
	$((Book;*;3;*;*;*;*),(*;*;*;*;*;0;*))$	27	0.9994
	$((Music;*;*;*;*;*;*),(Book;*;*;*;*;0;*))$	37	0.9989
Communication	$((*;*;0;*;*;*;*),(*;*;1;*;*;0;*))$	26	0.9998
	$((Guadeloupe;*;0;*;1;*;*),(Guadeloupe;*;*;*;*;*;*))$	25	0.9984
	$((Guyane;*;*;*;*;*;*),(Guyane;*;1;*;*;*;*))$	32	0.9958

conceptual links tend to connect nodes that belong to same communities. Moreover, this result demonstrates that a part of the intra-community links into a social network may be involved in a frequent conceptual link.

For example, the first line of the Table 4 provides relevant knowledge: First, $((*; 1; *; *; *; *), (*; 2; *; *; 3; *))$ is a frequent conceptual link, i.e. at least 10 % of the links of the network connect men $(*; 1; *; *; *; *)$ to women who have between 4 and 5 contacts $(*; 2; *; *; 3; *)$ (we remind that the minimum link support threshold was set at 0.1). Second, in each group, the percentage of nodes that belong to community 13 is exactly the same.

6 Conclusion

In this paper, we have addressed the problem of clustering from social networks. Unlike traditional approaches that focus separately on the design of new patterns of knowledge suited to the context or the optimization of existing algorithms, we have adopted in this work another point of view by focusing on the fusion of patterns and the useful knowledge emerging from such fusions. For this purpose, we have focused on two network clustering approaches extracting two kinds of knowledge: (i) the clustering of nodes through the identification of communities and (ii) the clustering of links through the search for frequent conceptual links. The main contributions of the paper can be summarized as follows.

- We have formally described the concepts of communities and frequent conceptual links and discussed both the problematic and the useful knowledge resulting from their fusion.
- We have proposed a set of measures, based on the notion of homogeneity, that aim to evaluate the amount of shared information between these patterns when they are extracted from a same network.

- Finally, we have applied these measures to three datasets: a proximity-based network, a product co-purchasing network and a phone call network. The results obtained have demonstrated the interest of the approach proposed since very interesting merged knowledge have been identified on each network.

This work demonstrates the interest to consider simultaneously several sources of knowledge. In future works we plan to extend the approach to other network mining methods.

More generally, this work also raises a variety of questions in terms of visualization, extraction algorithms and resulting meaning. For instance in our future works, we plan to propose more complete representations of networks combining into single visualizations several kinds of knowledge. Another interesting track should be to propose optimized algorithms able to extract in one run several kinds knowledge.

References

Agrawal, R., & Srikant, R. (1994). Fast algorithms for mining association rules in large databases. In *Proceedings of the 20th International Conference on Very Large Data Bases* (pp. 487–499).

Barrett, C. L., Bisset, K. R., Eubank, S. G., Feng, X., & Marathe, M. V. (2008). Episimdemics: An efficient algorithm for simulating the spread of infectious disease over large realistic social networks. In *Proceedings of the 2008 ACM/IEEE Conference on Supercomputing*.

Blondel, V., Guillaume, J. L., Lambiotte, R., & Lefebvre, E. (2008). Fast unfolding of communities in large networks. *Journal of Statistical Mechanics: Theory and Experiment, 2008*, P10008.

Croft, D. P., James, R., & Krause, J. (2008). *Exploring animals social networks*. Princeton: Princeton University Press.

El Gamal, A. & Kim, Y.-H. (2011). *Network information theory*. Cambridge: Cambridge University Press.

Fortunato, S. (2009). Community detection in graphs. *Physics Reports, 486*, 75–174.

Ganter, B., Stumme, G., & Wille, R. (2005). Formal concept analysis, foundations and applications. *Lecture Notes in computer science* (Vol. 3626).

Getoor, L., & Diehl, C. P. (2005). Link mining: A survey. *Physics Reports, 7*, 3–12.

Kumpula, J. M., Saramäki, J., Kaski, K., & Kertész, J. (2007). Limited resolution in complex network community detection with potts model approach. *Physics Reports, 56*(1), 41–45.

Lancichinetti, A., Fortunato, S., & Radicchi, F. (2008). Benchmark graphs for testing community detection algorithms. *Physical Review E, 78*, 046110.

Lehmann, S., & Hansen, L. K. (2007). Deterministic modularity optimization. *Physical Review E, 60*(1), 83–88.

Leskovec, J., Adamic, L. A., & Huberman, B. A. (2007). The dynamics of viral marketing. *ACM Transactions on the Web, 1*.

Mangiameli, P., Chen, S. K., & West, D. (1996). A comparison of som neural network and hierarchical clustering methods. *European Journal of Operational Research, 93*(2), 402–417.

Manyika, J., et al. (2011). Big data: The next frontier for innovation, competition, and productivity.

Newman, M. E. (2006). Modularity and community structure in networks. *Proceedings of the National Academy of Sciences, 103*(23), 8577–8582.

Radicchi, F., Castellano, C., Cecconi, F., Loreto, V., & Parisi, D. (2004). Defining and identifying communities in networks. *Proceedings of the National Academy of Sciences of the United States of America, 101*(9), 2658–2663.

Riadh, T., Le Grand, B., Aufaure, M., & Soto, M. (2009). Conceptual and statistical footprints for social networks' characterization. In *Proceedings of the 3rd Workshop on Social Network Mining and Analysis* (p. 8). ACM.

Scott, J. (2011). Social network analysis: Developments, advances, and prospects. *Proceedings of the National Academy of Sciences of the United States of America, 1*(1), 21–26.

Stattner, E. (2014). Link formation in a telecommunication network. In *2014 IEEE Eighth International Conference on Research Challenges in Information Science (RCIS)* (pp. 1–9). IEEE.

Stattner, E. & Collard, M. (2012a). Frequent links: An approach that combines attributes and structure for extracting frequent patterns in social networks. In *16th East-European Conference on Advances in Databases and Information Systems*.

Stattner, E. and Collard, M. (2012b). Social-based conceptual links: Conceptual analysis applied to social networks. In *International Conference on Advances in Social Networks Analysis and Mining*.

Stattner, E. and Collard, M. (2013). Towards a hybrid algorithm for extracting maximal frequent conceptual links in social networks. In *IEEE International Conference on Research Challenges in Information Science* (pp. 1–8).

Steinhaeuser, K., & Chawla, N. V. (2010). Identifying and evaluating community structure in complex networks. *Pattern Recognition Letters, 31*, 413–421.

Yang, J., McAuley, J., & Leskovec, J. (2013). Community detection in networks with node attributes. In *2013 IEEE 13th International Conference on Data Mining (ICDM)* (pp. 1151–1156). IEEE.

Yoon, S.-H., Song, S.-S., and Kim, S.-W. (2011). Efficient link-based clustering in a large scaled blog network. In *Proceedings of the 5th International Conference on Ubiquitous Information Management and Communication, ICUIMC 2011* (pp. 71:1–71:5). New York: ACM.

Zhou, Y., Cheng, H., & Yu, J. (2009). Graph clustering based on structural/attribute similarities. *Pattern Recognition Letters, 2*(1), 718–729.

Author Index

© Springer International Publishing Switzerland 2017
F. Guillet et al. (eds.), *Advances in Knowledge Discovery and Management*,
Studies in Computational Intelligence 665, DOI 10.1007/978-3-319-45763-5

Printed in the United States
By Bookmasters